Transactions on Computational Science and Computational Intelligence

Series Editor

Hamid Arabnia, Department of Computer Science, The University of Georgia, Athens, GA, USA

Computational Science (CS) and Computational Intelligence (CI) both share the same objective: finding solutions to difficult problems. However, the methods to the solutions are different. The main objective of this book series, "Transactions on Computational Science and Computational Intelligence", is to facilitate increased opportunities for cross-fertilization across CS and CI. This book series will publish monographs, professional books, contributed volumes, and textbooks in Computational Science and Computational Intelligence. Book proposals are solicited for consideration in all topics in CS and CI including, but not limited to, Pattern recognition applications; Machine vision; Brain-machine interface; Embodied robotics; Biometrics; Computational biology; Bioinformatics; Image and signal processing; Information mining and forecasting; Sensor networks; Information processing; Internet and multimedia; DNA computing; Machine learning applications; Multi-agent systems applications; Telecommunications; Transportation systems; Intrusion detection and fault diagnosis; Game technologies; Material sciences; Space, weather, climate systems, and global changes; Computational ocean and earth sciences; Combustion system simulation; Computational chemistry and biochemistry; Computational physics; Medical applications; Transportation systems and simulations; Structural engineering; Computational electro-magnetic; Computer graphics and multimedia; Face recognition; Semiconductor technology, electronic circuits, and system design; Dynamic systems; Computational finance; Information mining and applications; Astrophysics; Biometric modeling; Geology and geophysics; Nuclear physics; Computational journalism; Geographical Information Systems (GIS) and remote sensing; Military and defense related applications; Ubiquitous computing; Virtual reality; Agent-based modeling; Computational psychometrics; Affective computing; Computational economics; Computational statistics; and Emerging applications.

For further information, please contact Mary James, Senior Editor, Springer, mary.james@springer.com.

More information about this series at http://www.springer.com/series/11769

Maurizio Palesi • Ljiljana Trajkovic
J. Jayakumari • John Jose
Editors

Second International Conference on Networks and Advances in Computational Technologies

NetACT 19

Springer

Editors
Maurizio Palesi
Department of Electrical, Electronic and Computer Engineering
University of Catania
Catania, Italy

Ljiljana Trajkovic
School of Engineering Science
Simon Fraser University
Burnaby, BC, Canada

J. Jayakumari
Electronics & Communication Engineering
MBCET
Thiruvananthapuram, Kerala, India

John Jose
Indian Institute of Technology Guwahati
Guwahati, Assam, India

ISSN 2569-7072 ISSN 2569-7080 (electronic)
Transactions on Computational Science and Computational Intelligence
ISBN 978-3-030-49502-2 ISBN 978-3-030-49500-8 (eBook)
https://doi.org/10.1007/978-3-030-49500-8

This Springer imprint is published by the registered company Springer Nature Switzerland AG
The registered company address is: Gewerbestrasse 11, 6330 Cham, Switzerland

Preface

The main focus of the **NetACT** conference is to provide a forum for researchers, practitioners, and other stakeholders in infrastructure development to understand, assimilate, and interpret the latest innovations and, most importantly, their impact on the growth of the nation. This conference aims to discuss these aspects where researchers can share their concerns and achieve outputs that will help to widen their knowledge.

The 2nd International Conference on Networks and Advances in Computational Technologies **(NetACT19)** was held at Mar Baselios College of Engineering and Technology from July 23 to 25, 2019. The keynote addresses were given by **SriKrish Prasad**, *Senior Vice President and GM, Cloud Platform Business, VMWare, California, USA,* **Dr. Manoj B.S,** *Prof and Head, Dept of Avionics, Indian Institute of Space Science and Technology,* **Dr. John Jose,** *Asst. Professor, Dept of Computer Science and Engineering, IIT Guwahati,* **Dr. Samit Bhattacharya,** *Associate Professor, Dept of Computer Science and Engineering, IIT Guwahati*, and **Dr. B Valsa,** *Deputy Director, VSSC.*

We received 104 submissions this year. After a rigorous review check by the program committee, we decided to accept only 27 papers under the regular paper category. We are very thankful for Springer for supporting NetACT19, and Mrs. Suvira Srivastav and Nidhi Chandhoke for the continuous support and help.

We hope that the proceedings will be useful for all researchers working in the relevant areas.

Burnaby, BC, Canada — Ljiljana Trajkovic

Guwahati, India — John Jose

Trivandrum, India — Jisha John

Contents

Malware Attacks: A Survey on Mitigation Measures

Anna V. James and S. Sabitha

1 Introduction

Malicious software or Malware can be defined as a software that "deliberately fulfills the harmful intent of an attacker." Terms, such as "worm," "virus," "Trojan horse," "ransomware," etc., are the different classes of malware. Highlighting security vulnerabilities in the software or showoff of technical ability was the motivation for malware creators at the early times. Today, there is a flourishing underground economy based on malware. Now, it is no longer the fun factor, but the perspective of the money that can be made drives the development of such malware.

Introduction of new malware every day is a challenge to antivirus vendors. The main challenge of antivirus writers is the growing stream of obfuscated malware samples with several variations to avoid existing detection methodology. Very recent type of malware named as "ransomware" that extorts money from the victims became most popular with the cybercriminals. Compared to traditional malware, the attack pattern of ransomware is different. Traditional malware uses sophisticated coding techniques to steal the credential or to conduct any targeted attack. On the other hand, ransomware is designed purposefully to ask money from the victims by making the computer system unusable. In most of the cases, ransomware uses cryptographic technology to encrypt the user data.

It is critical to identify these types of malware, due to the fact that they cause lots of damage to large surface area. For detection, dynamic analysis is more popularly used so as to overcome the limitations of static analysis. Nowadays, Machine Learning techniques are also applied along with dynamic analysis [1, 2].

A. V. James (✉) · S. Sabitha
College of Engineering Trivandrum, Thiruvananthapuram, India
e-mail: annavjames@cet.ac.in; sabitha@cet.ac.in

M. Palesi et al. (eds.), *Second International Conference on Networks and Advances in Computational Technologies*, Transactions on Computational Science and Computational Intelligence, https://doi.org/10.1007/978-3-030-49500-8_1

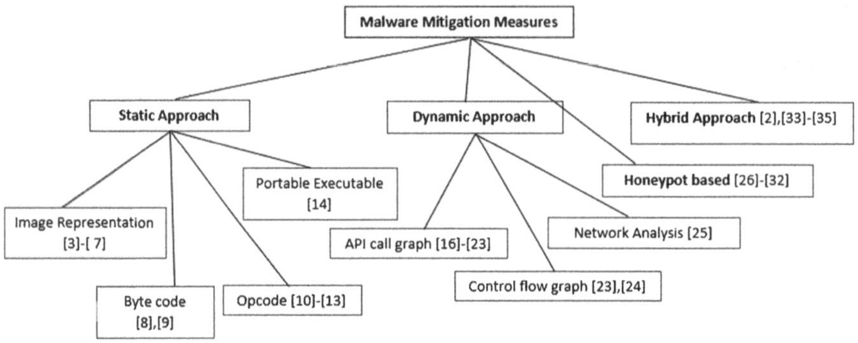

Fig. 1 Malware mitigation strategies

Antivirus (AV) vendors strive to keep the pace with sophisticated malware variants. The malware mitigation strategies used for detecting other malware variants can also be used in detection of ransomware. In this paper we tried to cover various best approaches used in detection of malwares, starting from traditional but effective signature (static) based approach to highly efficient hybrid approach. We also discuss Honeypot based approach that has been successful in detecting specific kinds of malwares like ransomwares. Figure 1 covers different techniques used in each approach. In fact, when new variants are spread, signature based mechanisms are easily deceived. Furthermore, very sophisticated packing techniques implemented by current ransomwares to evade detection, e.g. obfuscated API calls, delivering the static analysis useless.

Malware detection methods are fundamentally categorized into different categories from different points of view. They can be classified into four types as in Fig. 1: Signature based approach, Dynamic approach, Honeypot based approach, and Hybrid approach. The following sections discuss in detail about each of them.

The rest of the paper is organized as follows: In Sect. 2 we cover malware detection methods; Sect. 3 illustrates a comparison table; and finally in Sect. 4 the summary of the survey is discussed.

2 Mitigation Measures

2.1 Static Approach (Signature Based)

One of the traditional malware detection strategy is static approach. It examines the malware binaries without executing them. It is one of the fast and safe technique for malware detection. It mainly uses the hash signature, embedded strings, byte code distribution, etc. present in the malicious binary. This technique does not work well for sophisticated malware. Signatures are Short strings of bytes that are unique for each program. However, this signature based method is not effective against modified and unknown malicious executable.

Image Representation: Binary Texture Analysis Image Representation based technique makes use of pattern based method in identification of malware signature in an image. Image texture based features can be used for various applications such as image classification, image search, etc. Image representation of the malware binary is performed and texture based features are extracted from it as done by L. Nataraj [3]. The 2D matrix of grey scale image is generated by first converting malware binary into an array of 1D 8-bit integers as shown in Fig. 2. GIST, SIFT features are extracted from the image for further processing.

In [4], L. Nataraj use a traditional approach for classification of grey scale image. They use the idea that the variants of a malware have similar images and different malware have different images. In [5] S. Choi et al. propose a deep learning based method for malware classification by using the above method. The extracted feature set from malware is used to train kNN classifier. Nataraj [4] uses GIST to compute the mean value of magnitude of local feature and generate 320-dimensional GIST feature vector. Gibert [6] uses the binary image as input to CNN.

Other approaches [7] make use of deep Convolutional models to classify image based on LBP features. In addition to that, new variants of malware are created by changing only limited part of code hence, images can be used to detect slight changes by retaining the overall structure. Convolutional neural networks are found to be the best model used for image classification problems.

Byte Code Sequence Byte level details in a malware binary can be used for finding the relation with other binary files. Use of n-grams byte features for detecting malware has been done by E. Raff et al. in [8]. They treat the malicious binary as sequence of bytes and n consecutive bytes are considered as individual feature. It looks for the unique combination of n byte grams. Different works check for $n = 1$ to $n = 8$ bytes, while Tabish et al. [9] worked on byte level file content in a block-wise manner. Finally, the block-wise classification results of a given file are correlated to classify it as benign or malware. Byte code sequence requires no knowledge of format of file and is harmless in nature.

Opcode Sequence Opcodes and byte sequences can be interchangeably used for malware detection. But, the main advantage of opcode over byte code in malware

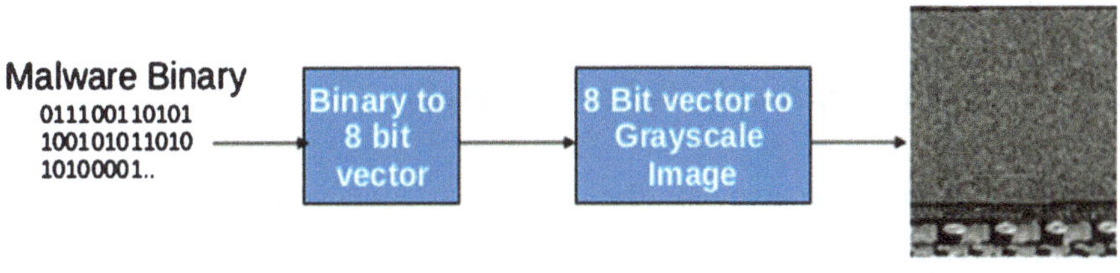

Fig. 2 Malware as 2D image

detection is that, opcode is efficient in detecting obfuscated and metamorphic malware. Malware executable and benign executable have different frequency distribution of opcodes (say mov, push). The similarity degree of two executable is compared based on these features. By comparing opcode distribution in malicious and non-malicious samples, detection and differentiation of advanced malware can be done.

Akkas et al. [10] performed assembly analysis on the ransomware samples and succeeded to figure out how the first files are created and how user's files are encrypted and also were able to identify the beginnings of the threads that encrypt the data.

Bilar [11] does the most significant research on OpCodes. His works proved that single OpCodes can be used as a feature in malware detection. For that, he analyzed the capability of single OpCodes statistically and demonstrated their reliability to determine the maliciousness of an executable. He also proved that OpCodes can be used as a powerful representation for executable files.

In [12] I. Santos et al. proposed a method which relies on the frequency of presence of opcode sequences. This approach is not effective for packed malware.

Runwal et al. [13] proposed another method that can be used for detecting unknown as well as metamorphic malware families by comparing closeness of simple graph. For that, they extracted OpCodes from malware and benign types, and occurrence frequency is noted for each pair opcode. This value is used to construct graph and is used to predict the maliciousness of a new executable by calculating the closeness of graph. Frequency distribution and graph construction from opcodes are two efficient approaches that can easily detect obfuscated malware samples.

Portable Executable Portable Executable (PE) is a file format of executable files in Windows Operating System. Most of the virus reported so far belong to PE type [14]. Viruses like CodeRed, Killonce, CIH, CodeBlue, Nimda, Sobig, Sircam, and Love Gate aim at PE files. The essential information used to load a PE file is contained in DOS MZ header and all PE files start with it. The PE file comprises of multiple sections and each section contains data with common attributes. PE parser extracts the APIs called by a PE file from the import table. For static extraction of API execution calls, PE parser extracts a 32-bit unique global API ID.

Ye et al. [14] proposed a system resting on the analysis of Windows APIs invoked by PE files, and using Objective-Oriented Association (OOA) mining developed an Intelligent Malware Detection System (IMDS) for classification.

2.2 *Dynamic Approach (Behavior Based)*

As static approach captures only the information available in the malware executable, it can be easily evaded using simple obfuscation techniques. Hence, Dynamic approach which analyses the malware at run time can be used to analyze the specific behavior of malware [1, 15]. Behavior based analysis can be efficiently used to detect several families of malware by inspecting what it does rather than what it says. It involves more complex tasks in acquiring and extraction of dynamic features from malware logs. But still dynamic approach is efficient in detection. These mechanisms help in detecting the program that generates new mutants continuously.

API Call Graph API, Application Programming Interface, is used by almost all programs to send requests to operating system. One of the most attractive way that represents the malware behavior is by API calls. Hofmeyr et al. were the first to use sequences of API call for constructing feature set of malware [16]. They used system call sequences for performing anomaly based detection. Short system call sequences were used to make the behavior profile of normal behavior. Hamming distance values above user-specified threshold are reported as anomalies. Afterward, an extensive research was made on using API calls by Bergeron et al. [17], Sekar et al. [18], and Sung et al. [19], etc. Even though it performed well in malware detection, there have been two main problems such as Handling of large set of rules for constructing the classifier and Finding effective rules to classify new file samples. By using post processing techniques of associative classification the above two problems were overcome by, Ye et al. [20]. In [21], Chi-squared testing and insignificant rule pruning are applied initially, followed by database coverage by rule ranking mechanism based on the Chi-square measure and Pessimistic error estimation. The best first rule is used for final prediction. CIDCPF is in-cooperated into existing IMDS system and generates CIMDS [20]. Post processing is used for the first time in malware detection for associative classification.

Code graph called topological graph is built from malicious and benign executables by Jeong and Lee [22] from API calls. The main drawback of this approach is that the code graph tends to be too large. Hence, the size of code graph is reduced by Lee et al. [23] by classifying API calls into 128 groups.

Control Flow Graph If a graph can be drawn representing the control flow of a program, it can be used for analyzing the behavior of the program. A directed graph called Control Flow Graph is used for the same. Each node in the graph represents the statement of the program and the edge between the nodes represents control flow between the statements. The statements can be either assignment statement in the program, copy statement, branches, etc. J. Lee et al. in [23] represent the malware detection as sub-graph isomorphism problem. A set of normalization operation is performed on the executable after disassembling it. It can reduce the effect of mutation techniques and can unveil the flow connections between malicious and benign code. As shown in Fig. 3 the corresponding CFG is generated. Newly generated CFG is compared against the CFG of a normalized malware to check the presence of sub-graph which is isomorphic to CFG of the normalized one.

Bonfante et al. [24] use CFG as signature for detecting malware. Each assembler is composed of four types of instructions, namely conditional jumps (jcc), function calls (call), non-conditional jumps (jmp), and function returns (ret). Any contiguous sequence of instructions is abstracted between nodes named "inst" and "end." So, they defined six types of node: inst, call, jmp, ret, jcc, and end. The CFG based on

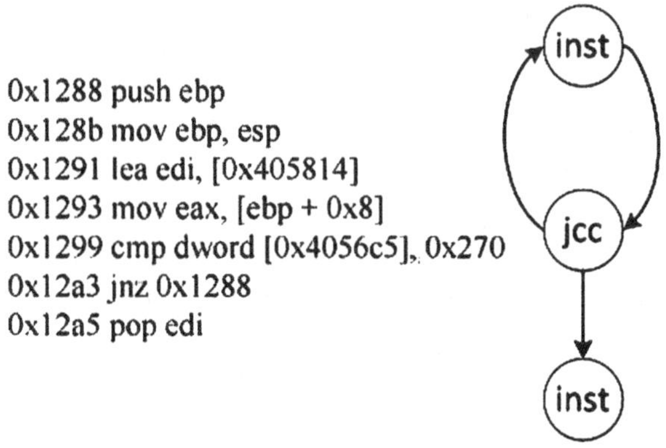

Fig. 3 CFG Extraction proposed by Bonfante et al. [24]

these types is constructed as illustrated in Fig. 3. Then, these nodes are reduced and are used as a signature for each file.

Network Analysis Dynamic malware analysis also monitors the network level activity of the malware. It tries to observe and capture the messages and packets send between the network and malware. Malware from ransomware families tends to communicate with the command and control server for its operation like encryption. This malicious traffic can be obtained using network analysis of malware. Network analysis operates over different OSI protocols like TCP/IP, HTTP, UDP, etc. Network analysis tools like Wireshark are used for this purpose. Some programs try to send and receive data requests from different IPs. These IPs would be TOR exists usually used in case of ransomwares. As specified in [25] malware analysis of ransomware includes this technique. They try to connect to several websites which are unsuccessful to connect from normal browsers. These anomalous behavior can be used to distinguish between the malware activities. Vigneswaran et al. [25] Several other intrusion detection systems try to find the anomalous behavior by using KDD data set for normal network operations and current values of these variables are used for classification.

2.3 Honeypot Based Approach

Honeyfiles are trap files employed in the deception environment in order to track and trap the malware with specific behavior.

Cryptostalker, a real time detection tool based on file system activity for windows and linux [26], when more than a specific number of files are within a time interval, it generates an advertisement. For earlier detection Pingree [27] proposes a technique. Another variant of this technique relies on the admission of the cyber kill-chain model by Hutchins et al. [28], in which an attack can occur if each step of the chain is executed sequentially.

In particular, the use of honeyfiles is an advisable mechanism in the phases of "permanence-exfiltration" and "lateral movement" in case of Ransomwares as in [29]. A real tool developed called Anti Ransom for Windows platforms [30]. In order to prevent data from ransomware and other malicious apps in Windows 10, Microsoft introduced a control folder access in [31]. R-Locker by Gomez-Hernndez et al. in [32] proposes a novel approach which in addition to detection it also thwarts the malicious activity and is specific for Unix platform. It deploys a FIFO like structure rather than normal file and can completely block the program accessing it. The cost and complexity of this solution is really low and does not impede with the normal operation of the environment. It does not require previous knowledge or training, and is efficient in fighting against unknown, zero-day attacks.

2.4 *Hybrid Approach*

Hybrid approach as the name suggests contains combination of several traditional approaches such as static, dynamic, etc. There are many works done on the grounds of ransomware with hybrid features in cooperating static and dynamic approaches. There have been several works done by using static and dynamic feature set. Ahmadian and Shahriari [33] in 2entFOX capture static and dynamic features of highly survivable ransomwares and designed a detection system using Bayesian belief network which uses statistical possibilities of the extracted features. The feature sets used in this system can be increased or decreased according to the security countermeasures and every module of this framework can be used in other driven systems too.

RansHunt proposed by Hasan and Rahman [34] is an Analysis Framework based on Support Vector Machines uses integrated feature set by integrating static and dynamic features. Another efficient layered approach developed by Shaukat and Ribeiro [2] is RansomWall, a Layered Defense System against Ransomware Attacks using Machine Learning. It employs a layered defense system which incorporates static analysis engine in the initial layers followed by honeyfiles and then dynamic analysis engine. It has a file backup layer to replace the modified files. It is one of the promising approaches for the early detection of ransomware.

3 Comparison

Tables 1 and 2 compare between different approaches in Static Malware analysis, Dynamic Analysis, Mitigation using Honeypot and Hybrid Approach. The comparison table is splitted as it does not fit on a single page. It also discusses the advantages and disadvantages associated with each methodology. Signature based solutions which uses opcodes, byte codes, PE headers, etc. can be used to detect malware. They are easy to implement with minimum cost and time but cannot overcome

Table 1 Comparison of different static approaches

Methodology	Refs.	Advantages	Disadvantages
Image representation	[3–7]	– Fast and Easy detection – Easily finds polymorphic code – Techniques like SIFT enable to efficiently find malware signature – Obtains large number of feature set	– Malware binary has to be converted to fixed sized images – Some portion of image would be pruned due to large size – Increases computational complexity
Byte code sequence	[8, 9]	– Easily obtained from malware executable – Feature set generation is also easy task	– Not efficient for obfuscated byte code – Large number of features
Opcode sequence	[10–13]	– Efficient than Byte code – Ability to detect obfuscated and metamorphic malwares – Different frequency distribution of opcodes enables detection easier – Reduces the false positive rate	– Takes into account only the opcode frequency – Information regarding malware behavior is not considered – It is difficult to evaluate. – Imbalance datasets
Portable executable	[14]	– PE contains implementation information of the executable – Can keep track of the API calls – Detects malware before their execution – Detects previously undetectable malicious executable – Detects borderline binaries	– Applicable only for windows executable

obfuscated malwares. In response to this, dynamic approaches were proposed, which analyze the program behavior during execution. They require a large amount of resources and have a substantial overhead on the system. Honeyfiles can be easily deployed and are cost efficient but can be applied to malware of specific types. Hybrid methodology combines the advantages of both static and dynamic approaches. Even though there are overhead on the system, the efficiency of these systems is very high.

Table 2 Comparison of dynamic, honeypot and hybrid approach

Methodology	Refs.	Advantages	Disadvantages
API call graph	[16–23]	– Detects polymorphic and unknown malware – Fewer false positives than other scanners – Outperforms other classification approaches in both detection ratio and accuracy – Detects metamorphic malware – Outperforms other classification methods in terms of performance and efficiency – Generates semantic signature	– Large set of generated rules for building classifier – Only provides binary predictions – Large size of graph for comparison
Control flow graph	[23, 24]	– Detects metamorphic malwares – High detection ratio – Low false positive rate	– Did not compare the efficiency of its algorithm with other techniques – Did not evaluate false negatives
Network analysis	[25, 25]	– Real time detection of network flow – Able to obtain specific malicious behavior – Enable to prevent malware propagation	– Cannot obtain information if highly sophisticated networks are used – Difficult in handling data of different formats
Honeypot method	[26–32]	– Simple implementation – Reduces time and space complexity – Requires no training – Detects and prevent malware	– Can only be used for specific malware types
Hybrid method	[2, 33, 34]	– Improves the accuracy of malware detection effectively – Low false positive ratio	– Increased time complexity

4 Conclusion

In this survey we discussed about several malware detection techniques employed so far especially for ransomware and proposed a unique classification scheme for malware detection techniques. The objective of the survey is to provide a procedure, which could be suitable for further studies and to develop malware detection techniques. Since there are traditional approaches that are employed still for malware detection, this survey focused on various other heuristic methods that are successfully applied for malware detection. It also provides an insight of different detection techniques that can be employed based on the application and behavior of malware under consideration.

References

1. Homayoun, S., Dehghantanha, A., Ahmadzadeh, M., Hashemi, S., Khayami, R.: Know abnormal, find evil: frequent pattern mining for ransomware threat hunting and intelligence. IEEE Trans. Emer. Topics Comput. **8**(2), 341–351 (2017). https://doi.org/10.1109/TETC.2017.2756908
2. Shaukat, S.K., Ribeiro, V.J.: RansomWall: a layered defense system against cryptographic ransomware attacks using machine learning. In: IEEE 10th International Conference on Communication Systems Networks (2018). https://doi.org/10.1109/COMSNETS.2018.8328219
3. Nataraj, L., Yegneswaran, V., Porras, P., Zhang, J.: A comparative assessment of malware classification using binary texture analysis and dynamic analysis. In: Proceedings of the 4th ACM Workshop on Security and Artificial Intelligence (2011). https://doi.org/10.1145/2046684.2046689
4. Nataraj, L.: Malware Images: Visualization and Automatic Classification. Vision Research Lab, University of California, Santa Barbara (2011). https://doi.org/10.1145/2016904.2016908
5. Choi, S., Jang, S., Kim, Y., Kim, J.: Malware detection using malware image and deep learning. In: 2017 International Conference on Information and Communication Technology Convergence (ICTC), pp. 1193–1195 (2017). https://doi.org/10.1109/ICTC.2017.8190895
6. Gibert, D.: Convolutional Neural Networks for Malware Classification. A thesis presented for the degree of Master in Artificial Intelligence, Universitat de Barcelona (UB) (2016)
7. Luo, J.-S., Lo, D.C.-T.: Binary malware image classification using machine learning with local binary pattern. In: IEEE International Conference on Big Data (BIGDATA) (2017). https://doi.org/10.1109/BigData.2017.8258512
8. Raff, E., Zak, R., Cox R., Sylvester, J., Yacci, P., Ward, R., Tracy, A., McLean, M., Nicholas, C.: An investigation of byte n-gram features for malware classification. J. Comput. Virol. Hacking Tech. **14**, 1–20 (2018). https://doi.org/10.1007/s11416-016-0283-1
9. Tabish, S.M., Shafiq, M.Z., Farooq, M.: Malware Detection using statistical analysis of byte-level file content. In: Conference Proceedings of the ACM SIGKDD Workshop on Cyber Security and Intelligence Informatics, Paris (2009). https://doi.org/10.1145/1599272.1599278
10. Akkas, A., Chachamis, C.N., Fetahu, L.: Malware Analysis of WanaCry Ransomware. https://courses.csail.mit.edu/6.857/2017/project/20.pdf
11. Bilar, D: Opcodes as predictor for malware. Int. J. Electron. Secur. Digit. Forensics **1**(2), 156–168 (2007). https://doi.org/10.1504/IJESDF.2007.016865
12. Santos, I., Brezo, F., Nieves, J., Penya, Y.K., Sanz, B., Laorden, C., Bringas, P.G.: Idea: opcode-sequence-based malware detection. In: International Symposium on Engineering Secure Software and Systems, pp. 35–43 (2010). https://doi.org/10.1007/978-3-642-11747-3_3
13. Runwal, N., Low, R.M., Stamp, M.: OpCode graph similarity and metamorphic detection. J. Comput. Virol. **8**(1–2), 37–52,(2012). https://doi.org/10.1007/s11416-012-0160-5
14. Ye, Y., Wang, D., Li, T., Ye, D., Jiang, Q.: An intelligent PE-malware detection system based on association mining. J. Comput. Virol. **4**(4), 323–334 (2008). https://doi.org/10.1007/s11416-008-0082-4
15. KALPA, Introduction to Malware. http://securityresearch.in/index.php/projects/malware_lab/introduction-to-malware/8/
16. Hofmeyr, S., Forrest, S., Somayaji, A.: Intrusion detection using sequences of system calls. ACM J. Comput. Secur. **6**(3), 151–180 (1998)

17. Bergeron, J., Debbabi, M., Desharnais, J., Erhioui, M.M., Tawbi, N.: Static detection of malicious code in executable programs. Int. J. Req. Eng. **2001**(184–189), 79 (2001)
18. Sekar, R., Bendre, M., Bollineni, P., Dhurjati, D.: A fast automaton-based approach for detecting anomalous program behaviors. In: Proceedings 2001 IEEE Symposium on Security and Privacy (2001). https://doi.org/10.1109/SECPRI.2001.924295
19. Sung, A.H., Xu, J., Chavez, P., Mukkamala, S.: Static analyzer of vicious executables. In: IEEE 20th Annual Computer Security Applications Conference, pp. 326–334 (2004). https://doi.org/10.1109/CSAC.2004.37
20. Ye, Y., Li, T., Jiang, Q., Wang, Y.: CIMDS: adapting postprocessing techniques of associative classification for malware detection. IEEE Trans. Syst. Man Cybern. C **40**(3), 298–307 (2010). https://doi.org/10.1109/TSMCC.2009.2037978
21. Snedecor, W., Cochran, W.: Statistical Methods, 8th edn. Iowa State University Press, Iowa City (1989)
22. Jeong, K., Lee, H.: Code graph for malware detection. In: Information Networking. ICOIN. International Conference, pp. 1–5 (2008). https://doi.org/10.1109/ICOIN.2008.4472801
23. Lee, J., Jeong, K., Lee, H.: Detecting metamorphic malwares using code graphs. In: Proceedings of the ACM Symposium on Applied Computing, pp. 1970–1977. ACM, New York (2010). https://doi.org/10.1145/1774088.1774505
24. Bonfante, G., Kaczmarek, M., Marion, J.Y.: Control Flow Graphs as Malware Signatures. WTCV (2007)
25. Vigneswaran, K.R., Vinayakumar, R., Soman, K.P., Poornachandran, P.: Evaluating shallow and deep neural networks for network intrusion detection systems in cyber security. In: Ninth International Conference on Computing, Communication and Networking Technologies (ICCCNT), Bengaluru (2018). https://doi.org/10.1109/ICCCNT.2018.8494096
26. Cryptostalker. https://github.com/unixist/randumb#cryptostalker-example
27. Pingree, L.: Emerging Technology Analysis: Deception Techniques and Technologies Create Security Technology Business Opportunities. https://www.gartner.com/doc/reprints?id=1-2LSQOX3&ct=150824&st=sb&aliId=87768
28. Hutchins, E.M., Cloppert, M.J., Amin, R.M.: Intelligence-driven computer network defense informed by analysis of adversary campaigns and intrusion kill chains. Lead Issues Inf Warf Secur. Res **1**, 1–14 (2011)
29. Moore, C.: Detecting ransomware with honeypot techniques. In: Cybersecurity and Cyberforensics Conference, pp. 77–81 (2016)
30. Yago, J.: Security Projects: Anti Ransom (2017). http://www.security-projects.com/?Anti_Ransom
31. GBH: Microsoft introduced a control folder access to prevent data from ransomware and other malicious apps and threats in Windows 10 insider release
32. J.A. Gmez-Hernndez, Alvarez-Gonzlez, L., Garca-Teodoro, P.: R-locker: thwarting ransomware action through a honeyfile-based approach. Comput. Secur. 73, 389–398 (2018)
33. Ahmadian, M.M., Shahriari, H.R.: 2entFOX: a framework for high survivable ransomwares detection. In: 13th International ISC Conference on Information Security and Cryptology (2016). https://doi.org/10.1109/ISCISC.2016.7736455
34. Hasan, M.M., Rahman, M.M. A support vector machines based ransomware analysis framework with integrated feature set. In: 20th International Conference of Computer and Information Technology (ICCIT) (2017). https://doi.org/10.1109/ICCITECHN.2017.8281835

A Survey on Online Review Spammer Group Detection

A. Thahira and S. Sabitha

1 Introduction

Nowadays, various e-commerce websites (e.g., Amazon, Flipkart, Myntra, and etc.) are available to people for purchasing a product. These sites provide customers with the opportunity to write their opinion about the product they purchased. These opinions are called reviews of a particular product. Various sites are also available to provide reviews about services (e.g., Yelp). In recent years, the customers are very much influenced by these online review systems for making purchasing decision, that is, whether to buy a particular product/service. By using these reviews, people evaluate the quality of the products/services. These review systems are also used by manufacturers to improve the quality of their product/service.

1.1 Online Review Structure

Typically, an online review contains the review title, review text, review star rating, review date and time, the product's unique identifier, the customer's unique identifier, a helpfulness vote, and a verified purchase tag. These components are called review metadata. Besides these components, an online review also contains the IP address or MAC address of the reviewer, which is private information not shown in the review system.

A. Thahira (✉) · S. Sabitha
College of Engineering Trivandrum, Thiruvananthapuram, India
e-mail: sabitha@cet.ac.in

M. Palesi et al. (eds.), *Second International Conference on Networks and Advances in Computational Technologies*, Transactions on Computational Science and Computational Intelligence, https://doi.org/10.1007/978-3-030-49500-8_2

1.2 Review Spam

Recently, this review system has been controlled by fraudulent reviewers. They may be individuals or organized groups (group spammers); they are selected by product manufacturers who want to promote their products or degrade competitors' products by writing positive reviews or negative reviews respectively.

1.3 Review Spam Detection

People are very much influenced by reviews for online purchasing. Therefore, detecting these spamming activities on review systems is of utmost importance in ensuring consumer trust in online purchasing, as well as retaining the authenticity of online reviews. Over the past 10 years, various approaches to review spam/review spammer detection have been developed. The various detection approaches can be categorized in two ways, **based on detection targets** and **based on spamming features used**. Based on detection targets, detection methods are categorized as detection of review spam, detection of review spammers, and detection of review spammer groups.

In the three detection targets, the most frequently explored topic is the detection of review spam written by fake reviewers. Based on textual details given by the reviewers, Liu et al. [1] categorized low-quality and high-quality reviews. However, they do not focus on detecting spam activity in reviews. Jindal et al. [2] first proposed an approach to review spam detection and using Amazon review datasets for analysis.

The detection of review spammers refers to the detection of reviewers who write fake reviews to deceive customers. Some work [3, 4] detected both review spam and review spammers on a unified framework. They used a graphical framework, which utilizes the relationship between reviews, reviewers, and products.

Some recent studies have focused on detecting review spammer groups. In order to make a great impact on reviews and earn more profit, review spammers work in collaboration. A review spammer group is a group of reviewers working collaboratively to post fake reviews from multiple entities. Mukherjee et al. [5] first introduced a review spammer group detection framework using frequent itemset mining (FIM). Most of the recent works are based on graph-based detection, because review spammers may leave behind more clues regarding their collaborative abnormal behavior in the targeted network [6–9].

In review spam detection, feature extraction is considered the most crucial step. An accurate and efficient detection framework needs a minimal and effective feature set that best describes the spamming characteristics of fake reviews. Based on spamming features used for the detection of review spam, methods are categorized as textually based, behaviorally based, and relationally based.

Textual features focus on a reviewer's writing styles and language to detect review spam. Behaviorally based features related to the review metadata, or features associated with the behavior of the reviewer. Spammers can sometimes easily imitate textual and behavioral patterns, but find it very hard to mimic the network structure of genuine reviewers. Recent studies [3, 4, 6–8, 10–12] focused on the relationally based approach to review spam detection. Relational features play a more important role in the detection of review spammer groups, because these are the only possible indicators to spot highly suspicious review spam groups. Most of the graph-based approach to review spam detection is based on relational characteristics. Few of the works [4] combined these three features to spot review spam more effectively.

In the review spam detection domain, review spammer group detection is not so extensively addressed. But it is the most frequently explored topic nowadays, because group review spammers occur more frequently and are more harmful than individual review spammers. This paper highlights the taxonomy of review spammer group detection techniques. The rest of the sections are designed as follows: Sect. 2 describes the taxonomy of review spammer group detection techniques. Section 3 describes the various group spam features used in existing detection approaches. Section 4 evaluates the performance of existing approaches. Section 5 highlights some future research directions, followed by the conclusion.

2 Taxonomy of Review Spammer Group Detection Techniques

In order to fully control the sentiment of the target products, split total effort, and camouflage (i.e., hiding their spamming behavior by arranging some group members to review irrelevant products or review normally to a mislead spam-detecting technique), review spammers (fake reviewers) often work collaboratively [7]. These review spammer groups are highly damaging compared with individual review spammers. The group of reviewers means a set of reviewer IDs. It may be a single a person with multiple IDs, multiple persons, or a combination of both. Studies have shown that individual review spammer detection is not adequate for review spammer group detection. Mukherjee et al. [5] first introduced review spammer group detection using the FIM technique.

Basically, the review spammer group detection technique contains two steps.

- Find the candidate review spammer groups from a review dataset
- Rank/classify the candidate review spammer groups to find real spammer groups

Figure 1 shows the taxonomy of review spammer group detection techniques.

The various pieces of research on review spammer group detection can be grouped, based on the method used for candidate spammer group detection. They include FIM, cosine pattern mining, and graph-based approaches.

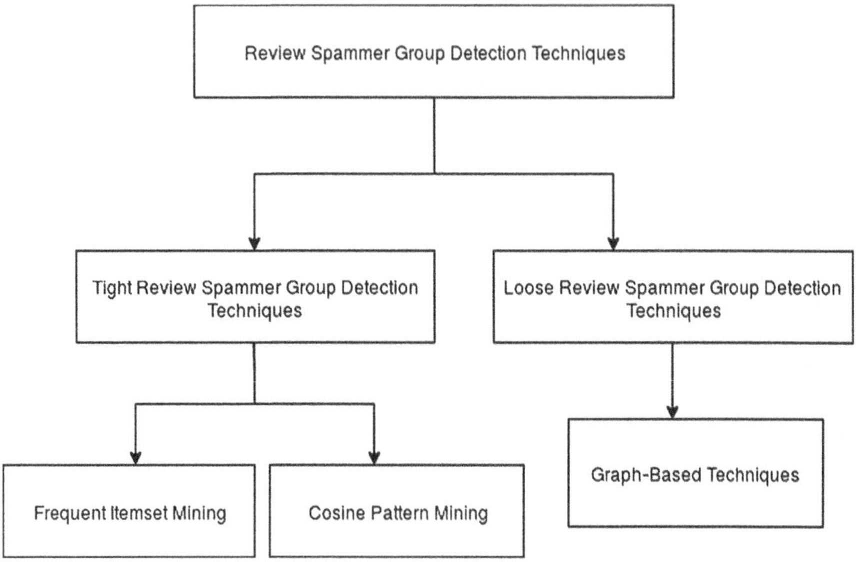

Fig. 1 Taxonomy of review spammer group detection techniques

2.1 Tight Spammer Group Detection Techniques

Tight spammer group refers to a group in which each group member has to review all the target products. In the existing approaches, they use FIM [5, 10, 13, 14] or cosine pattern mining [12] techniques to detect tight spammer groups.

Frequent Itemset Mining (FIM) Some of the existing works [5, 10, 13, 14] used FIM to detect review spammer groups. FIM can be used to find a group of reviewers working together on multiple target products. They first used FIM to detect a candidate spammer group and then based on group spam features ranked the candidate review spammer groups.

The concept of the detection of review group spamming was first introduced by Mukherjee et al. [5] in 2012. First, they used FIM to find candidate spammer groups. In this step, they produced a set of transactions from extracted review data. Each transaction represents a unique product and contains all reviewers (their IDs) who have reviewed that product. Based on all transactions, they performed FIM. The resulting frequent itemsets (also called patterns) are considered as candidate review spammer groups. Next, they computed spam indicator values to detect true review spammer groups from candidate review spammer groups. Finally, they ranked the detected candidate review spammer group based on the calculated spam indicator value. They used ICF to learn and produce the final ranking of the candidate groups.

Xu et al. [13] detected collusive spammers on Chinese review websites (amazon.cn). First, they used FIM to create a colluders' dataset. Then, they calculated textual and behavioral (individual and collusive) spam features. Finally, two approaches are used to detect colluders—the KNN-based method and the generalized graph-based method. In the KNN-based method, they computed the similarity of two reviewers based on the similarity of their corresponding groups and then k most similar reviewers are selected. In the graphical classification method, they used transaction correlations across reviewers (if a pair of reviewers reviewed at least one product within a predefined time interval), to detect colluders. The graph-based method is based on a pairwise Markov network and an approximate inference algorithm (ICA).

Zhang et al. [10] used a statistical model for collusive spammer detection. They argued that spamming features provided by Mukherjee et al. [5] are not efficient enough to detect collusive activities in smaller groups. Thus, they proposed measures such as rating consistency, consistency in targeted businesses, temporal synchronization, activity similarity, workload similarity, and first-review synchronization to identify smaller collusive spamming groups. First, they used FIM to detect a candidate spammer group. Then they used LCM in an unsupervised way to detect real spammer groups.

However, there are many drawbacks to using FIM to generate candidate groups. These include: (1) By FIM, only groups working on at least three products can be detected. However, groups that focus on promoting or demoting one or two products are also usual in a real scenario. (2) FIM-based methods do not consider a time window when creating candidate spammer groups. However, a time window is a more important factor in detecting spammer group activity because for maximum profit gain they need to finish their group spamming activity within a specific time limit. Thus, the candidate groups generated by FIM may be of lower quality. (3) FIM can only detect tight spammer groups. Nowadays, to avoid detection, the review spammers adjust their spamming activities very cleverly and carry out their spamming activities loosely. However, FIM cannot detect loose spammer groups.

Cosine Pattern Mining Zhang et al. [12] employed cosine pattern mining (CPM) to detect tight candidate spammer groups. Cosine threshold is used as the measure of tightness and less coincidentally generated groups are extracted. To mine the cosine pattern from the product–reviewer relationship, they used an frequent pattern (FP)-growth-like algorithm. Then they used a heterogeneous information network (HIN)-based classification method to detect real spammer groups from detected candidate spammer groups. The HIN-based approach is successfully used in [15] to classify reviews as spam or nonspam.

2.2 Loose Spammer Group Detection Techniques

Loose spammer groups refer to groups in which reviewers are not required to review each target product in some group spam campaigns. Existing approaches used graph-based algorithms [6–9] to detect loose spammer groups.

Fig. 2 The bipartite representation of the review dataset

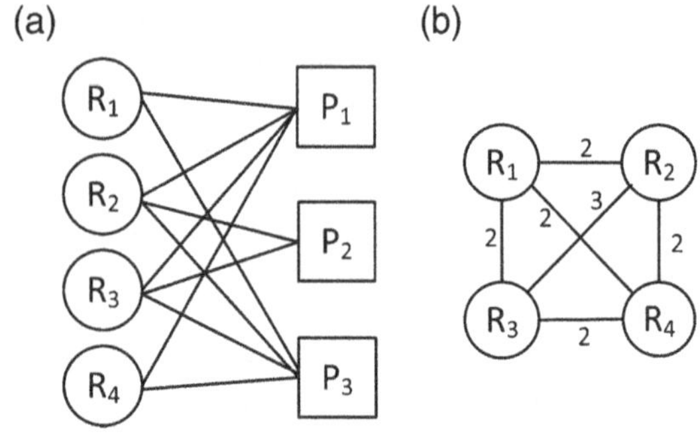

Using Graph-Based Methods Studies have shown that the detection of review spam (review, reviewer, or review spammer group) using textually or behaviorally based approaches is not as efficient as the relationally based approach because textual and behavioral characteristics are relatively easier to mimic by review spammers than network-based characteristics. Also some studies have shown that group spamming contains more spam clues than individual review spamming when considering relational characteristics; hence, group spamming is easier to identify with relational characteristics. In most of the graph-based approaches, the input is the bipartite (product, reviewer) or tripartite (product, reviewer and review) representation of the review data. Figure 2a shows the bipartite representation of the review data and Fig. 2b shows the corresponding reviewer graph.

In [3] and [4], the graph-based approach to reviewing spam detection was proposed. In [3], a Loopy Belief Propagation-based (MRF) inference algorithm was used, which is based on the network relationship. Rayana and Akoglu [4] was an extended version of [3], which included some meta information (e.g., star ratings, timestamps, and review content) and they showed better improvement. In their approach, we can detect spammer groups by performing a graph-clustering algorithm on a subgraph, which contains top-ranked reviewers and corresponding target products.

Ye et al. [6] proposed an unsupervised approach to spotting spammer groups. They used a reviewer–product bipartite graph for review spammer group detection and developed a two-step method in which first they identified target products and based on the top-ranked products, they next identified review spammer groups. For identifying target products, they quantified the extent to which network characteristics of a reviewer are manipulated by spamming activities by a measure called network footprint score. To calculated the network footprint score they considered neighbor diversity and self-similarity measures. After the computation of a network footprint score for products, they identified review spammer groups from induced subgraphs using a graph-clustering algorithm. To speed up the identification of similar clusters, they used locality-sensitive hashing (LSH).

Wang et al. [7] proposed a graph-based approach to detect loose review spammer groups via bipartite graph projection. They first represented the input review dataset as bipartite graph and then from this graph they created a reviewer graph. From the reviewer graph, they identified the review spammer group using divide and conquer strategy. But, this approach didn't consider rating score deviation between reviewers, which is one of the key factor for describing the collusiveness between two reviewers. Because, reviewers can't fake on rating score.

In [8], when creating a reviewer graph from a bipartite graph, the authors used the review time interval and the rating score deviation as the measures of collusiveness between the two reviewers. Then they used a divide and conquer algorithm to detect candidate review spammer groups. Finally, they ranked the candidate spammer groups based on the spamicity score.

Gu et al. [9] proposed an unsupervised LDA-based model for review spammer group detection. They used LDA in the product–cluster–reviewer concept. First, they created LDA clusters from review data. Then, reviewer graphs were created for each cluster and candidate review spammer groups were detected using the SCAN [16] algorithm. Finally, for each candidate group they computed the spamicity score and outputted review spammer groups based on this spamicity score value.

3 Group Spam Features Used in Existing Approaches

For the efficient detection of review spammer groups, a minimal and effective set of group spamming features are necessary. The various group spamming features used in the existing frameworks are the group time window (GTW), group deviation (GD), group early time frame (GET), group size ratio (GSR), group size (GS), group support count (GSC), group rating variance (GRV), group product reviewer ratio (GPR), group content similarity (GCS), and group member content similarity (GMCS). GTW, GD, GET, and GRV are behavioral features. GCS and GMCS are textual features. GSR, GS, and GPR are relational features.

GTW and GET are based on the review posting time. GD and GRV are based on the review rating score. The reviewing time- and rating score-related features play the most important roles in spotting spamicity compared with other features. Group spammers are short-term members and they post reviews early (at the time of the product launch) to make a great impact. Also, the group spammers either give a higher rating score for their product to promote or give a lower rating score for their competitors' product to demote. Thus, the spammers are not able to fake these two metadata. Recent works exclude textual features, because they often underperform in discriminating spammer/nonspammers. Behavioral features better reflect the group spamming activity.

4 Performance Analysis and Summary

Out of three approaches to review/reviewer spam detection (review, individual fake reviewer, or group review spammer), most of the works are focused on review spam detection. Group review spamming is least frequently explored. Studies have shown that, it is easier to detect review spammers than review spams. Because review spammers write multiple reviews, we get more information on spammer detection than the limited information from a review.

A main issue in the area of review spam detection is the absence of a huge and reliable ground-truth dataset. Based on practical experiences and the literature, the labeling of spam reviews is very difficult. However, the labeling of group review spammers is not very difficult when compared with review spams because of the clarity in the member's spamming behaviors. If an unsupervised method is used, human evaluation is also needed for the ranking or classification of spammer groups.

In spam feature-based approaches, text-based approaches are not accurate for current review spamming scenarios because the spammers are very clever at mimicking the textual characteristics of the original reviewer. Thus, in recent works, text-based features are not used for review spam detection. Studies have shown that a combination of behavioral and relational features outperforms well. Relational features are very helpful in the area of detection of group review spamming.

In a group review spamming scenario, existing approaches do not detect the type of group spamming (whether it is a single person with multiple IDs or multiple persons). It shows a research gap, but it is very hard to detect, because online purchasing sites provide only limited information about their reviewers. Based on the studies, the distribution of various research articles on review spammer group detection from 2012 to 2018 is shown below. Figure 3 shows the distribution of detection researches based on the techniques used for spammer group detection. Each slice represents the percentage of research focused on a particular detection technique. Earlier studies are mainly based on FIM. Very few work used cosine pattern mining technique for spammer group detection. In order to find review

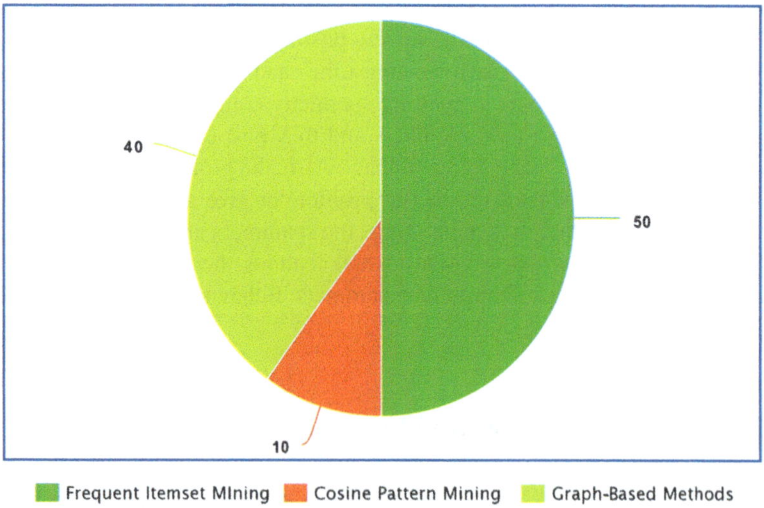

Fig. 3 Division of review spammer group detection research according to detection techniques

spammers in current review spamming scenarios, most of the recent research is based on the graph-based method. Compared with FIM-based approaches, graph-based approaches provide higher performance, but have higher computational complexity.

Table 1 summarizes the various research on review spammer group detection. It contains the methodology used by existing works, spamming features used, advantages, and disadvantages of the works. In most of the works, the performance is measured in terms of AUC, F1-score, and MCC. Studies have shown that the graph-based approach with time window- and rating score-related parameters as the measures of collusiveness give more accurate results in the current review spamming scenario. After the finding of candidate groups using any of the methodologies explained, feature extraction is the most important step and accuracy of the spammer group heavily depends on these features. Finally, the ranking or classification of the review spammer groups depend on reliable ground truth datasets, which also affect the accuracy of the spammer groups. Using graph-based algorithms and some relationally and behaviorally based features help in the efficient detection of these spammer groups from the highly suspicious, abnormal patterns of the graphs.

The spamming scenarios advance over the time; thus, a single approach to detecting all kinds of group review spammers with high precision is infeasible. Only a few works concentrated on group review spamming. Therefore, there is still a research gap, which is specified in Sect. 5.

5 Future Research Directions

- Develop a hybrid model to detect review spams and review spammer groups
- Combine various relationally based techniques in an appropriate way to improve the efficiency of detection
- Obtain an optimal set of group spamming features to improve the accuracy of detection
- Develop a distributed spam review detection approach to processing large review datasets efficiently.

6 Conclusion

This paper presents a taxonomy of review spammer group detection techniques and a performance analysis of these methods are included. Graph-based approaches show better performance than FIM and cosine pattern mining. The group spamming features used in several approaches are also discussed in this paper. Behavioral features especially the rating score and time-related features provide better detection accuracy. Textually based features do not contribute any improvement in the detection of review spammers in the current review spam scenarios.

Table 1 Comparison of existing approaches to review spammer group detection

Methodology	Refs.	Spamming features	Advantages	Disadvantages
FIM	[5, 10, 13, 14]	Textual and behavioral features	– Outperforms over supervised classification algorithms	– Low-quality candidate groups – Does not detect loose spammer groups – High false-positives – Does not consider time window for candidate spammer group detection
Cosine pattern mining	[12]	Behavioral group spam features	– Low false-positives – Low support count value needed compared with FIM	– Low-quality spammer groups – Does not detect loose spammer groups – Need for manually labeled dataset – High computational complexity
Graph-based approaches	[6–9]	Structural and behavioral group spam features	– High accuracy – Detects both loose and tight spammer groups – Outperforms other detection methods in terms of efficiency and precision	– High computational complexity

References

1. Liu, J, Cao, Y., Lin, C.Y., Huang, Y., Zhou, M.: Low-quality product review detection in opinion summarization. In: Proceedings of the 2007 Joint Conference on Empirical Methods in Natural Language Processing and Computational Natural Language Learning, Prague, pp. 334–342 (2007)
2. Jindal, N., Liu, B. Opinion spam and analysis. In: Proceedings of the 2008 International Conference on Web Search and Data Mining, pp. 219–230. ACM, New York (2008)
3. Akoglu, L., Chandy, R., Faloutsos, C.: Opinion fraud detection in online reviews by network effects. In: Proceedings of the Seventh International Conference on Weblogs and Social Media, ICWSM, Cambridge, 8–11 July 2013
4. Rayana, S., Akoglu, L.: Collective opinion spam detection: bridging review networks and metadata. In: Proceedings of the 21st ACM SIGKDD International Conference on Knowledge Discovery and Data Mining, pp. 985–994, Sydney, 10–13 August 2015
5. Mukherjee, A., Liu, B., Glance, N.: Spotting fake reviewer groups in consumer reviews. In: Proceedings of the 21st International Conference on World Wide Web. ACM, New York, pp. 191–200 (2012)
6. Ye, J., Akoglu, L.: Discovering opinion spammer groups by network footprints. In: Machine Learning and Knowledge Discovery in Databases. Lecture Notes in Computer Science, vol. 9284. Springer, Berlin, pp. 267–282 (2015)
7. Wang, Z., Hou, T., Song. D, Li, Z., Kong, T: Detecting review spammer groups via bipartite graph projection. Comput. J. **59**(6), 861–874 (2016)
8. Wang, Z., Gu, S., Zhao, X., Xu, X.: Graph-based review spammer group detection. In: Knowledge and Information Systems, pp. 1–27. Springer, Berlin (2017)
9. Gu, S., Wang, Z., Cao, J., Xu, X.: GSLDA: LDA-based group spamming detection in product reviews. Appl. Intell. **48**(9), 227–246 (2018)
10. Xu, C., Zhang, J.: Towards collusive fraud detection in online reviews. In: 2015 IEEE International Conference on Data Mining pp. 1051–1056, ICDM Atlantic City, 14–17 November 2015
11. Shehnepoor, S., Salehi, M., Farahbakhsh, R. NetSpam: a network-based spam detection framework for reviews in online social media. In: IEEE Transactions on Information Forensics and Security, pp. 1585–1595 (2017)
12. Zhang, L., He, G., Cao, J., Xu, B.: Spotting review spammer groups: a cosine pattern and network based method. Concurrency Comput. Practice Exp. **30**(20), e4686 (2018)
13. Xu, C., Zhang, J., Chang, K., Long, C.: Uncovering collusive spammers in Chinese review websites. In: Proceedings of the 22nd ACM International Conference on Conference on Information and Knowledge Management. ACM, New York, pp. 979–988 (2013)
14. Zhang, L., Wu, Z., Cao, J.: Detecting spammer groups from product reviews: a partially supervised learning model. In: IEEE Access, pp. 2559–2568 (2018)
15. Rastogi, A., Mehrotra, M. Opinion spam detection in online reviews. Interdiscip. J. Inf. Knowl. Manag. **16**(4), 1750036 (2017)
16. Xu, X., Yuruk, N., Feng, Z., Schweiger, T.: SCAN: a structural clustering algorithm for networks. In: Proceedings of the 13th ACM SIGKDD International Conference on Knowledge Discovery and Data Mining, San Jose, 12–25 August, pp. 824–833 (2007)

Detection of Deepfake Images Created Using Generative Adversarial Networks: A Review

K. Remya Revi, K. R. Vidya, and M. Wilscy

1 Introduction

Millions of digital images are being uploaded every day to the Internet due to the rising popularity of social medias like Facebook, Instagram, Twitter, etc. In most cases, the trustworthiness of these images is uncertain and hence determining the authenticity of images is a hot research topic. The communication of fake images in social medias may create undesirable issues among people. Most of fabricated news in social media contains fake or manipulated images. So it is very important to identify such image contents in order to stop propagation of false information.

The fake images can be created manually by humans using image manipulation operations like copy move [1] or image splicing [2] with the help of image editing tools. In copy move forgery, a part of an image is copied and pasted in another location of the same image and hence creating a forged image. In image splicing forgery, a composite image is generated using parts of one or more images and pasting the copied portions onto another image. Image forgery detection methods can be generalized into two: active and passive [3]. Pre-embedded details like watermark [4] or digital signature [5] are used in active detection technique. Passive techniques use image specific features to detect forged image without the help of pre-embedded details. Passive techniques are extensively explored as many of the photos circulated in the Internet not contain any pre-embedded details.

Many investigations have done in the area of image forgery detection using handcrafted feature engineering techniques. Some passive image forgery detection methods use handcrafted features like (a)Steerable Pyramid Transform (SPT) and Local Binary Pattern (LBP) [6], (b) Discrete Cosine Transform (DCT) [7],

K. Remya Revi (✉) · K. R. Vidya · M. Wilscy
Saintgits College of Engineering, APJ Abdul Kalam Technological University, Thiruvananthapuram, Kerala, India

M. Palesi et al. (eds.), *Second International Conference on Networks and Advances in Computational Technologies*, Transactions on Computational Science and Computational Intelligence, https://doi.org/10.1007/978-3-030-49500-8_3

Fig. 1 Deep fake images generated using PGGAN [17]

(c) combination of texture descriptors (LBP, Local Phase Quantization (LPQ), Binarized Statistical Image Features (BSIF), and Binary Gabor Pattern (BGP)) [8], (d) LBP and DCT [9], and (e) Rotation invariant Co-occurrences among adjacent LBPs [10]. These approaches attained a good classification performance when evaluated on image forgery datasets like CASIA v1.0, and CASIA v2.0 [11], etc.

Recently forgery detection approaches are reported which utilize data driven technique using convolutional neural network (CNN) [12]. Rao and Ni [13] designed a CNN for image forgery detection and then trained the network using patch samples of training images. Then patch based features of image are extracted from the pre-trained CNN and a fusion technique is used to combine the extracted features. Finally, the combined features are fed to a support vector machine (SVM) classifier for classification of images into original or forged. Rota et al. [14] trained a deep CNN using image blocks for image forgery classification. Zhou et al. [15] designed an image forgery detection technique using a rich model CNN (rCNN) architecture with a special block strategy.

Advancements in machine learning techniques help to create more realistic natural looking fake images and hence increased the difficulty level of fake image detection. Recently, the machine learning techniques like generative adversarial network (GAN) [16] is utilized to create deepfake images which may look legitimate and natural. Figure 1 shows the deepfake images generated using Progressive Growing GAN (PGGAN) [17].

The identification of deepfake images with human eyes is not easy. Anyone can purposefully fabricate fake news about government officials, politicians, and celebrities using deepfake images. For example, deepfake images or videos could show public officials or politicians taking bribes, showing racism or meetings with anti-national activists. Spreading of fake information may defame and hurt the sentiments of others and may lead to social and political unrest in a nation. Hence it is very important to develop deepfake image detection techniques which can validate the truthfulness of digital images shared via Internet. Therefore, the image forensic researchers are developing techniques to identify deepfake images. So in this paper, some relevant works related to deepfake image detection are investigated.

The rest of this paper is structured as follows. The general structure of generative adversarial network (GAN) is explained in Sect. 2 and the details of two commonly used GANs for generating deepfake images are given in Sect. 3. In Sect. 4, various deep fake detection methods are investigated and summarized. Lastly, the conclusions are drawn in Sect. 5.

2 Generative Adversarial Network

GAN [16] consists of two neural networks, i.e. a generator neural network (*G*) and a discriminator neural network (*D*). A general block diagram of GAN for generating fake images is shown in Fig. 2. Generator takes some random noise (n) as input and attempts to produce fake images *G*(*n*) which are similar to real image dataset (*x*), whereas the discriminator *D* aims to discriminate images generated by *G* from real images. The discriminator takes both real images and fake images as input and it estimates the probability of a sample coming from real image dataset rather than from fake images generated by *G*.

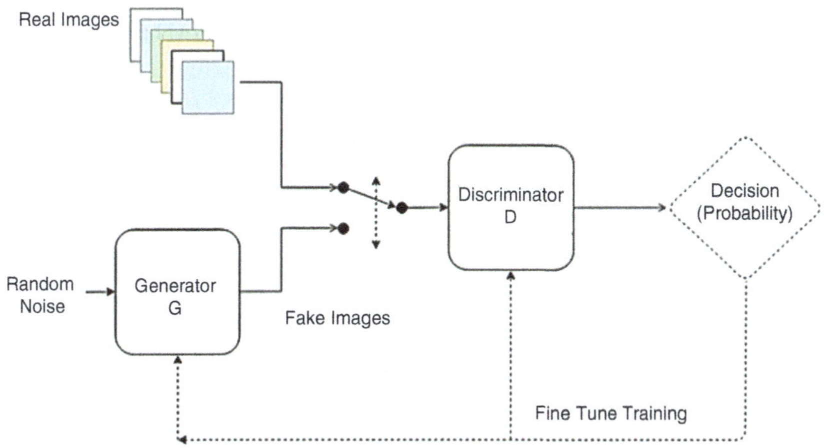

Fig. 2 General block diagram of generative adversarial network (GAN)

The discriminator will yield a probability value 1 when it is convinced an image is real and a 0 when it detects a fake image. The aim of the discriminator is to maximize the number of times it correctly classifies the type of image it receives as input; however, the generator is trying to make the discriminator less correct. Thus both networks are playing a game against each other, challenging to see who is superior at achieving their specific goal. So discriminator is trained to maximize the probability that it properly discriminates images into real or fake, while generator is trained to minimize the probability that fake images generated by it are determined by discriminator as fake images, i.e. to minimize $1 - D(G(n))$. Thus both the networks play a minimax game between them and it can be expressed mathematically as following value function as given in (1),

$$\min_{G}\max_{D} V(D, G) = \mathbb{E}_{x\sim p_{\text{data}}(x)}\left[\log D(x)\right] + \mathbb{E}_{n\sim p_n(n)}\left[\log(1 - D(G(z)))\right] \quad (1)$$

where $p_{\text{data}}(x)$ is data distribution of real images (x) and $p_n(n)$ is noise (n) distribution.

Once the necessary training is done, generator would be capable of producing natural and realistic looking fake images by using noise signals n, whereas the ability of D to differentiate deepfake images from real ones will also be improved. The various types of GANs used for generating deepfake images are discussed in the next section.

3 GANs Used for Generating Deep Fake Images

Two commonly employed GANs used for generating deepfake images are (1) Deep Convolutional GAN (DCGAN) and (2) Progressive Growing GAN (PGGAN).

3.1 Deep Convolutional GAN (DCGAN)

DCGAN is proposed by Radford et al. [18] and it is an enhancement of GAN. The architecture consists of convolution layers minus max pooling or fully connected layers and the generator network is shown as in Fig. 3. The down sampling and up sampling operations are done using convolutional stride and transposed convolution. The batch normalization is performed in generator network and discriminator network. This network generates good quality real looking deepfake images.

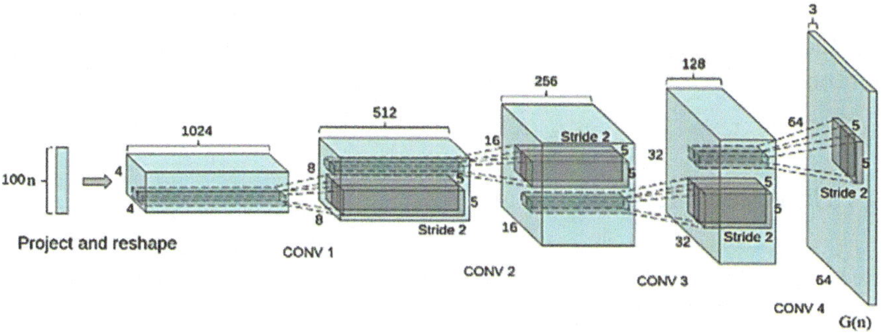

Fig. 3 Generator architecture of DCGAN [18]

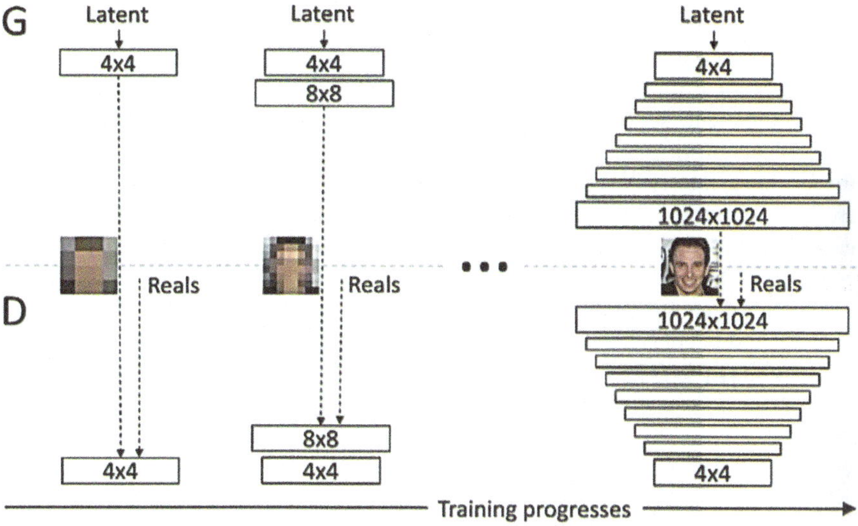

Fig. 4 PGGAN [17]

3.2 *Progressive Growing GAN (PGGAN)*

PGGAN is proposed by Karras et al. [17] and in this technique, initially a fewer number of layers are considered for training the generator and discriminator networks and as a result generator yields 4×4 resolution fake images. Then the discriminator is trained by taking these fake images along with real images which are scaled to same resolution. In training process, the number of layers is increased for both generator and discriminator networks as shown in Fig. 4 and hence progressively increasing the resolution of deepfake images.

Some other variants of GAN used for deepfake generation are Boundary Equilibrium Generative Adversarial Network (BEGAN) [19], Wasserstein Generative

Adversarial Network (WGAN) [20], Cycle GAN [21], etc. Different methods employed for detecting deepfake images from real ones are discussed in the coming section.

4 Detection Methods

Recently, the researchers started developing techniques to detect deepfake images and some relevant works are discussed in this section.

A method to identify Deep Network Generated (DNG) fake images is proposed by Li et al. [22]. DNG fake images are in Red, Green, and Blue (RGB) color space with no explicit associations among the color components and there are some clear differences between fake images and real images in other color spaces such as Hue, Saturation, and Value (HSV) and YCbCr. Also, these fake images are dissimilar from the camera captured images while considering red, green, and blue components together. Hence this method analyzes the differences in color components of images by separating image into R, G, and B components and also transforming into HSV and YCbCr color space. Then images in R, G, B, H, S, Cb are filtered using a high pass filter and the co-occurrence matrix is computed on each filter residuals. Finally, classifier is trained using a feature vector generated by concatenating the extracted co-occurrence matrixes. The GAN models used in this method for generating fake images are DCGAN, WGAN-GP, and PGGAN and real image datasets considered are CelebFaces Attributes (CelebA) and Labeled Faces in the Wild (LFW). The block diagram of this method is as shown in Fig. 5.

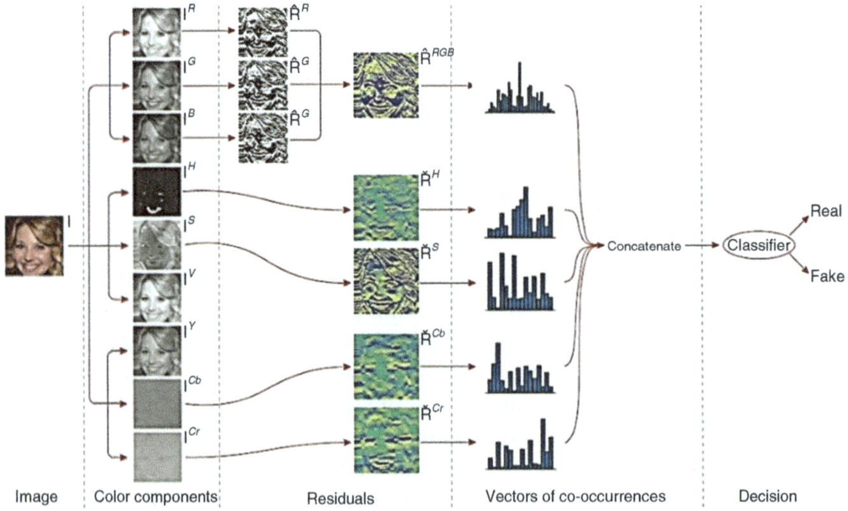

Fig. 5 Block diagram of the proposed method [22]

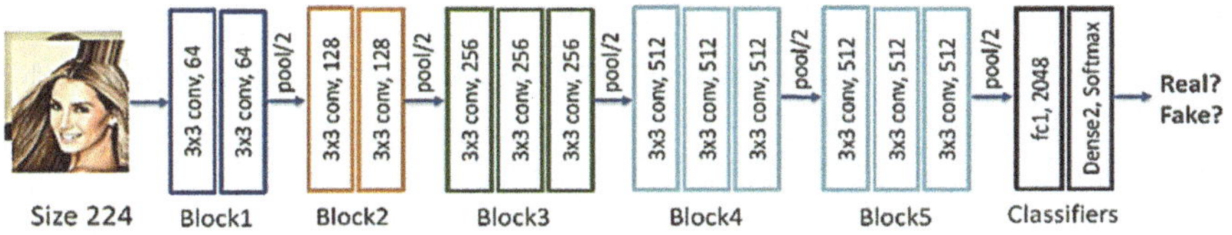

Fig. 6 Forensics face detection architecture [23]

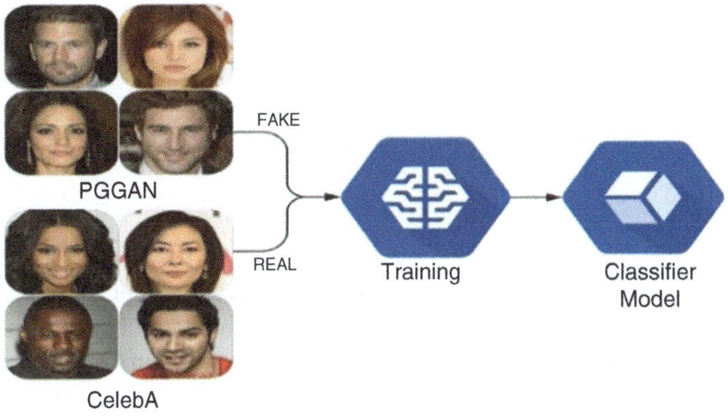

Fig. 7 Classification model [24]

Do et al. [23] proposed a deepfake image detection system using deep learning technique. They applied transfer learning technique and used pre-trained weights of VGG Face model to extract deep face features. VGG-Net as shown in Fig. 6 is used, which consists of convolutional layers, max pooling layers, a fully connected layer, and an output layer with 2-way softmax function for the classification of images. DCGAN and PGGAN models are used to generate deepfake images based on CelebA dataset. Finally, they evaluated performance of their method using data from the first mission of the AI Challenge contest.

Tariq et al. [24] proposed an ensemble classifier to detect deepfake faces generated by PGGAN. They designed ensembles of three different Shallow Convolutional Networks (ShallowNet) to detect deepfake human faces. The classification model is an ensemble of three ShallowNets, trained using “FAKE” and “REAL” images from PGGAN and CelebA dataset, respectively, as presented in Fig. 7.

McCloskey and Albright [25] proposed two methods based on Color Image Forensics and Saturation based Forensics for detecting GAN-generated images. The effectiveness of methods is evaluated with two benchmark datasets (GAN Crop image datasets and GAN Full image dataset) of Standards and Technology’s Media Forensics Challenge 2018. One method is based on Color Image Forensics where they used a pre-trained version of Intensity Noise Histograms (INH), and fine-tuned classifier with r and g chromaticity histograms. Area Under Curves (AUCs) of the classifier are 0.56 and 0.54 and hence it is clear that classifier not learned

Table 1 Layer architecture of CNN for CGFace model

Name of layer	Number and size of filters	Output dimension (height, width, channel)
Input	64 × 64	–
Convolution 1	8 filters with size 5 × 5	(60, 60, 8)
Convolution 2	8 filters with size 5 × 5	(56, 56, 8)
MaxPooling 1	2 × 2	(28, 28, 8)
Convolution 3	16 filters with size 3 × 3	(26, 26, 16)
MaxPooling 2	2 × 2	(13, 13, 16)
Convolution 4	16 filters with size 3 × 3	(11, 11, 32)
MaxPooling 3	2 × 2	(5, 5, 32)
Convolution 5	16 filters with size 3 × 3	(3, 3, 64)
Flatten	–	(576)
Fully connected layer	–	(256)
Fully connected layer	–	(2)

any relevant information regarding the color statistics of deepfake images versus real images. The second method is based on Saturation based Forensics where the measure of frequency of saturated and under exposed pixels in each image is taken as feature. The features are extracted from GAN-generated images and real camera images. These feature vectors are used for training an SVM. An AUC of 0.7 is obtained for both the evaluation datasets.

Dang et al. [26] proposed a Computer Generated Face Identification (CGFace) model based on deep learning technique to identify deepfake faces. A customized CNN is used for creating CGFace model and its architecture is as shown in Table 1. This model uses a dropout with probability of 0.2 to prevent overfitting of the model. The imbalanced data issue is addressed by creating an Imbalanced Framework (IF-CGFace) and which is done by training AdaBoost and eXtreme Gradient Boosting (XGB) using the features extracted from CGFace layers.

Various approaches for detecting deepfake images from real images are discussed and its performances are summarized in Table 2. It is evident that these techniques only perform well on deepfake images generated by any specific type of GANs, however, the performance of these methods is not evaluated on all types of deepfake images.

5 Conclusion

Deepfake images created by GANs are becoming more realistic due to the advancement of machine learning techniques. Thus identification of deepfake images from a real camera image is becoming a challenging task. Deepfake images can be purposefully used for propagating fake news as well as it can be used to create fake profiles in many social medias. Hence, it is very essential to develop accurate techniques for detecting deepfake images. In this survey, we investigated

Table 2 Summary of various methods to detect deepfake images generated by GAN

Methodology	Datasets	Performance	Limitations
Differences in color components of deepfakes and real images are analyzed for detecting deepfake images (Li et al. [22])	*Real image datasets*: Celeb A, HQ-CelebA, and LFW *GANs used for generating Deepfake images*: DCGAN, WGAN-GP, and PGGAN	Accuracy is >98%	Method is not evaluated on any practical case scenarios of deepfake images
Detection using convolutional neural network (Do et al. [23])	*Real image dataset*: Celeb A *GANs used for generating Deepfake images*: DCGAN and PGGAN *Evaluation dataset*: Images from AI Challenge Contest	Accuracy: 80% and Area under the ROC Curve (AUROC) is 0.807	Performance is not evaluated on deepfake images generated by WGAN-GP, BEGAN, etc.
Ensemble of neural network classifier (Tariq et al. [24])	*Real image dataset*: Celeb A *GANs used for generating Deepfake images*: PGGAN	Accuracy is 93.99% and 99.99% for small resolution images(64x64) and higher resolution images, respectively	Performance is not evaluated on deepfake images generated by DCGAN, WGAN-GP, BEGAN, etc.
Two methods based on (1) Color Image Forensics (2) Saturation based Forensics (McCloskey and Albright [25])	*Method 1*: *Real image dataset*: Celeb A *GANs used for generating Deepfake images*: PGGAN *Method 2*: *Real image dataset*: ImageNet dataset *Deepfake images*: LSUN dataset *Evaluation dataset*: GAN Crop image dataset and GAN Full image dataset of Standards and Technology's Media Forensics Challenge 2018	*Method 1*: AUROC 0.56 and 0.54 for GAN Crop image datasets and GAN Full image dataset, respectively *Method 2*: AUROC 0.7 for both the evaluation datasets	It is evident from AUROC that method gives comparatively a poor performance
Designed a Computer Generated Face Identification (CGFace) model based on customized CNN (Dang et al. [26])	*Real image dataset*: Celeb A *GANs used for generating Deepfake images*: PGGAN and BEGAN	Accuracy: 98% AUROC 0.81	Performance is not evaluated on deepfake images generated by DCGAN, WGAN-GP, etc.

generation of deepfake images by various GANs and reviewed different approaches for detecting deepfake images. However, these detection methods are only evaluated on deepfake images produced by a specific type of GANs and there arise a question of generalization capability of these methods.

References

1. Alahmadi, A.A., Hussain, M., Aboalsamh, H., Muhammad, G., Bebis, G.: Splicing image forgery detection based on DCT and Local Binary Pattern. In: 2013 IEEE Global Conference on Signal and Information Processing, pp. 253–256. IEEE, Austin (2013)
2. Ng, T.-T., Chang, S.-F.: A model for image splicing. In: 2004 International Conference on Image Processing (ICIP), pp. 1169–1172. IEEE, Singapore (2004)
3. Sutthiwan, P., Shi, Y.Q., Zhao, H., Ng, T.T., Su, W.: Markovian rake transform for digital image tampering detection. In: Transactions on Data Hiding and Multimedia Security VI, Lecture Notes in Computer Science, pp. 1–17. Springer, Berlin (2011)
4. Lu, C.S., Liao, H.Y.M.: Multipurpose watermarking for image authentication and protection. IEEE Trans. Image Process. **10**(10), 1579–1592 (2001)
5. Lu, C.S., Liao, H.Y.M.: Structural digital signature for image authentication: An incidental distortion resistant scheme. IEEE Trans. Multimedia. **5**(2), 161–173 (2003)
6. Muhammad, G., Al-Hammadi, M.H., Hussain, M., Bebis, G.: Image forgery detection using steerable pyramid transform and local binary pattern. Mach. Vis. Appl. **25**(4), 985–995 (2014)
7. El-Alfy, E.-S.M., Qureshi, M.A.: Combining spatial and DCT based Markov features for enhanced blind detection of image splicing. Pattern Anal Applic. **18**(3), 713–723 (2015)
8. Vidyadharan, D.S., Thampi, S.M.: Digital image forgery detection using compact multi-texture representation. J. Intell. Fuzzy Syst. **32**(4), 3177–3188 (2017)
9. Alahmadi, A., Hussain, M., Aboalsamh, H., Muhammad, G., Bebis, G., Mathkour, H.: Passive detection of image forgery using DCT and local binary pattern. Signal, Image Video Process. **11**(1), 81–88 (2017)
10. Isaac, M.M., Wilscy, M.: Image forgery detection using region–based Rotation Invariant Co-occurrences among adjacent LBPs. J. Intell. Fuzzy Syst. **34**(3), 1679–1690 (2018)
11. Dong, J., Wang, W., Tan, T.: CASIA image tampering detection evaluation database. In: 2013 IEEE China Summit and International Conference on Signal and Information Processing, pp. 422–426. IEEE, Beijing (2013)
12. Krizhevsky, A., Sutskever, I., Hinton, G.E.: ImageNet classification with deep convolutional neural networks. Commun ACM. **60**, 84–90 (2017)
13. Rao, Y., Ni, J.: A deep learning approach to detection of splicing and copy-move forgeries in images. In: 2016 IEEE International Workshop on Information Forensics and Security (WIFS), pp. 1–6. IEEE, Abu Dhabi (2017)
14. Rota, P., Sangineto, E., Conotter, V., Pramerdorfer, C.: Bad teacher or unruly student: Can deep learning say something in image forensics analysis? In: 2016 23rd International Conference on Pattern Recognition (ICPR), pp. 2503–2508. IEEE, Cancun (2017)
15. Zhou, J., Ni, J., Rao, Y.: Block-based convolutional neural network for image forgery detection. In: Digital Forensics and Watermarking. IWDW 2017, Lecture Notes in Computer Science, pp. 65–76. Springer, Cham (2017)
16. Goodfellow, I.J., Pouget-Abadie, J., Mirza, M., Xu, B., Warde-Farley, D., Ozair, S., Courville, A., Bengio, Y.: Generative adversarial nets. In: Advances in Neural Information Processing System 27, pp. 2672–2680 (2014)
17. Karras, T., Aila, T., Laine, S., Lehtinen, J.: Progressive growing of GANs for improved quality, stability, and variation. In: arXiv preprint arXiv:1710.10196, pp. 1–26 (2017)

18. Radford, A., Metz, L., Chintala, S.: Unsupervised representation learning with deep convolutional generative adversarial networks. In: arXiv preprint arXiv:1511.06434, pp. 1–16 (2015)
19. Berthelot, D., Schumm, T., Metz, L.: BEGAN: boundary equilibrium generative adversarial networks. In: arXiv preprint arXiv:1703.10717, pp. 1–10 (2017)
20. Gulrajani, I., Ahmed, F., Arjovsky, M., Dumoulin, V., Courville, A.: Improved training of Wasserstein GANs. In: arXiv:1704.00028v3, pp. 1–20 (2017)
21. Zhu, J., Park, T., Isola, P., Efros, A.A.: Unpaired image-to-image translation using cycle-consistent adversarial networks. In: Proceedings of the IEEE International Conference on Computer Vision, pp. 2223–2232 (2017)
22. Li, H., Li, B., Tan, S., Huang, J.: Detection of deep network generated images using disparities in color components. In: arXiv preprint arXiv:1808.07276, pp. 1–13 (2018)
23. Do, N.-T., Na, I.-S., Kim, S.-H.: Forensics face detection from GANs using convolutional neural network. In: 2018 International Symposium on Information Technology Convergence (ISITC 2018), South Korea (2018)
24. Tariq, S., Lee, S., Kim, H., Shin, Y., Woo, S.S.: Detecting both machine and human created fake face images in the wild. In: Proceedings of the 2nd International Workshop on Multimedia Privacy and Security, pp. 81–87. ACM, Toronto (2018)
25. Mccloskey, S., Albright, M.: Detecting GAN-generated imagery using color cues. In: arXiv preprint arXiv:1812.08247 (2018)
26. Dang, L.M., Hassan, S.I., Im, S., Lee, J., Lee, S., Moon, H.: Deep learning based computer generated face identification using convolutional neural network. Appl. Sci. **8**(12), 2610 (2018)

A Framework for Test Coverage of Safety and Mission Critical Software

P. Mithun, Anil Abraham Samuel, Prashant Ranjan, N. Jayalal, T. Gopalakrishnan, and B. Valsa

1 Introduction

Software testing is an important activity in the software development life cycle, and it becomes more important in the case of safety and mission critical software of satellite launch vehicles. Code coverage is an essential part in software testing, which directly gives an indication of effectiveness of test cases [1]. Code coverage can be described as the extent to which a certain part or structure of a code has been executed. It is a factor that can easily differentiate between a well-tested code and a poorly tested code. The coverage assessment is important to the qualification of the flight software of ISRO satellite launch vehicles.

Many software metrics exist that help in analyzing certain properties of a software quantitatively. These quantitative measures are further used for the assessment of software quality. Code coverage is one such metric that gives assurance on thoroughly tested code for minimum occurrence of bugs [1]. Code coverage additionally helps in finding the weakly covered code segments that can be used for designing efficient test cases, which in turn helps in reducing time needed for software testing [2]. This also gives details of frequently covered code segments, which further can be used for code optimization to minimize time complexity and thereby improve the real-time performance. The extent of functionality and code structure covered in testing is brought out by coverage analyzers.

Flight software of launch vehicles undergoes exhaustive verification and validation before it is inducted for the actual flight. The validation phase includes module-level tests and simulations at integrated level. As the dynamic testing of

P. Mithun (✉) · A. A. Samuel · P. Ranjan · N. Jayalal · T. Gopalakrishnan · B. Valsa
Vikram Sarabhai Space Centre, Trivandrum, Kerala, India
e-mail: mithun_p@vssc.gov.in; anil_abraham@vssc.gov.in; prashant_ranjan@vssc.gov.in; n_jayalal@vssc.gov.in; t_gopalakrishnan@vssc.gov.in; b_valsa@vssc.gov.in

M. Palesi et al. (eds.), *Second International Conference on Networks and Advances in Computational Technologies*, Transactions on Computational Science and Computational Intelligence, https://doi.org/10.1007/978-3-030-49500-8_4

software is conducted at this phase, code coverage proves to be an efficient metric to know whether all parts of software have been executed according to requirements or not.

2 Background

Flight software undergoes various phases of verification and validation before it is inducted for actual use in a mission. The major phases include code inspection, module-level tests, software fault injection tests, and different levels of integrated tests mostly through simulations. All these activities are done with different objectives. Code inspection is the verification of the code with respect to the design and brings out defects prior to testing and also brings out violations of many nonfunctional requirements. Module-level testing is efficient in bringing out coding-related and detailed design-related errors. Through integrated tests with flight software integrated in actual hardware packages and actual interconnections between them and mathematically based simulation of flight environment and flight conditions, the system-level requirements are validated. These tests are referred as integrated simulation tests. The coverage assessment in the simulations directly gives details of structural and functional coverage of the software under real flight environment, and this is very essential to bring the required confidence, as the software experiences the real possible input conditions.

In launch vehicle integrated tests, the test cases are designed by considering the mission requirements, software functional requirements, and software fault handling requirements. The integrated simulation tests are carried out in the actual target processors that are used in the flight. The processors used are indigenously developed and with specific instruction set architecture (ISA). The simulation runs are highly time sensitive. It has to adhere to a specified stringent cycle time to meet the integrated software test objective. Code instrumentation required for coverage analysis can affect the real-time performance of the flight software, which is not acceptable in integrated tests. The extra code inserted during instrumentation can conflict with the resource constraints in the target processor. Apart from that, the compiler/assembler are in-house developed based on a safe subset of Ada. Hence the direct coverage assessment of flight software in the integrated simulations is not attempted. These difficulties in in situ measurement of coverage (Method A) led to assessment of coverage by running the simulation test cases on an x86 host platform (Method B) using open-source tool suites described in Fig. 1.

3 Flight Software Coverage Assessment

In order to understand the test coverage of the flight software during the integrated simulation tests, a work around is evolved. The developments carried out to achieve

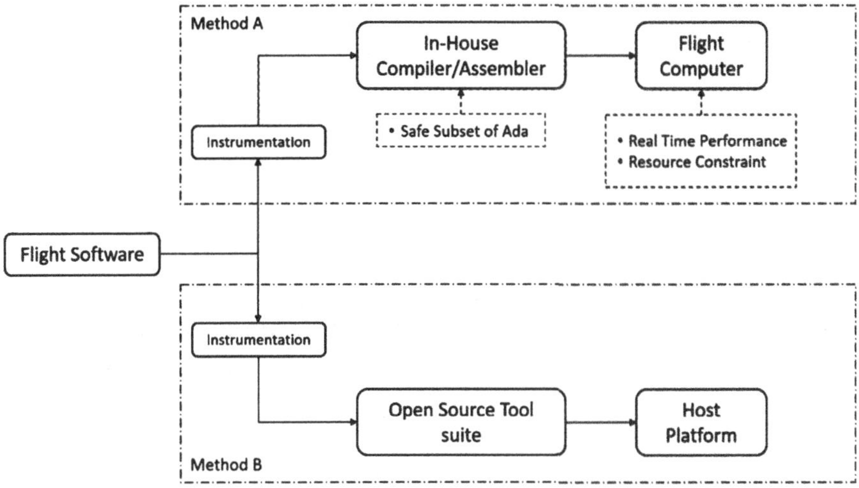

Fig. 1 Coverage analyzer option selection

the objective of retrieving the coverage information of the integrated test and its results are described in the paper.

The major steps in the software coverage analysis of integrated simulation tests are listed below and illustrated in Fig. 2:

- Flight profile generation
- Initialization data configuration
- Flight software instrumentation
- Scheduling execution of instrumented flight software
- Generation of the coverage data from execution trace
- Parsing of coverage data to derive coverage statistics
- Interpretation of coverage information for assessing the requirements covered

3.1 Flight Profile Generation

The simulations mimic various flight scenarios including propulsion dispersions, aero dispersions, separation disturbances, intercomputer communication failures, salvage schemes, sensor failures, and many other situations. All the integrated simulation runs taken with the flight hardware generate an immense amount of data. The data format of parameters follows well-defined format. A basis set of data required to execute the various flight software components is identified and extracted from this huge amount of data. This task is accomplished by developing a flight profile generator. This flight profile generator processes the data of avionics data bus of the flight computers acquired through a bus monitor. This data includes

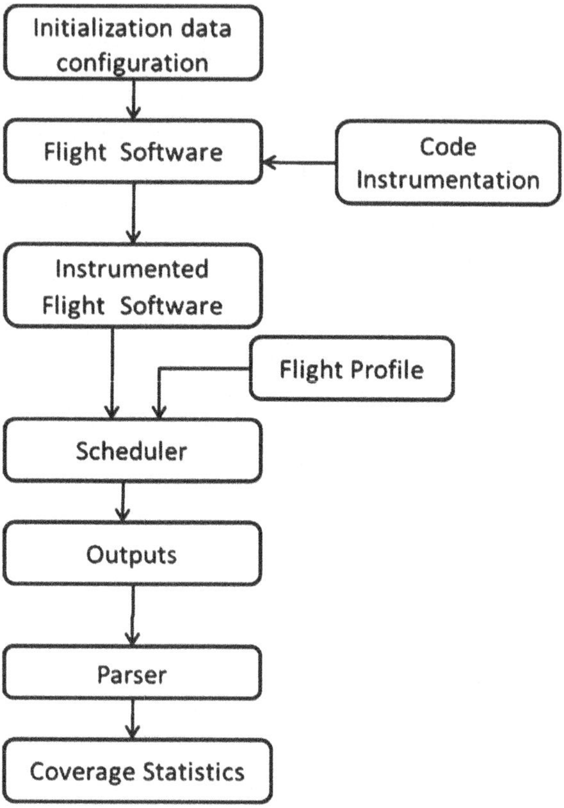

Fig. 2 Schematic of the coverage framework

the launch vehicle state vectors, attitude information, vehicle control, and mission sequencing and health parameters generated and exchanged by various computers in the flight system that are configured as bus controller and remote terminals in the avionics data bus. The flight profile generator extracts the relevant data and presents it in a well-defined format. The entire flow is illustrated in Fig. 3.

3.2 Initialization Data Configuration

The source code for each of the onboard software component contains a default set of initialization data. This initialization data has to be replaced with the set of initialization data used for the integrated simulation run. The initialization data sets containing various variables and its types and values are processed, and each of the variables in the source code is replaced with the new values automatically. The source code updated with the new set of initialization data shall be syntactically and semantically the same as the initial source code. Also, this should not create any error during compilation (Fig. 4).

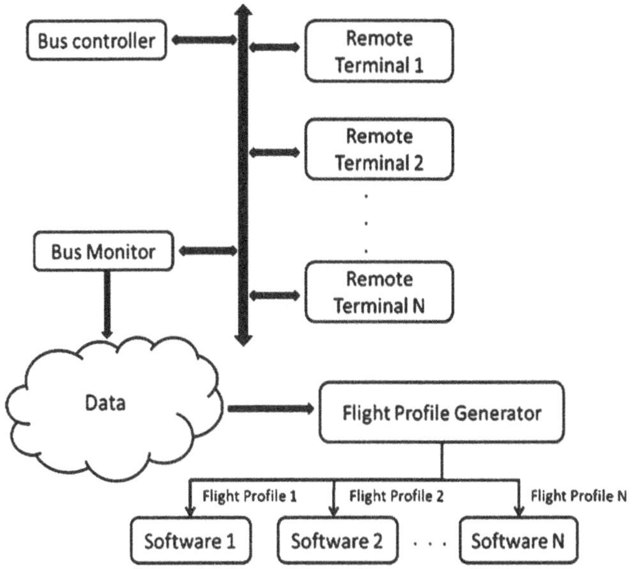

Fig. 3 Flight profile generator for generating the input profiles

Fig. 4 Configuring the software with initial conditions

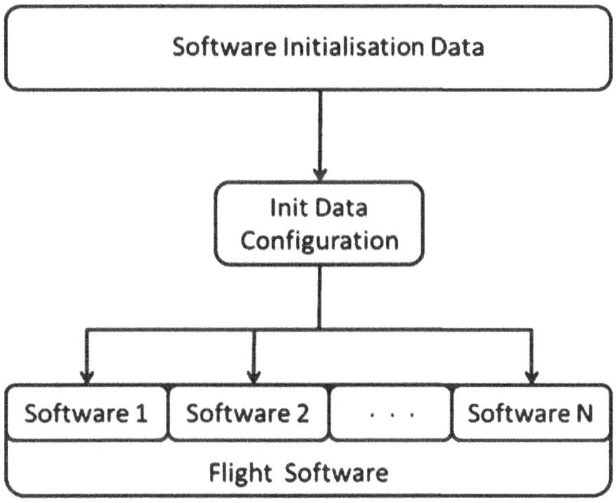

3.3 *Flight Software Instrumentation*

One of the major steps in coverage assessment is the code instrumentation. The software has to be instrumented such that execution trace can be extracted, which is the core information for deriving coverage statistics but keeping the logic of the software unaltered. It is necessary to ensure that after instrumentation, the real functionality of the software is not affected. Open-source tools [3] are used for instrumentation. This generates additional files during the software building and

Code (Ada and x86 assembly)	Instrumented Code (Ada and x86 assembly)

if (CurrSeqFlag < FDCCtrlIni)
then
.loc 1 2472 0 discriminator 1
movzwl _pdapmain__currseqflag,
%eax
 	
cmpw $2, %ax
jg L127
	if (CurrSeqFlag < FDCCtrlIni) then
	.loc 1 2472 0 discriminator 1
	movzwl _pdapmain__currseqflag, %eax
	cmpw $2, %ax
	jg L133
	movl _gcov0.pdapmain__rattrateerrfilter+104,
	%eax
	movl _gcov0.pdapmain__rattrateerrfilter+108,
	%edx
	addl $1, %eax
	adcl $0, %edx
	movl %eax,
	_gcov0.pdapmain__rattrateerrfilter+104
	movl %edx,
	_gcov0.pdapmain__rattrateerrfilter+108

Fig. 5 Code—with and without instrumentation

execution process, which holds the data for reconstructing the coverage information of the software. The difference in the host assembly code of an Ada comparison statement and its equivalent instrumented version is shown in Fig. 5. The additional information added enables easy extraction of coverage information.

3.4 Scheduled Flight Software Execution and Coverage Trace Generation

The execution of instrumented flight software is carried in an organized manner by a scheduler in a host machine with x86 processor. Because of the reasons mentioned earlier, this is not carried out in the actual target platform. The scheduler passes the flight profile required for the execution of each of the flight software component. The additional binary file created during execution contains arc transition counts of flow graph [4]. This information along with the flow graph information generated during compilation process is made use for mapping the execution trace at source code level and generating execution counts. These execution counts are obtained for each and every source file in all flight software components. Apart from the coverage information, software-related functional and performance parameters are extracted for guaranteeing the integrity of the code instrumentation and also for ensuring similarity in the control flow of the code execution in the actual target processor and the host processor. The coverage information can be accumulated for multiple simulation tests or can be interpreted run by run and analyzed in detail.

3.5 *Parsing Coverage Data for Derivation of Coverage Statistics*

After the execution of the instrumented code in a standalone mode in an x86 host platform, the coverage information is consolidated. The open-source tool [4] generates execution counts using the coverage information and the flow graph created at compile time. The tool provides information that distinguishes between blank lines, executable lines, and comments with certain prefixed patterns. The tool has the capability to generate information regarding decision coverage also but does not distinguish between branches in the actual code and branches included by compiler in object file. Thus the direct usage of the tool can cause incorrect indication of branch coverage.

To enable legible and understandable representation of the coverage information in the required format and also to avoid false findings, a parser was developed. The parser analyzes the coverage information to generate the statement and branch coverage. Specific signatures are looked during parsing to bring the correct statement and branch coverage details. In order to overcome the above-stated difficulty of identifying the branches in the source, logic was devised for assessing branch coverage from plain execution counts. The parser is built with the capability to bring coverage statistics at module level for procedures and functions, as well as source file level. The information gathered is further analyzed to find the weakly and frequently covered code segments. The entire flow of coverage information generation, parsing it, and formatted representation of the information is illustrated in Fig. 6.

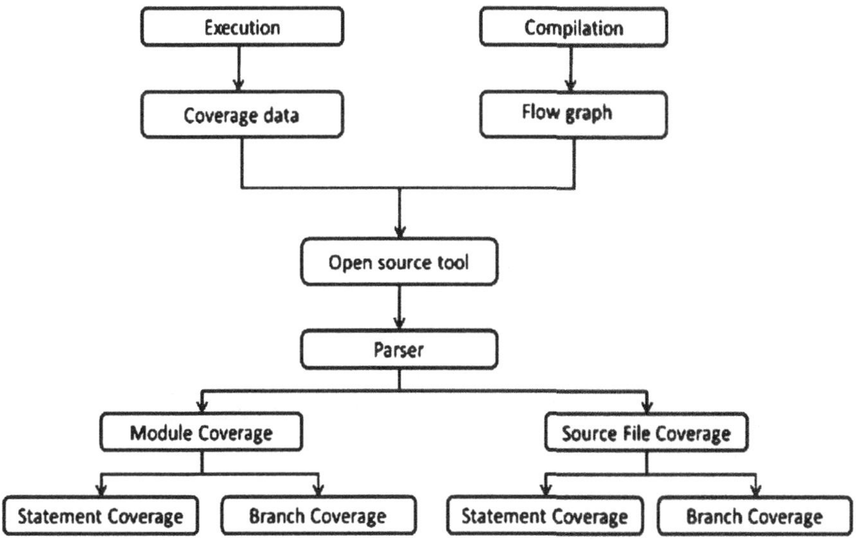

Fig. 6 Coverage statistics generation

4 Validation

One of the major steps in coverage assessment is flight software instrumentation. The software is modified to extract the execution trace. Also, the execution of the instrumented flight software is carried out in a host machine instead of the actual target machine. Thus, it is necessary to ensure that after instrumentation, the functionality of the software is not affected.

The integrated simulation test produces large set of data that are signatures of the flight software execution. The inputs used for the simulation tests are derived and formed as input profile for the execution of instrumented software in the host machine. The outputs produced by the instrumented software are compared and matched against the integrated simulation test outputs, which run in the target machines. In addition to this, original version of software is run in the host machine for comparison of the intermediate variables with instrumented version executed in host machine, as the target processor does not output all the intermediate variables of interest. This way it is ensured that the automatic code instrumentation has not affected the software functionality. The scheme is depicted in Fig. 7, and a typical result of comparison is shown in Fig. 8. It can be seen from the plot that the launch vehicle Pitch Error output parameter from simulation test matches with the output produced by the instrumented software running in the host.

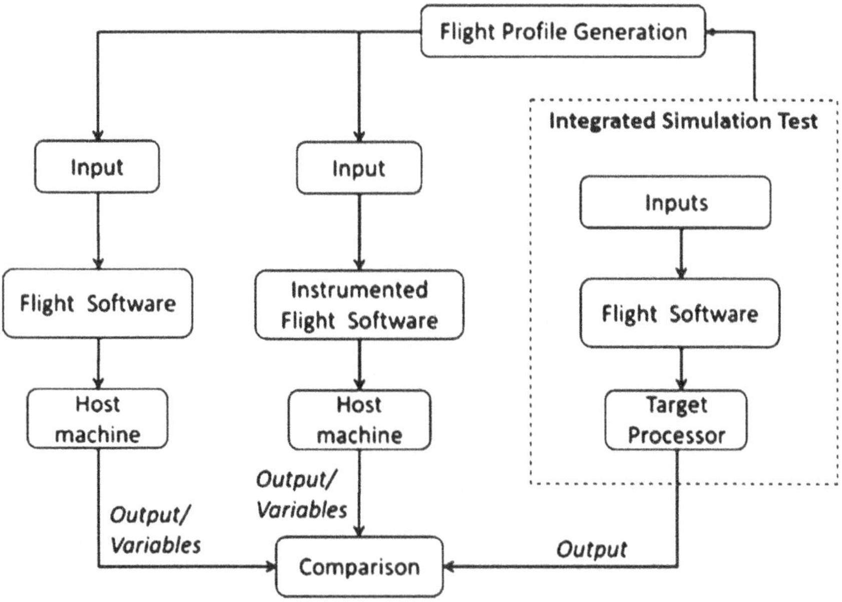

Fig. 7 Verification of instrumentation

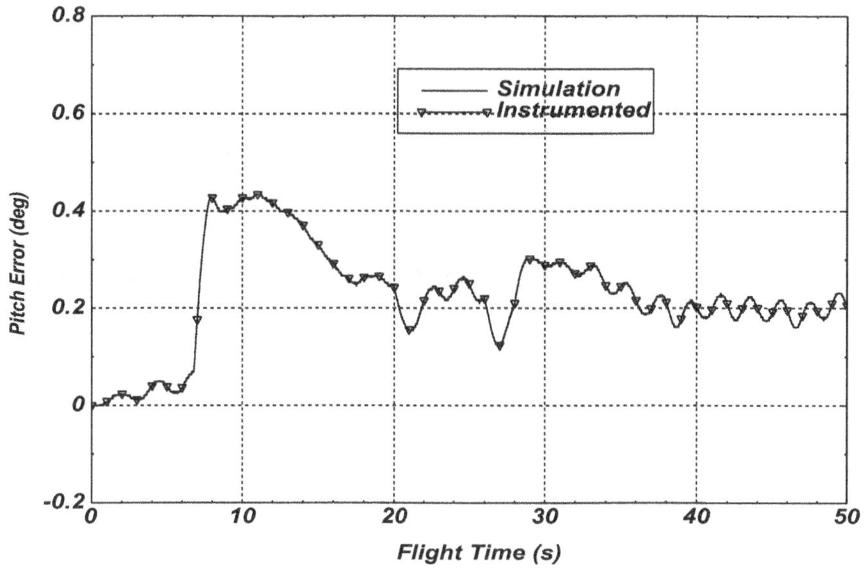

Fig. 8 Validation of the execution of instrumented software by matching the outputs

5 Results

Coverage analysis of launch vehicle flight software execution in Onboard computer In-Loop Simulation (OILS) test was carried out as part of Independent Verification and Validation (IV&V) activity of flight software. The coverage analysis was done in two parts. In first part, each case was separately executed using the coverage tool. In the second part, cumulative coverage information of all cases was analyzed, which gives the entire software coverage in simulations.

The coverage analysis indicates that 100% of statements and branches are covered in majority of the modules. However, the modules that are not covered 100% are specifically analyzed and ensured that all the required functionalities are validated. Some statements /branches are intentionally disabled by design or through initialization data, so that it will not be exercised in simulations.

The coverage analysis results of control software of launch vehicle are shown below in Fig. 9. Different color coding (green—100%, red—0%, orange—between 0% and 100%, and NA—modules with no branches) is used to interpret various levels of coverage. The bar chart showing the statement and branch coverage is given in Fig. 10.

Total Statement Coverage	91.25 %
Total Branch Coverage	79.59 %

Modules	Statement Coverage %	Branch Coverage %
rDAPCCompInit	100.00	NA
rRCSByteOnModComp	100.00	100.00
rIntgDeplete	00.00	00.00
rPropIntCntrl	85.25	68.75
rPh1DAPCComp	100.00	100.00
rPh2DAPCComp	100.00	83.33
rDAPCComp	92.00	75.00

Fig. 9 Coverage details

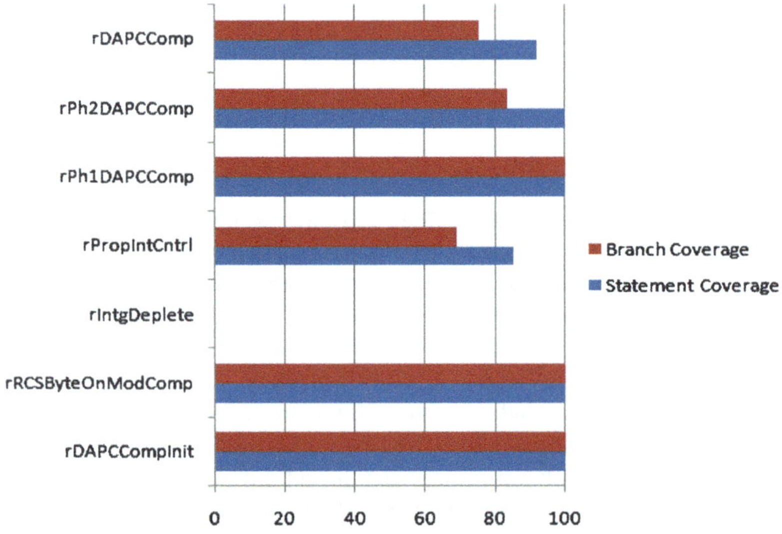

Fig. 10 Coverage metrics

6 Conclusion

The integrated system-level tests are often very important as the tests carried out are domain specific and will be closely matching with the operating conditions. The coverage analysis methodology demonstrated for the integrated simulation tests using this framework proves to be a good mechanism to build confidence on the extent to which software functionalities are covered in integrated system-level tests. All the software components analyzed show adequate structural and functional coverage. The explanation for the modules that have not achieved 100% coverage is consistent with respect to the mission and design requirements. However additional test cases are evolved and exercised to achieve satisfactory level of test coverage. Automated Coverage Analyzer is now inducted into the Independent Verification and Validation (IV&V) of new missions as this proves to be an efficient mechanism in addressing test adequacy, identifying weakly/frequently covered portions of software, and helps in efficient test case generation.

References

1. Hemmati, H.: How effective are code coverage criteria? In: IEEE International Conference on Software Quality, Reliability and Security (2015)
2. Shahid, M., Ibrahim, S., Mahrin, M.N.: A study on test coverage in software testing. In: International Conference on Telecommunication Technology and Applications (2011)
3. Gcc.gnu.org: Using the GNU Compiler Collection (GCC): Instrumentation Options (online). https://gcc.gnu.org/onlinedocs/gcc/Instrumentation-Options.html (2017). Accessed 10 Sep 2017
4. Gcc.gnu.org: Using the GNU Compiler Collection (GCC): Gcov (online). https://gcc.gnu.org/onlinedocs/gcc/Gcov.html (2017). Accessed 10 Sep 2017

Internet Censorship Based on Bayes Learning Model

Ajeesh Ramanujan and Blesson Andrews Varghese

1 Introduction

The Internet is a global computer network through which people around the globe communicate with each other, share files, learn and entertain themselves. It was originally intended for military and research purposes. There is no doubt that internet has brought together the entire world as a single global village. Internet, as a technology, has the enormous potential to be of benefit to human lives than any other invented technology in the world. However, a lot of people are using it to spread hatred, terrorism [18], and obscenity in the world. Therefore, it is necessary that the internet should be censored. Internet censorship refers to any process by which information that is publicized or viewed on the internet is controlled. Internet censorship in India typically occurs as DNS filtering which is often selective and is not entirely effective. The public often protests against censorship as they consider it as a violation of their freedom of speech. The lack of any common standards aggravates the issue. They support their claim by providing examples of authoritarian regimes which use censorship to suppress opposition parties and minority religious groups. However, censorship is the only tool available in the virtual world which can effectively counter social evils like terrorism, piracy, defamation, and fake news. National security can be assured only through censorship. It also protects each individual's right to be forgotten. Thus there is no denying that the internet should be censored to prevent unfortunate events.

The internet does not have any physical boundaries. Any precious information once posted online cannot be completely deleted due to replication. Restricting the

A. Ramanujan · B. A. Varghese (✉)
CSED, College of Engineering Trivandrum, Thiruvananthapuram, India
e-mail: ajeesh@cet.ac.in

M. Palesi et al. (eds.), *Second International Conference on Networks and Advances in Computational Technologies*, Transactions on Computational Science and Computational Intelligence, https://doi.org/10.1007/978-3-030-49500-8_5

posted information is also difficult as the internet is vast and provides anonymity. In this paper, we are giving an overview of current internet censorship methods in India. We also explain various circumvention tools used by citizens to overcome these sensors and the consequences of using the same. Later, we introduce a censorware having machine learning capabilities. It tries to identify whether a website should be filtered based on its content's degree of harmfulness. To the best of author's knowledge, such a study is not conducted earlier.

2 Internet Censorship

Internet censorship can be applied at national/ISP/institutional/user levels. It targets to block contents harmful/objectionable to authorities. It is implemented using several technologies like firewalls, proxies, and DNS filters. No single solution provides complete coverage. Therefore censoring organizations deploy several technologies together to achieve desired results. Censorship can be done technically as well as non-technically. Some of the most common technical methods are

1. **Internet Protocol (IP) Blocking**
   The most common internet filtering technique used by countries. Most of the censorware keep a regularly updated blacklist of IP addresses which are known to spread malicious/illegal content. All communications send to/received from such IP addresses are completely blocked. Usually, censorware employs a validation tool like VirusTotal to confirm that an IP address is a legitimate one to avoid the overhead of maintaining a blacklist. IP blocking only targets IP-based protocols like HTTP, FTP, and POP. IP blocking is usually circumvented by using a virtual private network (VPN) or finding uncensored proxies. Some websites own multiple IP addresses. In order to censor such websites, we need to block all such IPs. A major shortcoming of IP blocking is that, if the website to be censored is deployed on a shared web-server, all websites on the same server will be blocked. In India, most ISPs use IP blocking to block access to websites [2, 5–8]. All such blocks are not entirely effective as they do not prevent tech-savvy users from accessing such websites [9].
2. **Domain Name System (DNS) Filtering and Redirection**
   There will be a DNS authority in every country. Officials can deregister a domain that has illegal/harmful materials. Whenever a user tries to connect to such websites, DNS will not be resolved or incorrect IP address is returned using DNS hijacking or other means [3]. Users could circumvent DNS filtering easily by accessing foreign search engines and DNS servers.
3. **Uniform Resource Locator (URL) Filtering**
   URL filters verify that hyperlinks and URLs do not contain any malicious commands, keyword, or code [11]. URL filtering mainly targets HTTP based protocols. Attackers use encrypted protocols like VPN and TLS/SSL to circumvent such filters. Using escape characters in the URL will also confuse filters. Nowadays, URL filtering is used by web and email scanning engines to identify harmful emails and search results.

4. **Packet Filtering**
   All communication on the internet happens through packets. Packet filtering uses deep packet inspection to identify any forbidden content in the packet and drop it. Packet filtering targets all TCP-based protocols like HTTP, FTP, and POP. However, if the packet is encrypted, the filter will not identify any forbidden content.
5. **Network Disconnection**
   In highly sensitive situations, rather than trying to censor the network, authorities fully block the network with the help of network disconnection. They disconnect power to all routers or block communication to them. Highly privileged users can still use satellite ISPs to gain access to the internet in such scenarios.
6. **Network Attacks**
   Rather than fully blocking the network, we can target malicious websites and launch attacks like DoS to break it. Thus access to that website is prevented for a limited period.
7. **Search Result Removal**
   Government and legal authorities may force a major web portal or search engine to block a malicious website. Thus malicious website is excluded from their search results. As a result, the site is invisible to people who do not know where to find it. When a major website like Google does this, it has the same impact as censorship.

Internet content, like any other media, can be censored using nontechnical censorship methods like

1. **Legal Prosecution**
   Based on complaints received, Court will pass laws prohibiting various types of content and/or order the removal of content [13, 14, 16]. All publishers, authors, and ISPs are liable to remove, alter, or block access to those specific content. They may defend the judgment by going for an appeal and obtaining stay orders [10].
2. **Detention**
   Those publishers, authors, and ISPs who fail to remove illegal content will be arrested [12]. Later they may be punished with fines and imprisonment. As an author, he/she may be banned from further publishing for some duration. Businesses may be closed down by revoking their licenses.
3. **Blackmail and Other Criminal Practices**
   Publishers, authors, and ISPs who publish uncensored content may be threatened and attacked by people who were affected by this content. This may even lead to murder. People may employ hackers who will threaten ISPs and local authorities on behalf of them to work according to their interests.
4. **Bribes, Promotion, and Other Forms of Payment**
   Individuals/websites may be given incentives for supporting certain claims and viewpoints. They will be promoting articles and comments in support of one group or attacking opposition groups without notifying the readers.

5. **Controlling Network Access**
   Some social networking sites have mandated verifying phone numbers while registering. This has reduced anonymous attacks in social networks to some extent [15].

Internet censorship should be done in such a way that the internet remains a great source of reliable information. At the same time, we should protect those vulnerable to internet exploitation. As nobody has complete control over the internet, it is very difficult to punish a person for internet crimes like defamation, copyright infringement, and hate crimes. The uncensored internet can negatively affect the lives of several people. The censorship of internet can protect people from malware, ransomware received via internet making their internet life more safe and simple. Internet censorship prevents inappropriate information flow and ensures that critical information do not reach the wrong people. Internet censorship helps in preventing a large number of financial frauds and identity thefts

In recent years, internet bullying and violence has become a major concern [17]. Users can be anonymous on the internet and information spreads rapidly over the internet. Some users take advantages of such properties of the internet to create violence. The users may abuse, defame each other, and expose others privacy bringing great harm to them. Many celebrities are victims of such internet violence. There are incidents of internet users being cheated by other users through social networking sites. The occurrence of all these incidents and similar incidents make the internet censorship absolutely necessary and demanding.

Censoring the internet is not a simple process. Often, censorware suffers from several drawbacks like

1. **Overblocking/Over-Censoring**
   Overblocking refers to a scenario where legitimate content is getting blocked by censorware. For example, some health-related information may be censored unintentionally believing it to be porn material. Sometimes authorities prefer overblocking rather than risking allowing access to undesirable sites.
2. **Underblocking/Under-Censoring**
   Underblocking refers to a scenario where content that needs to be censored according to censorship policy is not censored properly using censorware. It happens when censorware fails to identify the content as undesirable. Whenever a new category of malicious information is uploaded to the internet, censorware will not censor the content unless updated quickly and accurately.
3. **Violation of Constitution**
   If any government try to censor a particular moral or political issue without valid reasons, it is considered as a violation of democracy and will be disapproved. Without adequate governmental supervision/permission, no censorware should be ideally deployed in a public network. Any form of internet censorship taking place should be informed to visitors using error 451.
4. **Legal Necessities**
   Internet censorship faces various legal actions in several countries. Censorship doesn't face many legal actions in aristocratic regimes like North Korea and

China whereas a large number of cases are filed against and file for censorship in democratic countries like India. In order to not face any legal actions, all censorware developers should ensure that their censorware does not suffer from overblocking and underblocking before mass distribution. They should document properly what all software standards were followed while development and testing. They should also mention any limitations identified in documentation.

Internet censorship circumvention refers to various processes of bypassing internet censorware and gaining access to censored materials. Usually, the common people lack expertise and knowledge to circumvent censorship but for most of the technologically savvy users, circumventing internet censorship is just a piece of cake. Circumventing works because censorship does not necessarily remove content from the internet but just makes it difficult to access it. Whenever a new blocking technology is introduced, anti-censorware developers reverse engineer it and find a new circumvention technology which can bypass it [4]. Different tools and strategies are used for internet censorship circumvention, including

1. **Cached Web Pages**
   Search engines like Google keep snapshots of web pages from an earlier point of time. Cached pages are identical to the original page in most cases. Even if the original website is blocked, cached web pages may still be accessible. The advantage of this technique is that no additional software needs to be installed
2. **Mirror Sites**
   Mirror websites or mirrors are replicas of other websites. Therefore even if the original website is blocked, copies of the website are still present at mirror sites which are not blocked. Using such sites, blocked content can still be accessed.
3. **Web to Email Services**
   Web to email services will return the contents of web pages with or without images as an email message. The content of a blocked web page can be accessed as an email using this service.
4. **Feed Aggregators**
   A feed is a Web document that is a shortened version of a Web page. Feed aggregator or RSS aggregator collects feeds from different web pages and shows it in a desktop window or web browser. Using such aggregators, blocked content can be retrieved directly.
5. **Direct IP Addresses**
   Several sites may own multiple domains. Only a few such domains or URLs may be blocked. Others will still be available. Trying to access an IP address directly will sometimes allow access to a blocked site. Some censorware can be fooled by entering the IP address in a base other than 10.
6. **Alternative DNS Servers**
   DNS server contains a database of public IP addresses and their associated hostnames. It helps in translating domain names to IP addresses as requested [1]. DNS servers are usually owned by ISPs and other private business organizations. Using DNS servers other than those supplied by default by an ISP may bypass DNS-based blocking.

7. **Proxy Websites**
   Proxy websites are the fastest way to circumvent censorship. They act as an intermediary between the user and the blocked website. User visits the proxy website and requests access to a blocked website by submitting the URL of the blocked website and initiating a connection. The proxy website will fetch the requested content and displays it.
8. **Reverse Proxy**
   A website may have several web servers behind a proxy. A reverse proxy server takes client requests from the internet and forwards it to one or more servers. These resources are then returned to the client as if they originated from a single server. Websites can avoid censorship by rerouting traffic using reverse proxies. Reverse proxies can also protect original characteristics and existence of actual web servers thereby making censorship difficult.
9. **Virtual Private Networks (VPNs)**
   Using VPN, censored users can create a secure connection to a country with relaxed censorship rules. Once connection is established, they can browse the internet as if they are in that country. Thus all blocked content can be accessed easily and safely.
10. **SSH Tunneling**
    SSH tunneling created an encrypted SSH connection. Users can transport all their traffic through this connection. Thus, both outgoing requests for blocked sites and the response from those sites are hidden from the censors.
11. **Sneakernets**
    A sneakernet refers to transferring information from one place to another by physically carrying electronic data on a storage media. Since we are not using any computer networks for transfer, no censorship is applicable to such transfer.
12. **Hybrid Combinations**
    Circumvention methods mentioned above can be combined to form hybrid methods which are more effective against censorship. For example, we can combine alternate DNS server technology together with VPN to create a smart DNS proxy server.

The above circumvention techniques differ in ease of use, speed, security, and risks. They target to achieve an uncensored internet connection. Rather than using the above techniques, using alternate protocols like FTP, telnet, or HTTPS will bypass some censorware. Some censorware can be fooled by conducting searches in a different language. Some countries have strict laws against circumvention. Yet, people are using several nonsecure ads based circumvention software. Internet censorship transparency is necessary to avoid confusions and negative attitude towards internet censorship. Only very few countries in the world openly admit that they practice internet censorship. Most of them would not even disclose censorship techniques employed, list of blocked websites, etc. leading to public protests against censorship.

The sensors may target nodes, users, or links. They may employ multiple strategies to filter malicious content. All censorware in the market should enforce censors at all scenarios without affecting performance. They must be scalable and cheap. They should provide accurate results with minimum false positives and false negatives. Every censorware should be capable of adapting against new circumvention techniques.

3 New Censorware Proposed

We are proposing a new censorware, Ever Learning Censorware, which learns continuously based on Naive Bayes learning technique. After every learning, it shall identify and filter harmful websites in a better way. The major components of the proposed censorware are

1. blacklist—A blacklist of domains or keywords and their degree of harm
2. classifier—A Bayesian classifier which will classify domains, keywords into harmful, moderate, and harmless
3. packet capture engine—It captures all network traffic. Whenever a site is accessed, it captures such packets, extracts domain/URL, and gives to a classifier for classification. If the packet is identified as harmful, communication is dropped.

In addition, we are storing recent harmful sites in a recent sites list to improve performance. It is periodically updated to remove entries older than 2 days. It will contain only one entry for each site with last accessed time. If a site is present in recent sites list as well as blacklist with a high degree of harm, it is simply returned as harmful.

The attacking mode chosen is to attack the harmful link in the network rather than targeting a particular node/user. The filtering approach used is as follows:

1. Drop all communications which cannot be analyzed at all.
2. IP Filtering—Drop packets send to and received from the website found on the blacklist with a high degree of harm.
3. Filtering based on classifier- Use classifier to find the nature of new domains not present in the blacklist. Add domain and degree of harm to the blacklist. If the domain is harmful, drop the communication. If moderate, do keyword filtering. If harmless, allow communication.
4. Keyword Filtering—Search for blacklisted keywords and blacklisted links in packet content. If the number of such keywords/links reaches a threshold (say 400), block communication, add/update the entry in blacklist with a high degree of harm. If all keywords have a low probability of being harmful (below a threshold), allow communication.
5. DNS Hijacking—In case of dropping communication, redirect the user to a block page confirming that content is being censored and asking for any suggestions. These suggestions can be stored in a central server and can be reviewed periodically.

Rather than completely blocking content which cannot be analyzed, we can log those domains and allow communication after getting user consent confirming communication is legal. An admin can periodically review non-analyzable sites and take remedial measures. In order to measure, the degree of harm of a site, a new censoring approach is identified based on Bayes theorem.

Particular words have particular probabilities of being harmful and getting censored. For example, keywords like *porn*, *drug*, *suicide*, *murder*, *book*, *pencil*, *movie*, *film*, *music* have respective probabilities (degree of harm) of 1, 0.9, 0.95, 0.93, 0, 0, 0.6, 0.7, 0.4. The filter will not know these probabilities without training. For manual training, the user must manually indicate a word and its degree of harm. If a new word is encountered after training, it is assigned a random degree of harm (say 0.4). This can be reviewed by a moderator initially. The degree of harm associated with every word will be continuously updated once the number of websites containing that word increases a threshold. It is calculated by dividing the total number of sites containing that word with the total number of sites inspected. After a significant amount of testing, the degree of harm associated with each word is expected to not deviate much. The probabilities associated with each word found on the website are used to calculate the probability that a website belongs to which category. Each word in a website (including domain name) contributes to the probability that the website is harmful. The website's probability of harm is computed and if it is greater than some threshold values (say 0.9, 0.5), websites are classified as harmful and moderate. Otherwise, it is classified as harmless

Let us assume that a website contains word "videos." It may be a benign website or a malicious website. Internet censorware will try to identify whether a website is harmful from this particular word. For that it uses the formula based on Bayes theorem

$$P_r(A|B) = \frac{P_r(B|A) \cdot P_r(A)}{P_r(B|A) \cdot P_r(A) + P_r(B|\neg A) \cdot P_r(\neg A)}$$

where

- $P_r(A|B)$ is the probability that accessed website is harmful knowing that it contains this keyword
- $P_r(B|A)$ is the probability of occurrence of this word in harmful websites. It is same as the degree of harm of word calculated by censorware
- $P_r(A)$ is the marginal probability that a website is harmful. It is calculated by dividing the total number of sites identified as harmful with the total number of inspected sites.
- $P_r(B|\neg A)$ is the probability of occurrence of word W in harmless websites.
- $P_r(\neg A)$ is the marginal probability that a website is harmless. It is calculated by dividing the total number of sites identified as harmless with the total number of inspected sites.

If we determine the harmness of a website only based on the presence of a single word, it is error-prone. We need to consider several words and combine their harm to determine a website's overall degree of harm. Combining Individual probabilities, we will get the following formula for Computing the probability that a website is harmful

$$P = \frac{P_1 \cdot P_2 \cdots P_N}{(P_1 \cdot P_2 \cdots P_N) + ((1 - P_1) \cdot (1 - P_2) \cdots (1 - P_N))}$$

where

- P is the probability that a suspected website is harmful.
- $P_1, P_2 \ldots P_N$ are the probabilities that a website is harmful knowing it contains words $w_1, w_2 \ldots w_N$
- N is the total number of valid words on the website

Rather than assigning a random degree of harm to new words not present in the blacklist, the classifier can also decide to discard such words for which there is no information available. Words like *the*, *a*, *some*, *is*, etc. for which degree of harm cannot be defined are ignored. It is obvious that we must not assign a degree of harm to numeric data, special symbols, and spaces. Even if we ignore such harmless components, there would not be much impact. We keep a list of such words to filter them. Even if censorware automatically adds a new such word to blacklist, it is removed later during periodic moderation. We can also try grouping words rather than a single word. Thus accuracy can be improved.

A background service should run continuously and stores the latest content of critical components like the blacklist, recent sites list, non-analyzable list to stable storage. This ensures that all the learned information are safe. Rather than trying to update all entries, we can recreate lists. Thus even if the censorware is terminated due to unforeseen consequences, all learnings are not lost. Censorware can be restarted and used based on these lists in stable storage. In such scenarios, we will only lose learnings that could have been conducted in a short interval from the last update time. Keeping backups of all these stable storage components is recommended to avoid data loss in case censorware is terminated while modifying the content in stable storage.

3.1 Implementation

The language used is Java 8. Java was used because of its support for network analysis and machine learning. Java applications can be modified, rewritten, or enhanced easily and can be run on almost all operating systems. An open-source Java library jnetpcap-1.4.r1425 was used to capture HTTP packets and get URL of visited websites. Jnetpcap is a java wrapper for popular libpcap and WinPcap libraries. In order to read the contents of a website, HtmlUnit was used. HtmlUnit is a headless web browser written in Java. It can simulate browsers like chrome and can extract data from websites. All lists (blacklist, recent sites list, non-analyzable list) were stored as simple Unicode files inside the project present in the file system. The entire project was built using Maven 3.5.3 to create our java app.

Implementation Modules

1. **Driver Engine**
   This module contains the main method. It initially calls init method to populate various components like the blacklist, recent sites list, etc. If required files are not available for populating the lists, initial training is carried out. Then Driver engine continuously spawns various threads for packet capturing, classification, automated component update, etc. after confirming they are not alive.
2. **Packet Capture Engine**
   It first gets a list of network devices on the running system. Second, it opens up the selected network device. Third, we create a packet handler which will receive packets from the libpcap loop. Fourth, we enter the loop and tell it to capture 10 packets. The loop method does a mapping of pcap.datalink() DLT value to JProtocol ID, which is needed by JScanner. The scanner scans the packet buffer and decodes the headers. The mapping is done automatically. If the header is of type HTTP, the URL of the accessed website is extracted. After confirming it is not yet tested recently, it is passed to the CheckDomain method of Bayes Classifier for analysis
3. **Bayes Classifier**
   The init method is located here. It also maintains array lists corresponding to blacklist, recent sites list, non-analyzable sites list, etc. It is implemented as a thread which periodically writes these array lists to their corresponding files (blocked, recent, non-analyzable) in the file system. When a site URL is passed to the CheckDomain method of Bayes Classifier, it tries to read the content of webpage using WebScraper. If the content is read, it applies Bayes classification method defined earlier to check the degree of harm of a website. If the website is harmful, it is added to the hosts file to block its further access. If the website is found to be not harmful, hosts file is updated to remove entries corresponding to this website. The website is added/updated in blacklist with a calculated probability
4. **Web Scraper**
   It accepts a URL and gets the content of corresponding web page using htmlunit library. The browser version was given as best_supported. If a website cannot be read, it returns 403 forbidden errors. Such websites are added to nan-analyzable lists.
5. **Word**
   It indicates an entry in the blacklist. It includes corresponding word/domain and its degree of harm.

6. **Site**
   It indicates an entry in the recent sites list. It includes site URL, last accessed time, and degree of harm.
7. **Manual Training Engine**
   It allows Admin to manually train the system. Admin can add a new entry to blacklist, search for any entry and can remove an entry from the blacklist. Admin can also write entire blacklist to blacklist.txt file.

3.2 Results

The initial training was carried out manually by populating blacklist with harmful, benign websites, words, and their degree of harm. Then Network data was sniffed continuously using packet capture engine module. If the system has multiple network devices, packets coming from any network device are captured. Whenever user tries to access a website, corresponding HTTP packets are captured successfully. The URL was extracted from those HTTP packets and was sent to analyze. As explained in earlier section, Bayes classifier analyzes degree of harmness of website. If the web page is analyzed and identified as not harmful, communication is allowed. Otherwise, communication is dropped and the domain name is added to the hosts file to prevent further access. In both cases, the blacklist is updated with non-redundant pairs of (words, the degree of harm) corresponding to website content and domain. When the user tries to access this site later, he would not be able to establish a connection. The degree of harm of each website is reevaluated periodically to avoid unnecessary long-term blocking errors. Thus censorware is successfully blocking websites identified as harmful. The success of censorware entirely depends on the blacklist and initial training should be carried out extensively to cover all domain types. As the training continues, size of blacklist is increasing exponentially resulting in performance degradation. As part of future research, blacklist may be changed to an indexed database to improve performance.

4 Conclusion

Internet censorship is really necessary for today's society. Since the internet is growing on a daily basis and has a wide range of applications, misuse of the internet can lead to drastic consequences. The censorware proposed in this project can overcome many limitations of the existing system and is far more efficient. It is also much transparent compared to existing leading to greater public support. The training can become a little cumbersome, but it can be managed. Manual intervention is needed only at the beginning. This design can be extended to implement similar censorware in routers and other internet endpoints. We can improve the proposed censorware by introducing a mechanism to analyze the https packet. One method suggested is reassembling TCP packets and analyzing the assembled packet. Another area which can be improved is the blocking policy. We can think of an intermediate DNS server or a Firewall rather than adding to the hosts file. As part of result analysis, we have identified that the size of the blacklist is increasing tremendously with each website tested. We can try to restrict size by grouping similar words, irrelevant words, etc.

References

1. Sarkar, P.K., Jain, A.: Intelligent Transport System. PHI Learning, 15 Nov 2017
2. Orlowski, A.: India Blocks Yahoo! Groups. The Register, 24 September 2003
3. Montieri, A., Pescape, A., Aceto, G.: Internet censorship in Italy: an analysis of 3G/4G networks. In: IEEE ICC 2017 Communications QoS, Reliability, and Modeling Symposium (2017)
4. Leberknight, C.S., Wong, F., Poor, H.V., Chiang, M.: A taxonomy of internet censorship and anti-censorship. In: Fifth International Conference on Fun with Algorithms (2010)
5. Raj, D.: BuyDomains.com Blocked in India for no Obvious Reason. TechBlogger (2012)
6. Government to block all porn sites in India, asks Internet providers to deny access to such websites. Mobiletor, 11 November 2014
7. India blocks 32 websites, including Vimeo and Github. India Today, 31 December 2014
8. India blocks Pakistani newspaper web site. Newswatch.in, 5 July 1999
9. Anwer, J.: Blocking Website in India: Reliance Communications Shows It Is Very Easy. Times of India, 24 December 2011
10. Pereira, L.: Singham Effect: File Sharing Sites Blocked. NDTV, 22 July 2011
11. Gupta, M., Verkamp, J.-P.: Inferring Mechanics of Web Censorship Around the World. FOCI '12, 2012
12. Pahwa, N., et al.: Updated: Indian Government Blocks Typepad, Mobango, and Clickatell. MediaNama, 4 March 2011
13. Pahwa, N., et al.: No more John Doe orders? Indian ISPs get court order for specificity in URL blocks. MediaNama, 20 June 2012
14. Pahwa, N., et al.: 219 Websites Blocked in India, after Sony Complaint. MediaNama, 7 July 2014
15. Orkut's tell-all pact with cops. The Economic Times, 1 May 2007
16. Ribeiro, J.: Delhi Court issues summons to Google, Facebook headquarters for objectionable content. PC World, 16 January 2012
17. Deibert, R.J., Palfrey, J.G., Rohozinski, R., Zittrain, J. (eds.): ONI Country Profile: India, Access Contested. OpenNet Initiative, MIT Press, November 2011, pp. 299–308
18. Sengupta, S.: India Blocks Blogs in Wake of Mumbai Bombings. The New York Times, 18 July 2006

Multiclass Sentiment Analysis in Text and Emoticons of Twitter Data: A Review

Nirmal Varghese Babu and Fabeela Ali Rawther

1 Introduction

1.1 Sentiment Analysis

Sentiment Analysis means the process of analyzing the views of various people in various situations based on polarity. It is aimed at users' attitudes toward various situations by investigating and extracting texts that involve the user's opinion, sentiments, etc. [12]. Nowadays, it is an emerging trend because many organizations or institutions follow this procedure to understand people's views and opinions. For example, the usage of a particular product can be analyzed by the way in which people respond to it [9]. This analysis also helps to find out the trends that people follow today. With the help of this analysis, the data or reviews posted by various users on social media, say, on Twitter, will be categorized using various classification algorithms. Sentiment analysis uses various natural language processing tools for the extraction and processing of data. These tools analyze the data and process based on the data they are handling.

One of the important features found in social media analysis states that Twitter analysis involves the usage of hashtags. As the maximum word count of data that can be posted online in Tweets is reduced to 280 characters; thus, it will be easy for the process of keyword matching. The data will be extracted using various keywords called hashtags, for example, #twitter. The data or Tweets matching with the keyword are extracted. After the extraction of data, various pre-processing [10] steps, including the removal of URLs, special symbols, full stops, stop words,

N. V. Babu · F. A. Rawther (✉)
Amal Jyothi College of Engineering, Kanjirapally, India
e-mail: nirmalvarghesebabu@cs.ajce.in; fabeelaalirawther@amaljyothi.ac.in

M. Palesi et al. (eds.), *Second International Conference on Networks and Advances in Computational Technologies*, Transactions on Computational Science and Computational Intelligence, https://doi.org/10.1007/978-3-030-49500-8_6

Fig. 1 Sentiment analysis steps

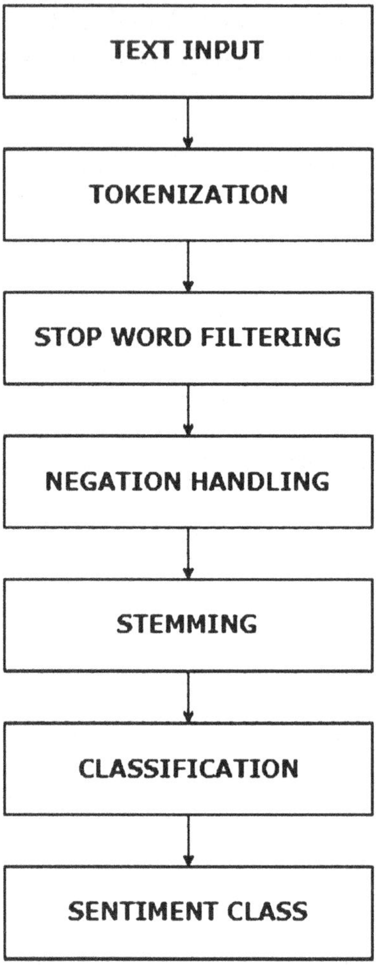

etc., are carried out, where the invalid or non-useful data are removed. These preprocessing helps to decrease the processing time of the available data, hence the expected output will be produced fast. Various natural language tools or packages are called for this particular procedure. Figure 1 shows the various steps in the sentiment analysis.

The next step is feature extraction [2], where the important features available in the data are extracted using various algorithms such as N-gram, term frequency, etc. Various features include sentiment features, unigram features, sarcastic features, and semantic features. The polarity or the score is calculated based on the extracted features available in a particular Tweet. After finding the sentiments of each data, various classification algorithms like Naive Bayes, K-NN, Random Forest etc. will be used to classify the tweets based in the various class or sentiments then the accuracy of each class will be calculated [11].

2 Preliminary Review

Based on the previous research, most of these have taken place in the binary and ternary classification of the texts. These reviews can be categorized mostly by:

- The classification algorithms used.
- The features used.
- The framework used.
- The emoticons used.

2.1 Classification Algorithms

In sentiment analysis, the classification of the sentiments or words is required to identify the user behavior or to perform the multiclass classification. There are several algorithms [15] that can be used to perform the classification, including Naive Bayes, K-NN, Random Forest, Support Vector Machine, etc.

Neural Networks The advantage of this technique [1, 4] is an ability to detect the possible interactions between the predictor variables. It could also perform complete detection without having any doubt in complex nonlinear relationship dependent and independent variables. Thus, it is selected as the best prediction method.

Naive Bayes The probabilistic classifiers [14, 16, 17] are based on applying Bayes' theorem with strong independence assumptions between features. The classifier was used to predict various values based on the features. Tirupati et al. [17] stated that the data collected from the Twitter streaming API and the sentiments were classified using Naive Bayes classifier and graphical user interface was there to observe the output [17]. Parveen and Pandey [14] stated that the Hadoop framework can be used for huge data storage and MapReduce can be used for data processing. Data can be extracted from the Twitter REST API and classified using the naive Bayes classification algorithm. The sentiments of Tweets with and without emotions were found [14]. Tago and Jin [16] considered neutral Tweets during the classification of sentiments and they were scored 0 [16]. Table 1 shows the confusion matrix of the Naive Bayes Classifier.

Table 1 Confusion matrix of Naive Bayes classifier

Prediction	Actual	
Negative	25	22
	4	5
Positive	11	10

Source: Ragupathy and Maguluri [15]

Table 2 Confusion matrix of the K-NN classifier

	Actual	
Prediction	Negative	Positive
Negative	0	0
Positive	50	51

Source: Ragupathy and Maguluri [15]

Table 3 Confusion matrix of the Support Vector Machine classifier

	Actual	
Prediction	Negative	Positive
Negative	0	0
Positive	150	401

Source: Ragupathy and Maguluri [15]

K-NN Stands for K-Nearest Neighbor algorithm, which is used to classify objects based on the neighbors near a particular class. Bouazizi and Ohtsuki [7] stated that the K-NN algorithm can be used to perform multiclass classification by categorizing various sentiment classes by identifying the neighbors nearby or satisfy a particular sentiment class [7]. Table 2 shows the confusion matrix of the K-NN classifier.

Random Forest Random forest operates by constructing decision trees at training time and the output is produced based on the class of the individual trees. Bouazizi and Ohtsuki [5] stated that the sentiment data belonging to the various sarcastic classes can be identified by finding the classes under the same tree or group by multiclass classification [5]. Also, in 2016, they noted that the total quantity of the features in the sentiment classes can be found out using the unigram score and pattern score [6].

Support Vector Support vector is a representation of points in space mapped so that they can be separated by a clear gap. It looks at extremes of datasets and draws a decision boundary, also known as a hyperplane, near the extreme points in the datasets. It is the widest margin that separates the two classes. Extreme data points are known as support vectors. Support vectors are most difficult to classify. The distance between the support vectors and the hyperplane is as wide as possible, which is another way of saying we maximized the margin. Baltas et al. [3] stated that the sentiments of the same sentiment class can be figured out using the classes that belong to the same type [3, 8]. Spark architecture has classifiers where various classification algorithms are there by default to perform the classification. Table 3 shows the confusion matrix of the support vector machine classifier (Table 4). Table 5 shows the experimental results using the naive Bayes and the support vector machine in Spark architecture.

Decision Tree A decision tree is a decision support tool that uses a tree-like model of decisions and their possible consequences can also be identified. The data with different numbers of levels, where the information gain in decision trees is biased

Table 4 Confusion matrix of decision tree classifier

	Actual			
Prediction	More negative	More positive	Negative	Positive
More negative	0	0	0	0
More positive	10	25	6	7
Positive	0	0	0	0
Negative	0	0	0	0

Source: Ragupathy and Maguluri [15]

Table 5 Experimental results using naive Bayes and support vector machine in Spark architecture

Classifier	Accuracy
Naive Bayes	85.4%
Support vector machine	86%

Source: Samar Al-Saqq et al. (2018)

Table 6 Binary classification F-Measure

Data set size	Naive Bayes	Decision tree
1,000	0.572	0.597
5,000	0.684	0.556
10,000	0.7	0.568
15,000	0.71	0.576
20,000	0.728	0.59
25,000	0.725	0.56

Source: Baltas et al. [3]

in support of those attributes with more levels and the calculations are very tough to resolve. Mainly, if many values are uncertain and if many outcomes are attached, greedy algorithms cannot give any assurance to return the globally optimal decision tree and can be decreased by training the multiple trees. Watanabe et al. [19] stated that hate speech sentiments can be identified by finding which sentiments fall under the corresponding sentiment class. Table 4 shows the confusion matrix of the decision tree classifier. Table 6 shows binary classification F-Measure.

2.2 *Features*

Various features [2, 3, 5–8, 14, 16, 17, 19] that are extracted during the feature extraction procedure include sentiment features, unigram features, punctuation features, sarcastic features, syntactic and stylistic features, top words, semantic features, pattern-related features, hate speech features. These features are extracted from the data to calculate the sentiments for performing the classification procedure. It is concluded that the features extracted are used to identify the polarity of the data.

2.3 Framework

Framework mainly focuses on the Hadoop and Spark framework. Hadoop [14, 17] and Spark [3, 8, 13] help in data processing where MapReduce takes an initiative. Hadoop is responsible for the linear processing of huge data sets, whereas Spark is responsible for the iterative processing, fast data processing, graph processing, real-time processing, etc.

2.4 Emoticons

Emoticons [18] can also be considered while performing sentiment analysis. Emoticons even have textual meaning from which the inner meaning can be extracted and a multiclass classification can be performed using the various types of emoticons that represent the exact textual data meaning. The sentiment analysis can be performed by collecting the data from various social media [20] and the emoticons can be processed the same way as the text.

3 Outcome of Survey

Using the Tweets collected from Twitter, the sentiments of the people under certain circumstances can be identified. Three types of classifications can be observed: Binary, Ternary and Multiclass classification. In binary, the sentiments classified are positive and negative, where both classes have a good level of accuracy, as there is no deep understanding of the words or sentiments, whereas in ternary classification, a new class, called neutral, where sentiments other than positive and negative are classified. Thus, a change, most likely less accuracy, can be seen. In multiclass classification, the sentiments are classified into many classes, where more precise and accurate classification will be achieved. Also, emoticons can be a part of the text, where they can be used to identify various sentiments. Mostly, the classifications are carried out using various classification algorithms such as Naive Bayes classifier, K-NN, Random Forest, Support Vector Machine, etc., using the features and patterns or the sentiments collected from the data. The efficiency of various classification algorithms using various sentiments can be verified using four performance indicators, i.e., accuracy, prediction, recall, and fmeasure.

A precise and accurate classification or analysis can be expected from the combination of multiclass classification in the Spark architecture using the K-NN classification algorithm or neural networks, as the Spark framework is best for parallel processing because of its in-memory database. Also, neural networks are considered to be the best prediction algorithms as the analysis mostly works at the level of prediction. Moreover, K-NN is also considered to be the best for figuring out the neighboring sentiments based on the sentiment classes.

4 Summary and Conclusions

Sentiment analysis is a wide area where a lot of research is going on nowadays, as it is an essential requirement for understanding how people behave in various situations under certain circumstances. Thus, understanding people's sentiments or emotions is not an enviable task. As the data are in the form of reviews or Tweets by various users online, a framework, for example, Hadoop or Spark, is essential for handling this large amount of data. After all the pre-processing and feature extraction steps, the data or sentiments are classified into various classes using various classification algorithms. Binary and ternary classifications are based on positive, negative, and neutral classes. In multiclass classification, the sentiments are classified into various sentiment classes; hence, a correct or accurate classification can be expected. As for future work, an algorithm can be developed to perform the sentiment analysis and classification of the data/Tweets that contain both the texts and emoticons by finding out the sentiments/polarity of both the texts and the emoticons together.

References

1. Ahan, M.R., Rohmetra, H., Mungad, A.: Social network analysis using data segmentation and neural networks. In: International Research Journal of Engineering and Technology (IRJET) (2018)
2. Alsaedi, N., Burnap, P.: Feature extraction and analysis for identifying disruptive events from social media. In: International Conference on Advances in Social Networks Analysis and Mining. IEEE/ACM, Piscataway/New York (2015)
3. Baltas, A., Kanavos, A., Tsakalidis, A.K.: An Apache Spark implementation for sentiment analysis on Twitter data. In: International Workshop of Algorithmic Aspects of Cloud Computing (2017)
4. Borele, P., Borikar, D.A.: An approach to sentiment analysis using artificial neural network with comparative analysis of different techniques. J. Comput. Eng. **18**, 64–69 (2016)
5. Bouazizi, M., Ohtsuki, T.: Sarcasm detection in Twitter. In: IEEE Global Communications Conference (GLOBECOM). IEEE, Piscataway (2015)
6. Bouazizi, M., Ohtsuki, T.: Sentiment analysis in Twitter: from classification to quantification of sentiments within Tweets. In: IEEE Global Communications Conference (GLOBECOM). IEEE, Piscataway (2016)
7. Bouazizi, M., Ohtsuki, T.: A pattern-based approach for multi-class sentiment analysis in Twitter. IEEE Access **5**, 20617–20639 (2017)
8. Bouazizi, M., Ohtsuki, T.: A large-scale sentiment data classification for online reviews under Apache Spark. In: 9th International Conference on Emerging Ubiquitous Systems and Pervasive Networks (2018)
9. Chauhan, C., Sehgal, S.: Sentiment analysis on product reviews. In: International Conference on Computing, Communication and Automation (2017)
10. Gokulakrishnan, B., Priyanthan, P., Ragavan, T., Prasath, N., Perera, A.: Opinion mining and sentiment analysis on a Twitter data stream. In: The International Conference on Advances in ICT for Emerging Regions (2012)
11. Kanavos, A., Nodarakis, N., Sioutas, S., Tsakalidis, A., Tsolis, D., Tzimas, G.: Large scale implementations for Twitter sentiment classification. MPDI Algorithms (2018)

12. Nanli, Z., Ping, Z., Weiguo, L., Meng, C.: Sentiment analysis: a literature review. In: International Symposium on Management of Technology (ISMOT). IEEE, Piscataway (2012)
13. Nodarakis, N., Sioutas, S., Tsakalidis, A., Tzimas, G.: Large scale sentiment analysis on Twitter with Spark. In: EDBT/ICDT Joint Conference (2016)
14. Parveen, H., Pandey, S.: Sentiment analysis on Twitter data-set using naive Bayes algorithm. In: 2nd International Conference on Applied and Theoretical Computing and Communication Technology (iCATccT). IEEE, Piscataway (2016)
15. Ragupathy, R., Maguluri, L.P.: Comparative analysis of machine learning algorithms on social media test. Int. J. Eng. Technol. **7**, 10425 (2018)
16. Tago, K., Jin, Q.: Analyzing influence of emotional Tweets on user relationships by naive Bayes classification and statistical tests. In: 10th International Conference on Service-Oriented Computing and Applications. IEEE, Piscataway (2017)
17. Tirupati, M., Pabboku, S., Narasimha, G.: Sentiment analysis on Twitter using streaming API. In: The International Advance Computing Conference (2017)
18. Wang, H., Castanon, J.A.: Sentiment expression via emoticons on social media. In: International Conference on Big Data. IEEE, Piscataway (2017)
19. Watanabe, H., Bouazizi, M., Ohtsuki, T.: Hate speech on Twitter: a pragmatic approach to collect hateful and offensive expressions and perform hate speech detection. IEEE Access **6**, 13825–13835 (2018)
20. Wolny, W.: Sentiment analysis of Twitter data using emoticons and Emoji ideograms. In: Information Systems Development: Complexity in Information Systems Development, Katowice (2016)

A Survey on DDoS Prevention, Detection, and Traceback in Cloud

Ajeesh Ramanujan and Blesson Andrews Varghese

1 Introduction

The World Wide Web Security forum defines Distributed Denial of Service (DDoS) as a coordinated attack launched by many users against one or more targets. The main actors in a DDoS attack are handler(s), agents, and victim(s). What makes DDoS attacks so dangerous is that the attacker is using client/server technology thereby multiplying the effectiveness of attack significantly using the computational power of compromised hosts which serve as attack platforms. During the initial phase of an attack, the attacker sets up a DDoS army by installing a malicious DDoS master program in one or more computers using a stolen account or id. Whenever any computer or device comes in contact with this compromised device, this worm is copied to those systems and are enslaved. These enslaved systems may include servers, handheld devices, personal computers, etc. Some of these enslaved systems become handlers and others will act as bots. Handlers will contain a malicious program to issue commands to bots. Attacker contacts handlers to communicate with agents. The attacker can launch hundreds to thousands of slave programs within seconds with the help of handlers. The second phase is to plan the attack. The attacker decides the targets, time of the attack, duration of attack, and bots to be used. The final phase is to launch the attack. When handlers receive the command from the attacker, they command bots controlled by them to launch the attack. The victim stands no chance against the aggregated computational power of these zombies and crashes. The attacker may use an additional central command and control server to manage all handlers and coordinate attack (Fig. 1).

A. Ramanujan (✉) · B. A. Varghese
CSED, College of Engineering Trivandrum, Thiruvananthapuram, India
e-mail: ajeesh@cet.ac.in

M. Palesi et al. (eds.), *Second International Conference on Networks and Advances in Computational Technologies*, Transactions on Computational Science and Computational Intelligence, https://doi.org/10.1007/978-3-030-49500-8_7

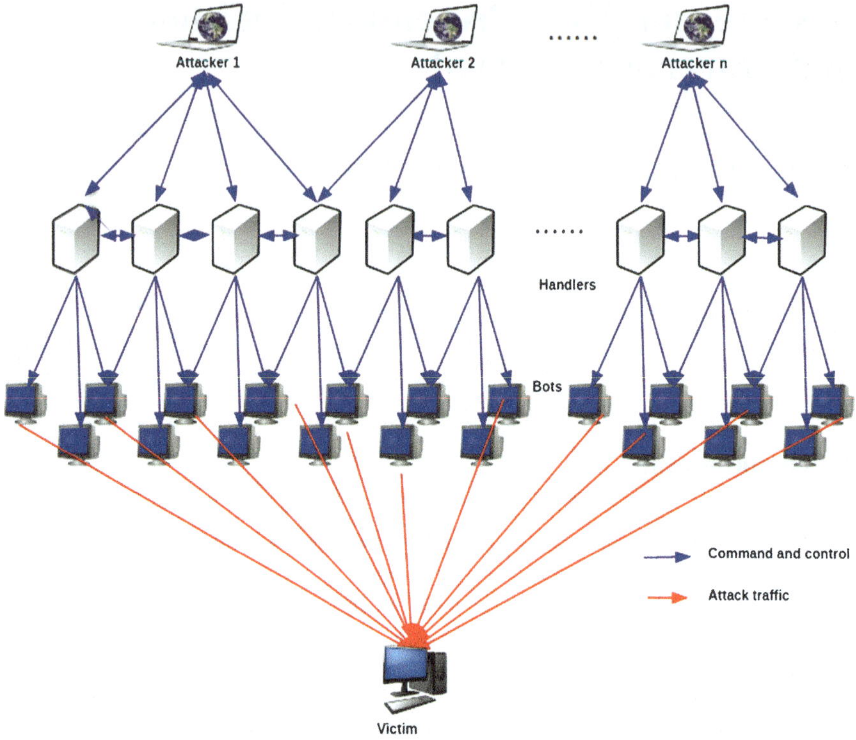

Fig. 1 DDoS attack model

In the main variants of DDoS attacks, Distributed Reflection Denial of Service (DRDoS) attacks are the most similar. DDoS and DrDos only differ in the way in which the attack is launched. In case of direct DDoS attack, bot directly sends service request packets to victim posing as a legitimate user. In the DrDoS attack, the attacker poses as the victim and sends spoofed packets to several unsuspecting nodes called reflectors. These spoofed packets contain the victim's address as source address and demand a reply from the reflector. The reflectors on receiving these spoofed packets send replies to the victim believing it to be the sender. Since the number of reflector and attack packets is large, the victim will receive a mass number of junk packets in a single moment. The victim needs to waste a sufficient amount of memory and computational power to store, read, and process these packets. The continuous arrival of packets will overwhelm the victim in a short period and the victim will not be able to service legitimate requests. Since DDoS defense applies to DrDoS too, DDoS defense measures apply to DrDos too.

DDoS attacks are becoming more sophisticated every year. Though small businesses are more prone to DDoS attacks, DDoS victims include major businesses like Yahoo, eBay, Microsoft too. Attackers are figuring out novel ways in which DDoS attack can be launched. It is important that we study all these attack types and design new preventive measures. We must be ready to detect, block, and traceback any new unforeseen attack type. Section 2 covers all known DDoS attack types. Section 3 lists major DDoS attacks in the past. We have included the most simple DDoS preventive measures in Sect. 3.1. Sophisticated preventive measures can be designed for each business separately in addition to the same. Sections 3.1 and 4 explain different DDoS detection and Traceback schemes.

2 DDoS Classification

DDoS attacks can be broadly classified as bandwidth-based and resource-based attacks. Both types can be further divided into different subtypes based on the vulnerability they are exploiting and method used.

2.1 Bandwidth Depletion Attack

These kinds of attacks mainly consume victim's bandwidth by flooding the target system with unwanted useless traffic. As a result, legitimate traffic is prevented from reaching the victim network. Based on the way in which attack is launched, bandwidth attacks are classified as flooding attacks and amplification attacks

- **Flood Attacks**
  Zombies send a huge volume of attack traffic to the intended victim. This attack traffic per second will be usually much greater than the victim's network bandwidth. As a result, the network is clogged and the victim's bandwidth is saturated eventually. Thus legitimate traffic cannot reach the victim and no legitimate service requests reach the victim. There are mainly 2 kinds of flooding attacks:
  - **UDP Flood Attack**
    UDP is a stateless transmission protocol. It allows packets to be sent to a network without connection establishment. Attacker sends several UDP packets to random or specified ports of the victim system. On receiving these packets, the victim checks for applications associated with these packets. On finding no such application, the victim replies back with a "Destination Unreachable" packet. Attacker sends more and more packets continuously overwhelming the victim. As a result, the victim can no longer respond to legitimate requests
  - **Ping(ICMP) Flood Attack**
    ICMP is an internet protocol used by network devices to communicate with each other. ICMP request needs the receiving server resources to process each request send a response. Attacker knowing about this feature misuses it and sends a large number of ping requests (ICMP_ECHO_REPLY) to the

victim. The victim is overwhelmed as it must send responses to all the packets received. As a large number of request and reply packets are sent in both direction at the same time, the victim's network is clogged.

- **Amplification Attacks**
  Amplification attacks exploit the broadcasting feature of the internet. The attacker spoofs itself as the victim and sends thousands of packets to an IP range. These packets demand a reply. As a result, all the systems which received the attacker's packet send a reply to the victim system believing it to be the sender. A large number of systems sending reply cause malicious traffic in the victim's network. Amplification attacks mainly occur in the following forms
  - **Smurf Attack**
    Attacker posing as victim sends packets to systems (amplifier) that support broadcasting. All these packets have victim's IP as the return address. The network amplifier on receiving the request sends ICMP_ECHO_RESPONSE packet to all systems in the address range. This packet demands all receiving hosts to respond with ICMP_ECHO_REPLY. All receiving systems in address range believe it to be a legitimate request from the victim and send ICMP_ECHO_REPLY message to the victim. Though the victim simply ignores these messages, network is flooded with useless traffic causing loss/delay of legitimate requests
  - **Fraggle Attack**
    It is a variation of the Smurf attack. Similar to smurf, the attacker spoofs as the victim and sends UDP packets to a character generating port. All these packets contain the victim's address as the return address. The return port of victim will be a character generating port too. All systems which received above packets echoes back to the character generator port in the victim. Victim on receiving does the same as UDP echo packets are used. This process continues endlessly until the victim is dead

2.2 Resource Depletion Attack

These attacks target the most limited and precious resources of the network like memory, processing time, etc. Victim services request using these resources. Thus if these resources are exhausted, legitimate users cannot be serviced anymore. The following are major types of resource depletion attacks

- **Protocol Exploit Attacks**
  Attacker consumes a lot of resources by exploiting any specific feature of one or more protocols used/installed by the victim. Examples of protocol exploitation include TCP SYN attacks, authentication server attacks, and PUSH + ACK attacks.

- **Malformed Packet Attacks**
  A malformed packet has malicious information embedded in it along with data. This malicious information may mean nothing or may include flags which increase processing time of that packet. These packets take more processing time thereby increasing the workload of victims. The victim will ultimately crash or reset when a ton of malicious packets is received. Malformed packet attack mainly occurs in two forms.
  - IP Address attack: The source address and destination address of the attack packet will be the same. As a result, the victim is unable to judge how to handle this packet. It may abruptly stop execution or may give an incorrect response after some time.
  - IP packet options attack: The optional fields in a packet are present to specify additional information about the processing of a packet different from the standard way of execution. IP packet options attack makes use of optional fields to create a malformed packet. For example, if Quality of Service bit is set to 1, the victim takes additional time to process the packet.

3 Major DDoS Attacks in Past

Year	Description
1998	DDoS discovered, but has not gained popularity
1999	Trinoo network floods the University of Minnesota, Massive attack using shaft
2000	Michael Calce (Mafiaboy) launched an attack on Yahoo's website and several sites affecting stock market
2001	Attack size ranges over Gbps
2002	DrDoS discovered. A 1-h long attack affected all 13 DNS root name servers
2003	DoS against the SCO Group Inc., and eBay
2004	US credit card processing firm Authorize.Net fights DDoS
2005	Hamburg-based gambling site jaxx.de was blackmailed to pay 40,000 Euros to stop an ongoing DDoS attack
2006	DNS reflector attacks were discovered. DDoS attacks targeted the blog of Michelle Malkin for over a week
2007	Estonia was hit with a massive DDoS attack targeted at government services as well as financial institutions and media outlets
2008	Anonymous launched DDoS against the Church of Scientology. Amazon.com was also targeted. Conficker worm used a vulnerability in MS Windows to create botnets
2009	Series of coordinated cyber attacks against major government, news and financial websites in South Korea and the United States
2010	Operation payback conducted by anonymous brought down websites of MasterCard, PayPal and Visa
2011	Three members of LulzSec hacktivist group claim CIA website shutdown

(continued)

2012	Leading US banks targeted in DDoS attacks. Targeted banks include Bank of America, JPMorgan Chase, Citigroup, U.S. Bank, Wells Fargo and PNC
2013	150 Gbps DDoS attacks against non-profit anti-spam organization Spamhaus
2014	An attack reaching 500 Gbps hit Hong Kong Democracy voting website
2015	GitHub and BBC were affected by 500 Gbps DDoS attacks
2016	Series of DDoS was launched against DNS provider Dyn affecting internet services in Europe and North America
2017	Mirai Botnet which use IoT devices to create Botnet was discovered
2018	GitHub hit with 1.35 Tbps attack. DDoS-for-hire service is gaining popularity
2019	Number of DDoS attacks falls but sophistication improves

3.1 DDoS Prevention Techniques

These methods try to prevent DDoS from happening by safeguarding the network and blocking the attacker. Major DDoS prevention techniques include

- **Ingress Filtering**
  This kind of filtering blocks all incoming packets not having a legitimate source IP address. Ingress filtering is effective against IP spoofing
- **Egress Filtering**
  This method uses an outbound filter. It only allows packets having valid IP address range to leave the network
- **Secure Overlay Services (SOS)**
  This method initially defines a list of legitimate servers. An incoming packet is accepted only if it is from a known legitimate source. An overlay network placed before the victim will filter all other packets.

4 DDoS Detection Techniques

Detecting the attacker earlier reduces the impact of the attack by preventing its propagation. This helps in preventing complete system failure. DDoS detection methods are mainly categorized as

- **Anomaly Detection**
  This method recognizes any abnormal behavior or anomalies in the performance of the system. If the behavior deviates considerably from normal system behavior, it is considered an attack. This method suffers from a lot of false positives during flash crowds, high traffic scenarios, etc [7, 9].

- **Misuse Detection**
  Victim or a central monitoring system keeps a database of DDoS attacks. Attack database contains an attack signature or pattern of well-known DDoS attacks. When a similar pattern is observed, DDoS attacks are confirmed.

When a DDoS attack is confirmed, we should immediately block the attacker to prevent our system from crashing. This is done either manually or automatically using access lists. Then, to identify people responsible for such attacks and ensure this incident never repeats, we should track the attacker and bring him under the law.

5 DDoS Traceback Techniques

DDoS Traceback methods try to locate the attacker accurately using minimal resources in a short period. Let there be an attackers launching malicious traffic against a victim V through links connecting each other. Any traceback scheme should at least identify those routers directly connected to the victim through which malicious traffic arrives. Some traceback schemes also report route between these routers and victim. An ideal traceback scheme should completely identify all attackers in a single traceback process conducted from same traceback point. Trying to correlate the data from different traceback processes is extremely difficult as well as meaningless for a time-dependent event. The number of false positives should be minimum and network performance should not be affected. A practical traceback system should track 100 attack sources from a network of 1 million devices using a minimum number of packets. Most of the practical traceback systems do not trace complete attack routes as attack packets from a single source may not have same attack path due to the internet's dynamic nature.

DDoS Traceback mechanisms can be classified as Intra-AS and Inter-AS. If traceback occurs within a network, it is called Intra-AS. If traceback ranges across various networks, it is called Inter-AS. Obviously, Intra-AS traceback cannot be used in real life as an attacker may not be in the same network. Most popular inter-AS traceback mechanisms are explained below.

5.1 Link Testing Traceback

Link test traceback process begins with the victim. At every moment, link test traceback traces one step closer to attack source via upstream link. It identifies parent router from which attack path arrived using intrusion detection system installed on every router. This scheme is not suitable if an attack is not continuous. Link test traceback is further classified as input debugging and controlled flooding. In input debugging technique, the victim detects DDoS attacks and builds attacks

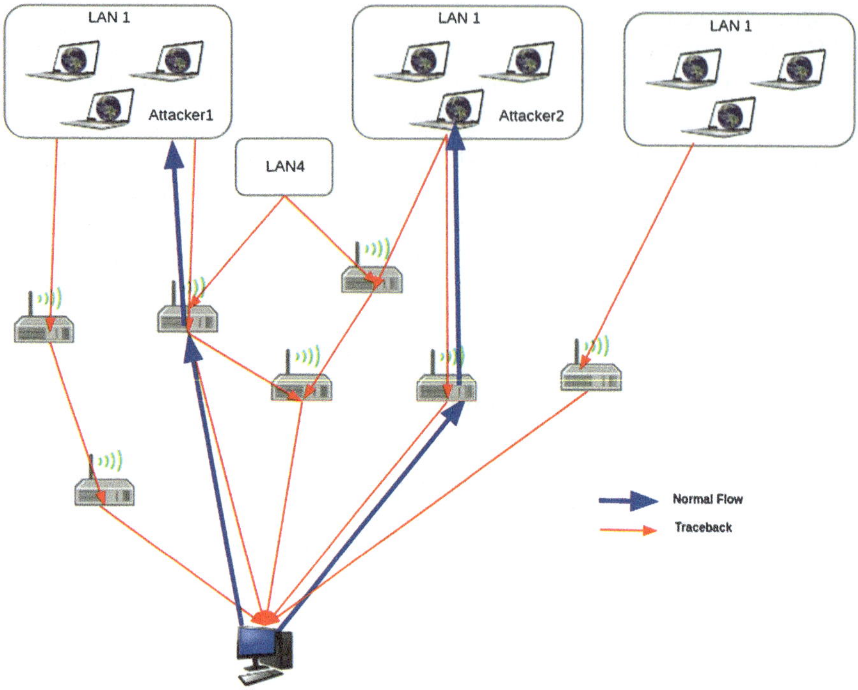

Fig. 2 Link test traceback

signatures (or attack patterns) based on that. The victim then recognizes subsequent attack packets based on the attack signature developed. This recognition happens at every router in this level and corresponding upstream routers of the affected router in this level and so on. This continues until the attacker is identified. We need to regularly update the attack signature for detection to be proper. This detection overhead is improved in control flooding. In this scheme, DDoS is detected by iteratively flooding each incoming connection of the victim with mass volumes of junk traffic. Victim identifies the attack link by noticing the change of rate of packets. It will be drastic in case of attack links. This process is continued in the next immediate upstream router of the attack path identified. This continues until the attacker is reached. Link test traceback mechanism is very old and is no longer used in industry. It utilizes a lot of limited resources and introduces additional traffic to the network thereby boosting DDoS. Link testing traceback is depicted in Fig. 2.

5.2 Messaging

This traceback scheme utilizes ICMP traceback message or iTrace. ICMP message packet has the same destination as a suspicious packet and takes the same route as a

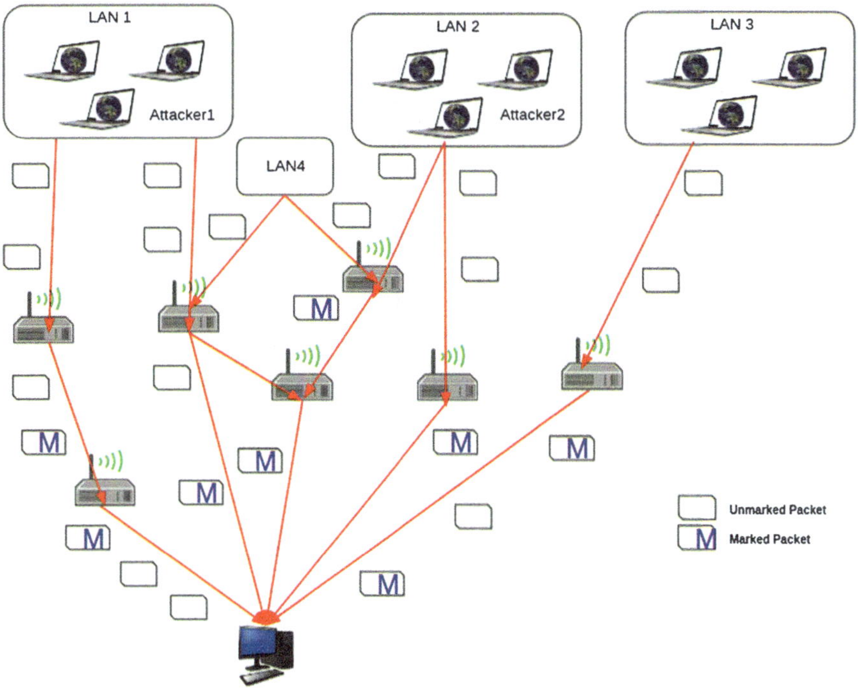

Fig. 3 ICMP based traceback

suspicious packet. When a suspicious packet is received at a router, an ICMP packet is generated with a random probability. ICMP message contains the information of neighboring routers on the path to the destination and source along with the original packet. In order to prevent spoofing, the ICMP packet has an authentication field. When the traffic is high, the victim will eventually receive an ICMP packet from most routers. The schematic representation of the scheme is shown in Fig. 3.

5.3 *Logging*

In this scheme, a router will store digest, signature, and IP header field of all incoming/outgoing packets in local storage. The overview of logging scheme is shown in Fig. 4. When a DDoS attack is confirmed, victim asks all upstream routers to share details of the attack packet. Those routers which have details will be included in attack path. This process continues from identified routers until the attack path is fully traced. The major advantage of Logging scheme over previous traceback schemes is that tracing can be done even if the attack completed long before. However, logging scheme requires large storage on routers compared to

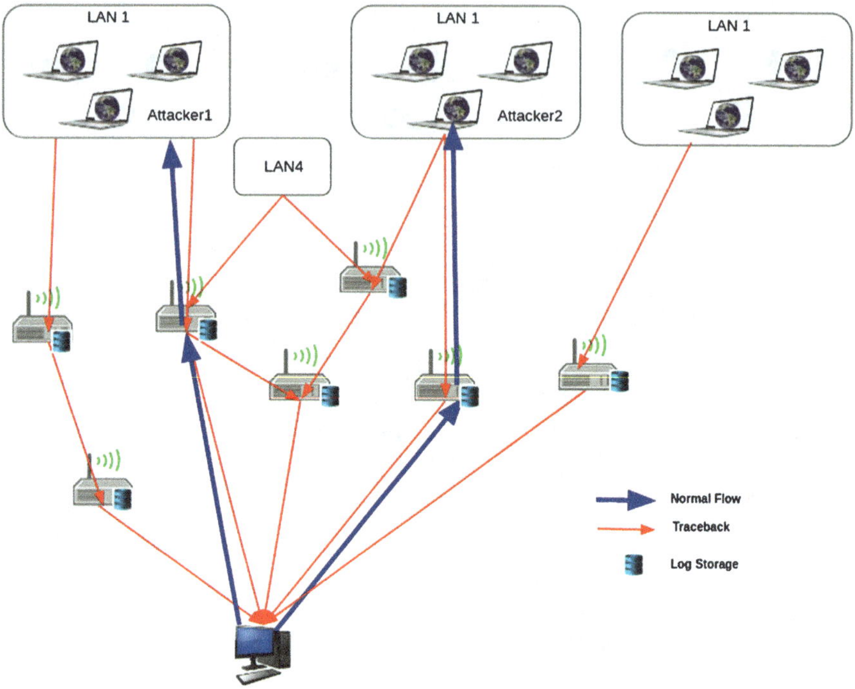

Fig. 4 Logging scheme

other schemes. Researchers have proposed the use of a hash function or bloom filter to reduce the data stored in routers [8].

5.4 *Packet Marking Techniques*

This technique is one of the most common DDoS traceback scheme. In this scheme, the router will insert unique marks into packets forwarded by them. Each mark will identify the corresponding inserted router. When a DDoS attack is confirmed, the victim can traceback attack using marks in attack packets. Two basic forms of Packet Marking are Probabilistic Packet Marking and Deterministic Packet Marking.

5.5 *Probabilistic Packet Marking (PPM)*

Savage et al. originally introduced Probabilistic packet marking (PPM) by defining effective methods to embed partial route path information and incorporate traceback information in packets. The identification field in the IP header is used to store

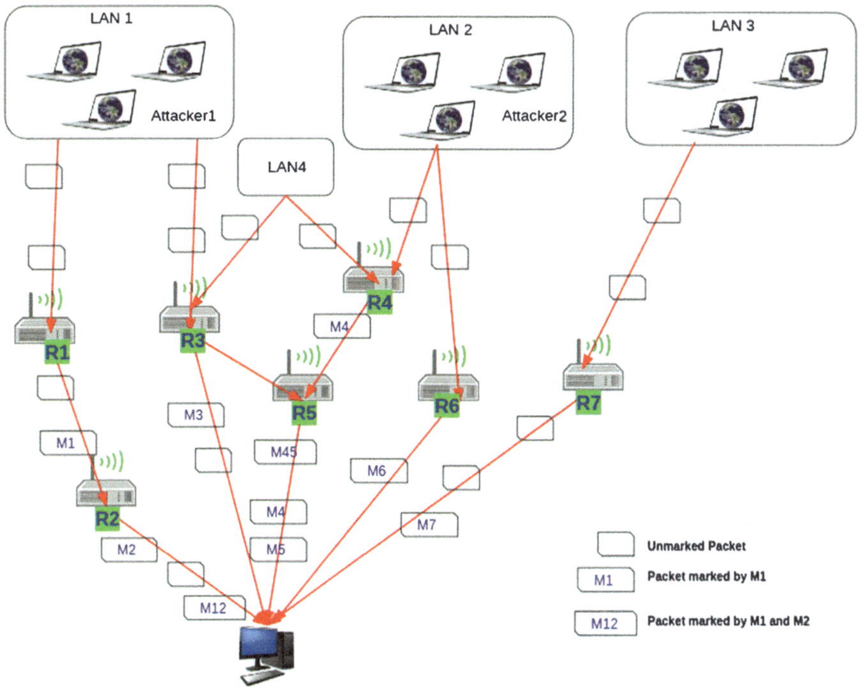

Fig. 5 PPM

marking information. Each mark contains hop count (5 bits) and router information. Hop count is necessary during reconstructing attack path. If router information is too large, it is split into multiple fragments and marked in multiple packets with fragmentation index. The victim will be able to trace back only after it receives multiple marked packets. Probabilistic packet marking technique works irrespective of network topology. Similar to the messaging scheme, it suffers from false positives. Another major drawback of PPM is the loss of marking information due to overwriting in marking space. Rather than marking every packet, packets can be marked based on some random probability say $p = 1/10$. Thereby network overhead can be reduced (Fig. 5).

5.6 Deterministic Packet Marking (DPM)

Deterministic Packet Marking overcomes disadvantages of Probabilistic Packet Marking. Whenever an ingress router encounters an unmarked packet, it marks the packet with its IP address based on a fixed probability. If marking space is not empty, it does not mark the packet (Fig. 6). Marking space include 1 bit Reserved flag in addition to the 16-bit Identification flag. The 32-bit IP address of

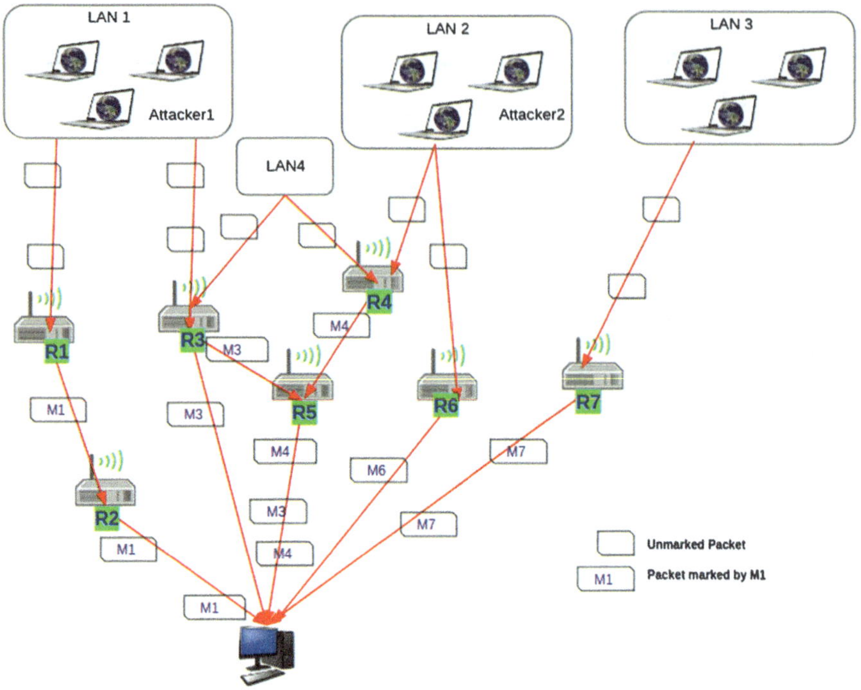

Fig. 6 DPM

ingress router is broken into 2 halves. Each fragment is stored in the identification field of the outgoing packet along with fragment index in reserved flag area. A victim will be able to recover the entire IP address of the ingress router when it receives both fragments. However, if the source address is spoofed, this scheme fails. Deterministic packet marking has less memory requirement compared to probabilistic packet marking. It will not be able to reconstruct the attack path if marking space is overwritten as in the earlier case. The enhanced DPM schemes split IP address into more fragments and use a hash digest to store the address of the marked router to avoid packet tampering [3].

Deterministic packet marking using redundant decomposition was introduced by two researchers Jin and Yang [5]. They used a technique called redundant decomposition to improve identity field coding of deterministic packet marking scheme. The ingress router's IP address is split into three redundant segments, 0–13 bits, 9–22 bits, and 18–31 bits. After that, five different hash results of these three segments are created by applying five different hash functions. Then, these three segments together with five hash results are recorded into eight outgoing packets randomly. When the victim receives most of these packets, it could reconstruct the source router IP. Even if one or two packets have tampered, it can be identified and rectified using other packets. Compared to DPM, packet space can be used more efficiently in DPM-RD, as we do not need to store hash digests for verification.

5.7 *Flexible Deterministic Packet Marking (FDPM)*

Flexible Deterministic Packet Marking (FDPM) is a more flexible and optimized version of Deterministic Packet Marking. It has better tracing capability compared to the probabilistic packet marking and deterministic packet marking. Depending on the load on the router, marking length and marking rate can be varied. However, it cannot fall below a certain requirement. In this scheme, in addition to the identity field and reserved flag, the type of service field is also used to store marks. Based on availability of Type of Service field, FDPM is categorized as FDPM-16, FDPM-19, and FDPM-24 (marking space = 24 bits). The FDPM reconstruction process is much simpler compared to previous schemes. FDPM recognizes marks and recovers marked router's address. FDPM needs only a few numbers of packets to reconstruct attack path with low false positives and false negatives [4]. If processing speed is lower than packet arrival rate, FDPM stores packets in a cache to process later.

5.8 *Hybrid Schemes*

Hybrid schemes [2, 10] combine marking and logging to improve individual marking and logging scheme. They partially record network information at routers as well as packets. There are mainly two hybrid schemes—Distributed Linked List Traceback (DLLT) and Probabilistic Pipeline Packet Marking (PPPM). The first scheme uses core routers to store marking information safely to collect later. The second scheme embeds the IP address of routers as the mark in suspicious packets going to the same destination. Thus the second scheme avoids storage in routers for a long time.

5.9 *Entropy Variation Schemes [1]*

This scheme is different from all previous traceback schemes and is relatively a new concept. In this scheme, we define a metric to measure the randomness of flows at a point in a network called entropy. Entropy drops quickly when there is a major change in network traffic. This entropy variation is a sign of DDoS flooding attack [6]. When a DDoS attack is detected, the victim launches pushback process to trace the location of bots. Pushback process initially identifies the immediate upstream routers of victim present in attack tree using flow entropy variations it has accumulated. Then the victim asks its immediate upstream routers to trace attack based on local entropy variations they have monitored. When those routers identify attack flows, they forward request to their immediate upstream routers to trace attack. This process repeats in a parallel and distributed manner until the attacker is identified or maximum traceable limit has reached.

6 Conclusion

This paper describes the detailed survey of DDoS attacks and measures to prevent, detect, and traceback it. We believe that this survey will be helpful for future researchers in this area. DDoS Attacks are becoming more frequent (7.5 million) and sophisticated (1.35 TBps). Attackers use vulnerabilities present in the system to launch DDoS. Cloud consists of new concepts, vulnerabilities that can be misused. Through our discussions, we have shown that DDoS can take several forms. Further research must be done for designing censorware that can adapt themselves to these new forms discovered daily.

References

1. Yu, S., Zhou, W. Doss, R., Jia, W.: Traceback of DDoS attacks using entropy variations. IEEE Trans. Parallel Distrib. Syst. **22**(3), 412–425 (2011). https://doi.org/10.1109/TPDS.2010.97
2. Al-Duwairi, B., Govindarasu, M.: Novel hybrid schemes employing packet marking and logging for IP traceback. IEEE Trans. Parallel Distrib. Syst. **17**(5), 403–418 (2006).
3. Yu, S., Zhou, W. Guo, S., Guo, M.: A feasible IP traceback framework through dynamic deterministic packet marking. IEEE Trans. Comput. **65**(5), 1418–1427 (2016). https://doi.org/10.1109/TC.2015.2439287
4. Xiang, Y., Zhou, W., Guo, M.: Flexible deterministic packet marking: an IP traceback system to find the real source of attacks. IEEE Trans. Parallel Distrib. Syst. **20**(4), 567–580 (2009)
5. Jin, G., Yang, J.: Deterministic packet marking based on redundant decomposition for IP traceback. IEEE Commun. Lett. **10**(3), 204–206 (2006). https://doi.org/10.1109/LCOMM.2006.1603385
6. Yu, S., Zhou, W., Doss, R.: Information theory based detection against network behavior mimicking DDoS attacks. IEEE Commun. Lett. **12**(4), 318–321 (2008). https://doi.org/10.1109/LCOMM.2008.072049
7. Yu, S., Guo, S., Stojmenovic, I.: Fool me if you can: mimicking attacks and anti-attacks in cyberspace. Comput. IEEE Trans. **64**(1), 139–151 (2015)
8. Tseung, C.Y., Chow, K.P., Zhang, X.: Extended abstract: anti-DDoS technique using self-learning bloom filter. In: 2017 IEEE International Conference on Intelligence and Security Informatics (ISI), pp. 204–204. Beijing (2017). https://doi.org/10.1109/ISI.2017.8004917
9. Lonea, A.M., Popescu, D.E., Tianfield, H.: Detecting DDoS attacks in cloud computing environment. Int. J. Comput. Commun. Control **8**(1), 70–78. https://doi.org/10.15837/ijccc.2013.1.170.
10. Sung, M., Xu, J.: IP traceback-based intelligent packet filtering: a novel technique for defending against Internet DDoS attacks. In: Proceedings of the Tenth IEEE International Conference on Network Protocols, 2002. Paris, pp. 302–311. https://doi.org/10.1109/ICNP.2002.1181417

Network Approach for Inventor Collaboration Recommendation System

Susan George, Hiran H. Lathabai, Thara Prabhakaran, and Manoj Changat

1 Introduction

Network models have been used since a long time to explain real life problems ranging from the topology of the worldwide web, citation networks, protein interaction networks and various other areas. Analysing problems related to patents and their co-inventors using patent-inventor network is a novel method for learning the internal structural aspects and their inter relationships. The tools and theory of networks provide us with an all-new perspective that allows us to study technological growth and its trajectories. In this study we use patent data using network approach to analyse the collaboration among inventors and to recommend suitable collaborations that can be forged. This might be helpful for inventors who are in various phases of their career to advance their career. For instance, a novice inventor might be able to know and plan which skills to acquire for collaborating with an inventor who is in a higher stage of career or the one who works in a different but related technological area. Sometimes, the collaboration recommendation may help an inventor to decide even on things such as migration to another organization.

Research on collaboration recommendation system in science and technology is predominantly found in scholarly collaboration recommendation for authors. One such important related works on collaboration recommendation system design is [3] in which link semantics is relied to recommend academic collaboration. Novelty of this research lies partly in the fact that a collaboration recommendation system for inventors is attempted for the first time, at least according to our knowledge. Another novelty is in the reliance of sheer network approach than semantics and thereby more simplicity in implementation and operation can be

S. George · H. H. Lathabai (✉) · T. Prabhakaran · M. Changat
Department of Futures Studies, University of Kerala, Thiruvananthapuram, Kerala, India

M. Palesi et al. (eds.), *Second International Conference on Networks and Advances in Computational Technologies*, Transactions on Computational Science and Computational Intelligence, https://doi.org/10.1007/978-3-030-49500-8_8

achieved. The core of the recommendation system is a kind of link prediction method which identifies the important non-existing linkages using the inventors separated by distance=2 (distance=1 indicates an existing collaboration) in the inventor collaboration network (II). Such indirect collaborations can be retrieved as a derived network II^2 of II and non-existing linkages can be identified. Direction of recommendation, i.e., which inventor is to be recommended to whom is an important factor. This can be identified using the metric receptivity index introduced in [4], which reflects the receptivity of the inventions co-invented by an inventor with respect to a particular field (indicated by citations from other patents in that research context). The norm followed here is that the inventor with high receptivity in a recommendation pair is to be recommended to the inventor with less receptivity. Once such non-existing relationships with highest weights qualified for the recommendation are identified, a threshold Th can be set to filter the most important relationships for recommendation. Detailed discussion of methodology for the design of the recommendation system can be found in Sect. 3.

Before that important concepts used in this work are discussed in next section.

2 Networks and Concepts

Important types of networks used are affiliation networks (Patent-Inventor affiliation), citation network (Patent citation network), collaboration network (Inventor–Inventor collaboration), and the derived collaboration network (Inventor–Inventor network with distance 2 collaborations).

2.1 Affiliation Networks

A network consists of a graph (which consists of vertices V and lines $L \subseteq V \times V$) and additional information about its vertices or the edges. A network can be termed as a structure, $N = (V, L, F, W)$ where $L = E \cup A$ is the set of lines, $E \cap A = \Phi$. Undirected lines E are called edges, and directed lines A are called arcs. F is the set of vertex value functions or properties and W is the set of line value weights.

In unweighted graphs, weightage is not significant. A network is edge-weighted, if each of its edge $l \in L$ has a weight w. So, the adjacency matrix of an unweighted graph will have binary values, i.e., 0 or 1, where 1 indicates the presence of an edge and 0 indicates the absence of an edge. In weighted graph, each edge is having a numerical weight. Depending on the type of nodes and the relations in a network, a network can be a one mode (1-mode) or a two mode (2-mode) network. All of the vertices in an one mode network will be of the same type. A two mode network (also called bipartite network) contains two types of vertices.

Affiliation networks are 2 mode networks that represent the relationships or affiliations of one kind of actors (say scholarly papers, managers, patents, etc.) to other kind of actors (say authors, organizations, journals, inventors, countries, etc.). Formally, an affiliation network of patents and inventors, namely the PI network[2], is a structure $PI = (P, I, L, W)$ where P is the set of patents that forms the first mode and I is the set of inventors that forms the second mode and L is the set of arcs which originates from P and terminates at I, W is the weight of lines/arcs. A PI affiliation network can be represented by $W = [w_{pi}]_{PI}$

$$W_{p,i} = \begin{cases} w(p,i), & if\ p,i \in \mathcal{A} \\ 0, & otherwise \end{cases} \tag{1}$$

where $\mathcal{A}$ is the adjacency matrix of PI.

2.2 Collaboration Networks

As our interest lies in inventor collaboration, the collaboration network discussed here will be inventor collaboration networks. These can also be termed as the co-inventorship networks, where vertices are inventors and edges are links of co-inventorship. A co-inventorship link indicates that the pair of inventors have co-invented at least one patent. Usually, the weight of a collaborative link indicates the number of times they have collaborated (i.e., the number of patents they have co-invented). These are usually undirected 1 mode networks. It can be formed in PAJEK [1] as discussed in [2] as:

$$PI^T \times PI = II$$

where PI^T is the transpose of PI network, which represents the affiliation of inventors to the patents. This network contains self-loops which reflects the total number of invented by an inventor as edge weight, which is usually removed unless required for special purposes.

Distance between two vertices/nodes is dependent on the number of edges to be traversed to reach from one vertex to other. If there is more than one such traversal paths, the shortest distance will be the one with minimum number of edges. If two vertices are connected by an edge (direct relation), then distance between the pair=1. If there are two edges between a pair of vertices (as the case of s and u, where there is no edge between s and u, but there is an edge between s and t and there is also an edge between t and u), then the distance between the pair=2. If there is also edge between s and u along with an indirect but redundant path through t, then the shortest distance is 1.

2.3 Inventor–Inventor Derived (Indirect) Network

The inventors separated by distance=2 in the inventor—inventor collaboration network (II) is very crucial for our recommendation system design and in order to focus on such non-existent collaborations, we derive another network from II as II^2 using :

$$II^2 = II \times II$$

Here also, self-loops will be there, which is conveniently eliminated. In II, if there are $m + 1$ paths of distance ≤ 2 where m paths are through m common co-inventors shared by an inventor pair s, u and the remaining one is an existing edge of collaboration between s, u. Let t be one such arbitrary inventor who has collaborated with both s and u, then edge weight w_{su} between s and u signifies the number of works co-invented by s with u and w_{tu} is the edge weight for edge between t and u, which reflects the number of patents jointly invented by t and u then the weight of edge between t and u in II^2 network will be

$$w'_{su} = w_{su} + \sum_{t=1}^{m} w_{st} \times w_{tu}$$

If there is no collaboration between u and w, then there will be no edge between u and w in II and hence $w_{uw} = 0$, therefore the weight of edge between v and w in II^2 will be

$$w'_{su} = \sum_{t=1}^{m} w_{st} \times w_{tu}$$

In this case, the edge with weight w'_{su} represents a truly non-existent collaboration. More this weight, more significant and strong the ties of inventors s and u to their common collaborators and it will be advantageous for s and u to collaborate if such a collaboration is feasible. Our task is, however, to identify the truly non-existent collaborations of significant strengths. The details of same can be found in Sect. 3.

3 Methodology

The flowchart shown in Fig. 1 gives a clear layout of the proposed recommendation system. The detailed information about each patent and their citation details are available for public from USPTO database. Patent citation network is constructed by using the patent-patent citation information which can be extracted from this patent data. Simultaneously, a two mode network (i.e., a network having two types

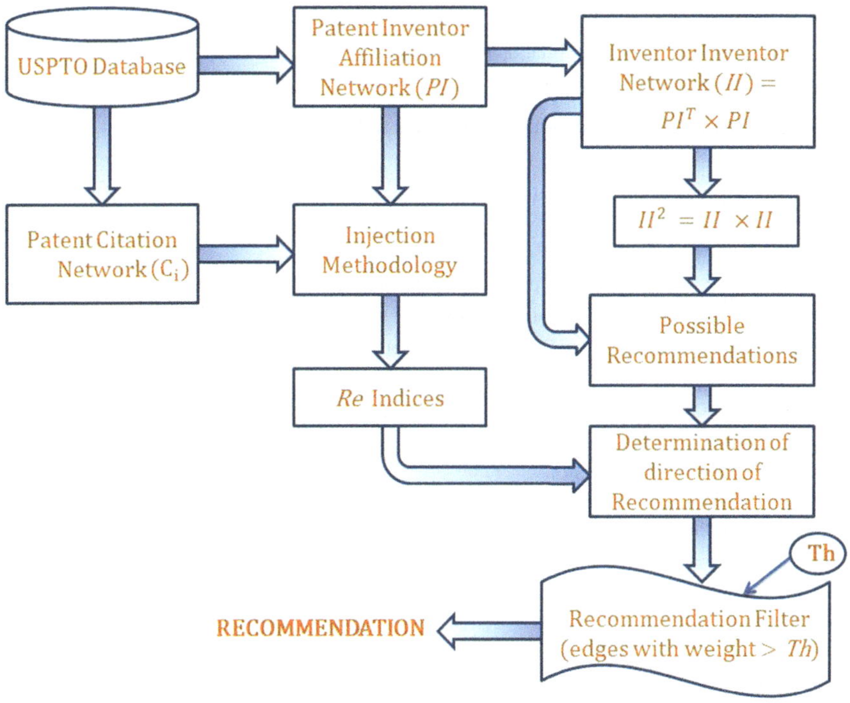

Fig. 1 Methodology for collaboration recommendation system design

of nodes) is created using patent and their inventors which gives a patent-inventor (PI) network. Typically both networks are unweighted. When indegree of patents in the patent citation network (indegree signifies the number of citations received by a patent from the patents in the network) is injected to the unweighted affiliations network, each arc (of affiliation) carries weight equal to the indegree of patent (source node) and this methodology for creating affiliation networks with important attributes of patents as arc weights is termed in [4] as the *Injection methodology*.

As injection methodology is applied here with indegree or citation counts as attribute, *Re* index of any inventor, Re_i can be calculated as the weighted indegree of the indegree-weighted PI affiliation network. This represents the overall productivity of an inventor in terms of the volume of receptivity of his patents.

$$Re_i = \sum_{k=1}^{n} indeg_k = wt.Indeg_i \tag{2}$$

This metric is to be reserved for determination of direction of the recommendation. Now, proceed with the creation of inventor–inventor collaboration network II as per the algorithm shown in Table 1 using the initial unweighted (non-injected) PI. Self-loops present in II has to be removed and using this loop-free II network,

Table 1 Algorithm for inventor collaboration recommendation

Input: PI, the Patent-Inventor affiliation network and
Ci, the patent citation network

Output : Arc set ER (Relationships for Recommendation)

1. Compute Re indices for all inventors using PI and Ci with the help of Injection methodology [4].
2. Obtain II network using $II = PI^T \times PI$
3. $II := II \setminus e_{vv}$, where $e_{vv} \in E$ represent the self-loops and E the set of edges in II network.
4. Obtain II^2, the distance 2 relationship network
5. $II^2 := II^2 \setminus e^2_{vv}$ where $e^2_{vv} \in E^2$ represent the self-loops and E^2 the set of arcs in II^2 network.
6. Possible relations for recommendation are $Er = E^{2*} \setminus (E^{2*} \bigcap E^*)$, where $E^{2*} = E^2 \setminus e^2_{vv}$ and $E^* = E \setminus e_{vv}$.
7. For $\forall e_{r_{vw}} \in Er$,

    if $Re_v > Re_w$

        Convert edge $e_{r_{vw}}$ as arc $e_{r_{wv}}$

    else if $Re_v = Re_w$

        Retain edge $e_{r_{vw}}$ as such

    else

        Convert edge $e_{r_{vw}}$ as arc $e_{r_{vw}}$

8. $ER = \phi$ and set value for Th.
9. For $\forall$ links $e_r \in Er$

    if $w_{e_r} \geq Th$

        $ER := ER \cup e_{r_{vw}}$

form II^2, the distance-2 relationship network. After the elimination of self-loops in II^2, edges which were commonly found in II^2 and II are removed from the II^2 network (see step 6 of the algorithm). This is because, such edges which signify already existing collaborations are redundant. Now the remaining edges form the set of inventor pairs for possible recommendation.

For identification of the direction of recommendation, the stored receptivity indices can be used. For a pair u and w, if Re index of u is found to be greater than that of w, inventor u is to be recommended to w as a collaboration suggestion. If both are found to be equally receptive, both can be recommended to each other. Practically this can be done by converting the undirected II^2 network as a mixed one (links as edges and arcs) with help of step 7. The recommendation level, i.e., the number of recommendations to be made can be controlled by using a threshold Th. Arcs in the processed II^2 whose weights (significance of collaborations of intermediary inventors) $> Th$ can be retained and others have to be filtered out.

Now the case study of 'Wireless power transmission' is discussed.

4 Analysis, Results, and Discussions

In this work, we have done network analysis for the field 'wireless power transmission', and a brief description of the field is discussed as follows.

4.1 Wireless Power Transmission

'Wireless Power Transmission' (WPT) is a method which is used to signify wireless power transference (transmission and distribution) at any level. The founder of AC electricity, Nikola Tesla, was first to conduct experiments dealing with WPT in 1899 [5–7]. Wireless power transmission(WPT) is reliable, efficient, fast, and have low maintenance cost. WPT is useful for both short range or long range. Conventionally, electricity is manufactured from different sources and is delivered through wires and cables to various destinations. Wired transmission (WT) is prone to my disadvantages such as (1) heavy transmission loss, (2) immense cost of the system infrastructure that includes high-tension transmission line cables, substations, and towers and diminished economic feasibility in electrifying remote areas, and (3) the impact on environment due to the disposal of used wires. At the moment, Wireless Power Transmission is in the forefront of electronics research. Our current study started as an initial attempt to investigate WPT inventor collaboration and to suggest recommendation for forging useful collaborations among inventors within this fast-growing industry. In this work, data for creating patent-inventor network is collected from USPTO (United States Patent and Trademark Office) database using the phrase 'wireless power transmission'.

PI network consists of x vertices and y arcs. Out of the x, i patents form the first mode and j inventors form the second mode. The patent citation networks consist of i patents (obviously) and n arcs. The collaboration network derived from PI network consists of j vertices and m edges (after self-loop removal). The distance 2 inventor–inventor network consists of j vertices and p edges (after self-loop removal). After identification of direction of recommendation using the step 7 of algorithm depicted in Table 1, we proceeded to filtering of most significant recommendations. Out of these p edges, the ones with highest weights are retained and others are filtered out using threshold, Th which is set as 20. Resulting relationships and recommendations that can be made are shown in Table 2. Analysis of these results and possible implications are discussed next.

Table 2 Final results for recommendation when Th is set as 20

Recommendation (Th = 20)						
Link weight > Th	Inventor to be communicated			Inventor to be recommended		
	Inventor ID	Re index	Inventor name	Inventor ID	Re index	Inventor name
87	6989664-5	10	Young-Tack Hong	8129322-3	11	Young Taek Hong
83	5801974-1	5	Byung-Chul Park	7053730-1	72	Yun-Kwon Park
83	6055233-1	5	Jae-Hyun Park	7053730-1	72	Yun-Kwon Park
83	8558644-6	5	Jung Hae Lee	7053730-1	72	Yun-Kwon Park
64	5801974-1	5	Byung-Chul Park	6265261-1	6	Ki-Young Kim
64	5801974-1	5	Byung-Chul Park	8730697-1	6	Dong Zo Kim
64	6055233-1	5	Jae-Hyun Park	6265261-1	6	Ki-Young Kim
64	6055233-1	5	Jae-Hyun Park	8730697-1	6	Dong Zo Kim
64	8558644-6	5	Jung Hae Lee	6265261-1	6	Ki-Young Kim
64	8558644-6	5	Jung Hae Lee	8730697-1	6	Dong Zo Kim
54[a]	5801974-1	5	Byung-Chul Park	6149435-2	5	Jin Sung Choi
54[a]	6055233-1	5	Jae-Hyun Park	6149435-2	5	Jin Sung Choi
54[a]	6149435-2	5	Jin Sung Choi	8558644-6	5	Jung Hae Lee
53	9272630-8	1	Jin Sung Chol	7642882-5	28	Eun-Seok Park
50[a]	5801974-1	5	Byung-Chul Park	8730697-10	5	Chang Wook Yoon
50[a]	6055233-1	5	Jae-Hyun Park	8730697-10	5	Chang Wook Yoon
50[a]	8558644-6	5	Jung Hae Lee	8730697-10	5	Chang Wook Yoon
43	6160435-1	7	Min-Gyu Kim	7864775-2	24	Yunjung Yi
39	7720503-3	7	Han-Byul Seo	7349374-5	34	Dong Youn Seo
33	5504635-1	1	Chang-Ho Lee	7642882-5	28	Eun-Seok Park
32	6389058-2	4	Byoung-Hoon Kim	7349374-5	34	Dong Youn Seo
30	5504635-1	1	Chang-Ho Lee	8120534-6	12	Young-ho Ryu
30	6103432-3	15	Ki-jun Kim	7864775-2	24	Yunjung Yi
30	9272630-8	1	Jin Sung Chol	6149435-2	5	Jin Sung Choi
30	7308624-2	22	Daewon Lee	7864775-2	24	Yunjung Yi
28	5801974-1	5	Byung-Chul Park	6989664-5	10	Young-Tack Hong
28	6055233-1	5	Jae-Hyun Park	6989664-5	10	Young-Tack Hong
28	8558644-6	5	Jung Hae Lee	6989664-5	10	Young-Tack Hong
26	6006539-2	3	Jung-Hoon Lee	7864775-2	24	Yunjung Yi
25	6587697-1	15	Stephen E. Terry	5534734-2	22	John M. McNally
24	7460504-5	3	Pablo Anigstein	5388124-1	45	Rajiv Laroia
22	7240502-8	5	Young-Seob Choi	8050208-5	36	Suck-Chel Yang

[a]This pair is an edge and two way recommendation of inventors is possible

4.2 Inventor Recommendations and Implications

Thirteen of the inventors in the recommendation table are working for Samsung Electronics. They are Young-Tack Hong, Byung-Chul Park, Jae-Hyun Park, Jung

Hae Lee, Jin Sung Choi, Jin Sung Chol, Chang-Ho Lee, Young Taek Hong, Yun-Kwon Park, Ki-Young Kim, Dong Zo Kim, Eun-Seok Park, and Chang Wook Yoon.

In further discussions, the receptivity indices of each inventor are shown beside him/her in braces. Inventor Yun-Kwon Park (72) with the highest Re in the list can be recommended to Byung-Chul Park (5), Jae-Hyun Park (5) and Jung Hae Lee (5). Rajiv Laroia (45) who have the second highest Re can be recommended to Pablo Anigstein (3). Suck-Chel Yang who have 36 as receptivity is recommendable to Young-Seob Choi (5). Dong Youn Seo (34) has to be recommended to two inventors- Han-Byul Seo (7), Byoung-Hoon Kim (4). Eun-Seok Park (28) can be recommended to Jin Sung Chol (1) and Chang-Ho Lee (1), two less receptive inventors.

Yunjung Yi (24) is found recommendable to Min-Gyu Kim (7), Ki-jun Kim (15), Daewon Lee (22) and Jung-Hoon Lee (3). John M. McNally (22) can be recommended to Stephen E. Terry (15). Young-ho Ryu (12) can be recommended to Chang-Ho Lee (1) again. Young Taek Hong (11) can be recommended to Young-Tack Hong (10).

Dong Zo Kim (6) and Ki-Young Kim (6) has to be recommended to the trio Byung-Chul Park (5), Jae-Hyun Park (5), and Jung Hae Lee (5). Several authors with Re=5 can be mutually recommended as shown in Table 2.

Most of the inventor recommendations are found to be intra-organizational within Samsung Electronics. These inventors may be working in different but related kind of R & D projects. The inventors for whom the recommendations are communicated can plan their career by identifying the skills to be acquired to migrate to the other kind of R & D projects.

Yunjung Yi (24) from 'Honeywell' is found to be recommendable to Min-Gyu Kim (7) from 'MagnaChip Semiconductors', Daewon Lee (22) from Polaris Innovations, and to both Ki-jun Kim (15), Jung-Hoon Lee (3) from Samsung Electronics. These kind of intra-organizational recommendations are also found with the usage of Th= 20. As the scope of this work is fixed to a demonstrative level, Th value is not lowered much. Lowering of Th (to 15 or 10) might have resulted in more recommendations and thereby more inter-organizational recommendations.

5 Conclusion and Possible Future Works

A preliminary design for an inventor collaboration recommendation system is conceived for the first time according to the best of our knowledge. Unlike the existing recommendation systems for scholarly collaboration recommendation, our design is purely network based one and no semantics is used. This ensures relative simplicity in implementation and operation. Little emphasis on usage of semantics may raise the concern of ending up with unacceptable results. For instance, an inventor working in one specialized technological area may receive a suggestion to collaborate with one working in other technological area. Mostly, for such a collaboration to materialize that inventor has to attain some new technical skills and knowledge about the new technology area. However, the career prospects of inventors always lie with their ability to acquire technical skills both in their own specific areas as well as other areas that are more or less related to their areas. Due to this fact, there is always an opportunity associated with the risk of cross-technological suggestions of inventors. This risk can be addressed to an extent by integrating semantic parameters to the present design. Semantic validation of the present method can determine whether there is a need for such an exercise, which is beyond the scope of this work. Such an investigation is intended as a future research pursuit. However, an advantage of the present design is that enough flexibility is there to incorporate semantic parameters to improve the system if such requirements surface. Alternatively network approach for identification of extent of technological area relatedness of pair of inventors in a recommendation through the usage of derived networks of inventor's technological area affiliation is investigation worthy. Such an investigation might improve the possibility of keeping our framework purely network based one. The system is tested using a case study from the research field 'Wireless power transmission'. Validation using different case studies that represent different fields is beyond the scope of this work and may be attempted in future. With a threshold set at moderately high value, the system could identify some of the key intra-organizational recommendations within Samsung electronics and most suitable inter-organizational recommendations. More recommendations can be retrieved with the adjustment of threshold to much lower values. Thus, some of the other suitable modifications for the betterment of this design are discussed next.

Choice of an optimum threshold level has to be addressed. May be the examination of distribution of arc/edge weights in processed II^2 network could achieve that. Another important betterment needed is related to the optimization of recommendations, i.e., the most suitable data organization of recommendations for the mailing/communicating system. May be the adjacency list representations could address this concern. These are the major possible future endeavours other than the likely incorporation of semantic parameters in the design if there is a requirement.

Acknowledgments This work used the facility provided by 'Scientometric lab' (order No.Pl.A1/Annual plan 16-17/Imp.plan/16 dtd. 29/11/2016), Department of Futures Studies, University of Kerala.

References

1. Batagelj, V., Mrvar, A.: Pajek-program for large network analysis. Connections **21**(2), 47–57 (1998)
2. Batagelj, V., Cerinšek, M.: On bibliographic networks. Scientometrics **96**(3), 845–864 (2013)
3. Brandão, M.A., et al.: Using link semantics to recommend collaborations in academic social networks. In: Proceedings of the 22nd International Conference on World Wide Web. ACM, New York (2013)

4. Lathabai, H.H., Prabhakaran, T., Changat, M.: Contextual productivity assessment of authors and journals: a network scientometric approach. Scientometrics **110**(2), 711–737 (2017)
5. Moore, D.M.: Peaked-wave wireless transmission. U.S. Patent No. 755,305. 22 March (1904)
6. Nikola, T.: Apparatus for transmitting electrical energy. U.S. Patent No. 1,119,732. 1 December (1914)
7. Tesla, N.: The future of the wireless art. Wireless Telegraphy Telephony, 67–71 (1908)

Architecture of a Semantic WordCloud Visualization

Vinitha M. Rajan and Ajeesh Ramanujan

1 Introduction

There has been a monumental increase in the amount of raw text being generated from every realm in the form of online articles, journals, research works, educational materials, news feeds, blogs, celebrity talks, speeches, or interviews.... Such an upward surge in information explosion leads to a proportional increase in the demand for analysing the core content from huge repositories of heterogeneous text data easily and efficiently. This led to the emergence of "WordClouds" [4, 11, 30, 32] designed as a "Text Visualization" [12, 16] technique to obtain the most prominent keywords from a piece of text or document.

The origin of wordcloud visualization [29] dates back at least 100 years, while designing the political posters in Soviet Union. It was further reformed for numerous real-life applications in the 1990s, notably displaying the top organizations as quoted by Fortune magazine [29], analysing the search phrases leading to a website and plotting the most significant landmarks in the city of Paris [29]. The initial representations and format of wordclouds, as formal methods for text (automatic) summarization, were often termed as *TagClouds* [26, 28, 29]. *TagClouds* have the same visual encoding for words as that of wordclouds. The words are ranked according to their frequency of occurrence ("Term-Frequency") [24] and proportionately assigned with the font sizes. All the keywords with varying sizes are arranged sequentially along multiple horizontal lines. *TagClouds* had gained much popularity for being a naive user-friendly visualization [31] providing easier analysis of raw text documents.

V. M. Rajan (✉) · A. Ramanujan
College of Engineering Trivandrum, APJ Abdul Kalam Technological University, Thiruvananathapuram, Kerala, India
https://www.cet.ac.in/

M. Palesi et al. (eds.), *Second International Conference on Networks and Advances in Computational Technologies*, Transactions on Computational Science and Computational Intelligence, https://doi.org/10.1007/978-3-030-49500-8_9

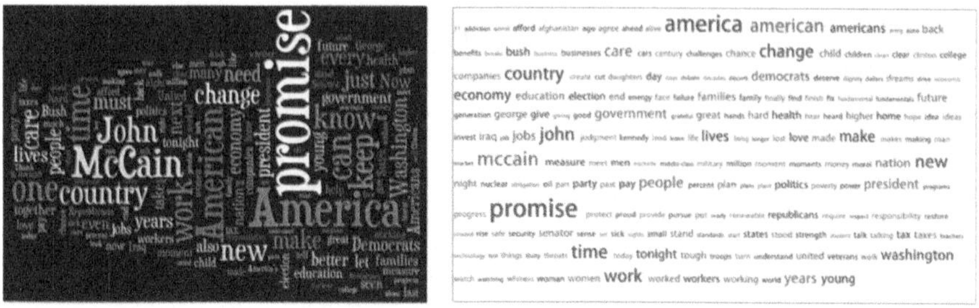

Fig. 1 WordCloud (left) and TagCloud (right) representations for the same textual content [30]

WordClouds of the form today were primarily designed and developed to improve the visual features of tagclouds (Fig. 1). Wordles [5, 8, 11, 30, 32] as they are popularly called are web applications to generate random wordclouds, where words are placed at random positions with "maximum compactness" and "minimum overlapping". Due to the enormous popularity of Wordles, many of its successors followed of which the most prominent are (1) *ManiWordle* [11]: "Manipulable Wordles" for providing flexible word manipulations in the output wordcloud visualizations; (2) *WordlePlus* [8]: extending *ManiWordle* to touch enabled devices; (3) *EdWordle* [32] : to "Edit" wordclouds preserving location consistency; and (4) *MetroWordle* [13]: a urban application of *Wordle* to visualize text along with geographic information to identify the "Place Of Interests (POI)".

The common design technique of wordles (random wordclouds) can be identified by the greedy approach being used for its word placement algorithm [30] generating the initial layout ("blueprint") for the wordcloud. Thus the primary layout construction of wordles can be closely related to the NP-Hard "Bin-packing" [3] problems. Such two-dimensional packing algorithms strive to "pack" the words to the closest space available, without considering the "semantic" relationships among them. From the perspective of naive users, in natural language visualization techniques such as wordclouds, the words which are physically closer are often expected to be "semantically related" to each other. In this regard, "Semantic WordClouds" [2, 4, 22, 27, 33, 34] are becoming prominent every day, which try to attain ("realize") the highest possible "word semantics" in the generated visualizations. The enormous popularity of semantic wordclouds is quite evident from the massive research undergoing in the domain. Analysing the various related works in the literature [1, 2, 4, 22, 25, 27, 33, 34], it can be observed that semantic wordcloud visualizations unanimously follow a generic pipeline for their architectural framework. Many approaches are available for each stage of the pipeline and multiple combinations of them can be developed to generate the visualizations.

In this paper, we provide a detailed account on the architecture and the design pipeline stages of a semantic wordcloud visualization. Such a tutorial can be extremely helpful in developing the prerequisite basis for many of the upcoming researches in semantic wordclouds. The paper has been structured as follows: Sect. 2

gives an overview of the various research works which have been referred and studied upon in the domain of wordclouds; Sect. 3 gives an in-depth description of the semantic wordcloud architecture and the constituent design components followed by a theoretical conclusion in Sect. 4.

2 Related Work

The significance and a brief overview of wordcloud functionalities can be obtained from the survey article [12] "Overview Of Text Visualization Techniques" by Kostiantyn Kucher and Andreas Kerren. The review article [12] accounts wordclouds as one of the accepted standards of text visualization to obtain the core contents of a text document at the "word level". "TagClouds" are the pioneers in the evolution path of wordclouds as formal techniques of text summarization. F. B. Viegas and M. Watterberg of IBM Research, in their paper [29], give an introduction to "TagClouds", overview of its history, evolution timeline and its core visualization features. Various algorithms for generating tagclouds from the user generated tags can be obtained from the research article [26] published by the *Institute for Knowledge Management* team. A similar set of algorithms and its associated analysis have also been attempted by Owen Kaser and Daniel Lemire [10]. IBM Research has further proposed a protocol [23] to quantitatively evaluate tagclouds with sufficient guidelines and datasets.

IBM Research continued with their study on wordclouds with "Wordle" [5, 30] being first in the series of "Random WordClouds". The primary motivation for the development of wordle was to improve the "ransom note" [21] appearance of tagclouds to the "cloudy" representation of today. The visual features and a case study of its unanimous popularity, the "participatory culture" and the evaluation results have been presented by F. B. Viegas, M. Wattenberg, and J. Feinberg in [30]. The article also throws light on the "Classic Wordle Algorithm" [30] which forms the backbone of many of its successors [8, 11, 32] and even other variants such as semantic wordclouds [4] in building the initial wordcloud layout. The first successor to Wordle, "ManiWordle" [11, 32] was developed by the research team of Seoul National University to make Wordle more flexible, by supporting user interactions to control the layout and manipulate individual words. Bongshin Lee of Microsoft Research, Jaemin Jo and Jinwook Seo of Seoul National University extended ManiWordle to "WordlePlus" [8] to incorporate human touch interactions. "EdWordle" [32], being one of the novel developments in the series, was developed by a joint team of various universities to provide consistency and maintain compactness while editing wordclouds. The latest entry in the series, "MetroWordle" [13] is a heavy-weight, real-life application used in the lines of "Google Maps" to identify the most important landmarks.

Since wordles unanimously follow a "randomized greedy" approach, "Semantic WordClouds" have become the present flavour especially due to their close relationship with the human thought processes of pattern identification. The unanimous popularity is quite evident from many of the recent research works [2, 4, 22, 27, 33, 34]. Initial path breaking results in the development of semantic wordclouds were provided by the research team from Arizona University [2]. They proposed and evaluated multiple algorithms to develop the visualizations such as "Context Preserving WordCloud Visualization", "Seam Carving", "Inflate & Push", "Star Forest", and "Cycle Cover". Performance analysis of the algorithms were done against the randomized versions (wordles) using the metrics : "Realized Adjacencies", "Distortion", "Compactness" , "Uniform Area Utilization", "Aspect Ratio", and "Running Time". Y. Wu and team [33] further elaborated and evaluated the framework for "seam carving" algorithm. These algorithms, metrics, and the associated datasets formed the basis for many variants in the domain of wordclouds for newer developments and performance analysis.

One method inspired by the above work is the implementation and evaluation of a "Fully Dynamic Semantic WordCloud" [4]. While the wordcloud layouts were developed using the algorithms above, they improvised the visualization to make it "dynamic" (evolving). The visualization evolves by accommodating the changes corresponding to an evolving input source text. Each wordcloud gets generated after a specific time interval such that the visualization is a coherent combination of the changes in the wordcloud output from the previous time interval and the present input text content. The application of semantic wordclouds in document summarization can be obtained from [22, 34] with the Zhejiang University [34] giving the architectural diagram (Fig. 3) which is an almost abstract representation of all the algorithms for semantic wordcloud generation. L. Barth and team [1] developed a formal representation "Contact Representation of Word Networks (CROWN)" [1] for the problem of semantic wordcloud generation, with the visualization being modelled as a "Word Graph" having the nodes as "axis-aligned" word rectangles. The problem tries to optimize the position of the rectangles in a given two-dimensional area with "maximum compactness" and "minimum overlapping".

One of the most recent research works in semantic wordclouds is provided by M. A. Hearst and team [7] which provided a standardized approach to evaluate the layouts using a series of experiments. They even provided a standard dataset for future analyses in semantic wordclouds. E. Schubert and team of Heidelberg University [25] attempted to bring in novelty to semantic wordcloud generation enhancing the standard "t-distributed stochastic neighbor embedding (t-SNE)" method. The team was successful in proving that the proposed system is more "memory efficient" than the above methods. With the reference to all the above articles, generating a semantic wordcloud visualization can be approximated to a pipelined architecture (Fig. 3) with multiple stages of various components, irrespective of the metrics and design choices used for each stage.

3 Semantic WordCloud Architecture

As mentioned before, a "Semantic WordCloud" [2, 4, 22, 27, 32–34] has a geometric layout where the words are arranged on a visual canvas by maintaining the semantic relationships among their positions. Figure 2 shows an example of a perfectly positioned semantic wordcloud formed with the words from a BBC news feed [33]. Each word is assigned to its respective "semantic cluster" which is a representative of the individual news items from the feed. The "quality" of a semantic wordcloud is usually adjudged by the total "realized" semantic relationships, clustering efficiency, compactness of words, and the amount of overlapping. The algorithmic framework (Fig. 3) for developing a semantic wordcloud is a pipelined execution of various stages such as (1) Natural Language Preprocessing [17, 18]; (2) Semantic Similarity Generation [2, 4, 22, 33, 34]; (3) Words Clustering [2, 4, 22, 34]; (4) Word Graph Visualization [1, 2, 4, 34]; and (5) WordCloud Layout Optimization [2, 19]. Following sections brief upon each of the stages (modules) and its respective components with reference to Fig. 3.

3.1 NLP Preprocessing

The first stage of the architecture is "Natural Language Processing" (NLP) [17, 18, 24], which is essentially a preprocessing step of the input raw text to extract its most significant keywords. The stage comprises the following components:

1. *Text Extraction:* [24] Exclusively extracting the raw text T from the given input document, irrespective of its format and discarding other components such as figures, font colours, and table structures.
2. *Keywords extraction:* [24] The extracted raw text T is being subjected to various natural language processing techniques [24] to generate the keywords such as: Dividing T into multiple sentences $S_e = \{s_i\}$ using a predefined "stop symbol" set; Decomposing each sentence s_i into its component phrases $P = \{p_i\}$ by

Fig. 2 Semantic WordCloud Visualization of a BBC news feed with the perfectly formed semantic clusters for each news category [32]

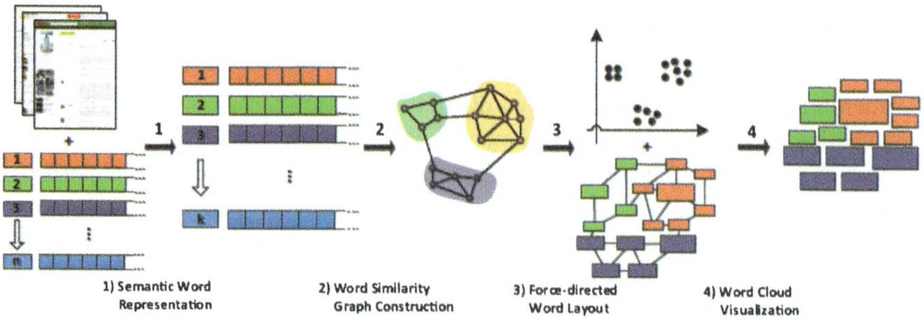

Fig. 3 Architectural overview [34] for generating a semantic WordCloud visualization

removing the "stop words" and other extraneous symbols; Finally filtering each phrase p_i to form the set of keywords $K = \{k_i\}$ using the appropriate "regular expressions".

3. *Keywords Ranking:* [24] The extracted keywords K are ranked using the metric "Term-Frequency" [24], wherein the keywords k_i are counted for their frequency of occurrence f_i in the extracted raw text T and sorted in reverse using the respective frequency values $\langle k_i, f_i \rangle$. This results in the set of ranked keywords $W = \{W_i\}$ from T.

3.2 Semantic Similarity Evaluation

Once the ranked keywords are obtained from the previous stage, the "semantic similarity" [2, 4, 22, 33, 34] measure can be evaluated between every pair of them $\langle W_i, W_j \rangle$ using the state-of-the-art metrics such as Cosine Similarity [4], Jaccard Similarity [4], or user-defined Ad hoc metrics [22, 34]. This results in a semantic similarity matrix S; mathematically represented as $S_{ij} = S\ (W_i, W_j)$ indicating the normalized similarity value within the range [0, 1], between words W_i and W_j. Cosine and Jaccard similarities require the generation of corresponding word vectors [4, 15], say $\langle C_i, C_j \rangle$ using standard tools such as wor2vec [15], Glove [20] in prior and the values are computed as

1. *Cosine Similarity:* [4] an approximate measure of the "cosine of the angle" formed by the word vectors is given by

$$S_{ij} = \frac{C_i \cdot C_j}{||C_i|| \times ||C_j||} \ [4]$$

where numerator represents the dot product of the two vectors and the denominator being the product of the vector's respective magnitudes.

2. *Jaccard Similarity:* [4] indicating the degree by which I_{ij}, the number of sentences having both W_i and W_j over U_{ij}, the total number of sentences having either W_i or W_j given by

$$S_{ij} = \frac{I_{ij}}{U_{ij}} \text{ [4]}$$

3. *Extended Jaccard Similarity:* [4] If the word vectors C_i and C_j are binary, then an extended measure can be used as

$$S_{ij} = \frac{C_i \cdot C_j}{||C_i||^2 + ||C_j||^2 - ||C_i|| \times ||C_j||} \text{ [4]}$$

4. *Ad hoc metrics:* [22, 34] Depending on the context in which the similarity measure has been defined, ad hoc (customized) similarity metrics can be defined such as

   (a)

$$S_{ij} = \sum_{W_i, W_j} e^{-\alpha.l} \times \frac{e^{\beta.d} - e^{-\beta.d}}{e^{\beta.d} + e^{-\beta.d}} \text{ [34]}$$

   defined as combination of both "path length" l between the words W_i and W_j, the "depth" d of their the "least common ancestor" in the WordNet knowledge base; α, β being heuristically determined parameters

   (b) *Word Co-occurrence:* the optimal distance between the two words W_i and W_j as indicative of the minimal or most frequent distance d between them in the underlying text.

3.3 Words Clustering

Once the semantic similarity matrix S is formed, it forms the basis for all the remaining stages of the pipeline architecture. Among them comes the clustering [2, 4, 22, 34] of the words to their most appropriate semantic clusters. The clusters are formed using the standard partitioning mechanisms such as K-Means [9] or K-Medoids [14] to form an ideal number of K clusters. K can be heuristically determined to an optimistic value, typically "$\sqrt{\frac{N}{2}}$" [4] where N is the number of nodes or words. Within each cluster, words can be positioned by utilizing the "Classic Wordle Algorithm" [30], to keep them compact without any overlapping. Further all the words of each cluster can be unanimously and uniquely assigned with suitable font colours. The "clustering efficiency" [7] can be analysed using the methods of "Newman-Girvan's modularity" algorithm [6].

3.4 Word Graph Visualization

The problem of semantic wordcloud generation can be mathematically modelled as generating a word graph G [1, 2, 4, 34] from the ranked keywords W, similarity matrix S, and visualizing G on a two-dimensional canvas. With the clustered keywords in hand, this stage aims at executing a predefined word placement algorithm wherein the words are assigned with their positions on the canvas. The following steps need to be executed in this regard:

1. *Words Visualization:* [8] The visualization step must be invoked at the word level before visualizing the wordcloud in its entirety. Visualizing a word on a two-dimensional canvas usually requires assigning the graphical data structure "bounding box" to each of them. A "bounding box" can be defined as a two-dimensional rectangle which maximally encloses a word of given font size (Fig. 4). The bounding box B_i for each ranked keyword w_i is usually defined with the respective lower left co-ordinates (x_i, y_i) on the canvas, width w_i, and height h_i , all measured in terms of pixel values. (x_i, y_i) is usually assigned during the execution of the geometric layout algorithm for word placement. w_i and h_i are determined by manipulating the approximate aspect ratio for the word W_i,

$$A_i = \frac{w_i}{h_i}$$

   proportional to its ranking in the NLP preprocessing step. The co-ordinates can be initially assigned using a randomized allocation and further changed with the proceedings of the word layout algorithm.
2. *Word Graph Visualization:* [1, 2, 4, 34] "Word Graph" G is defined as $\langle V, E\rangle$ where V is the set of vertices representing the keywords K, and E the set of edges between them with the edge weights corresponding to the respective similarity measures from S. Such a representation (Fig. 5) tries to "*realize*" maximum possible edges as defined by E and S preventing word intersections (overlapping of word rectangles).

   Once the keywords are clustered and coloured, the respective bounding boxes are updated with the new co-ordinates, possibly different from their

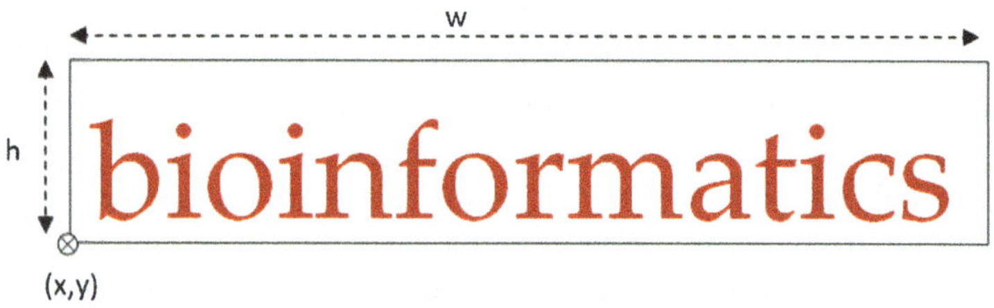

Fig. 4 Bounding box structure for a keyword

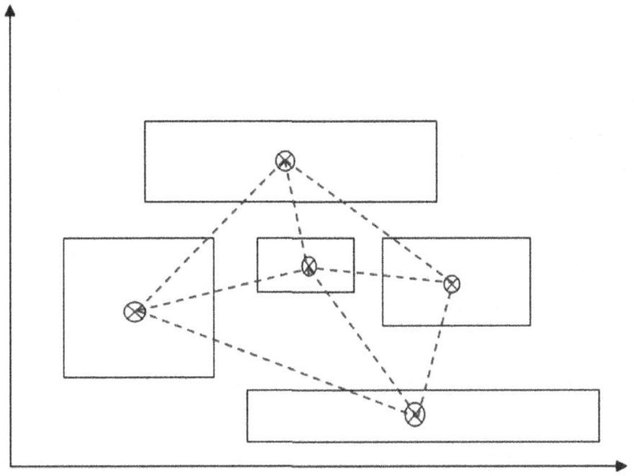

Fig. 5 Word graph visualization

previous assignments while realizing the word graph G. With this updated set of "bounding box" attributes $\langle (x_i, y_i), w_i, h_i \rangle$, each word W_i gets positioned on the canvas resulting in the final visualization output.

3.5 *WordCloud Layout Optimization*

The wordcloud layout algorithm in the previous step realizes the edge weights as distances measured in pixels, between the midpoints of word's bounding boxes $\langle B_i, B_j \rangle$. While the algorithm converges, the words w_i could potentially be scattered across the canvas while "realizing" the maximum number of edge weights. Since a WordCloud is qualitatively judged by its compactness and word overlapping, it would be ideal to apply "Rigid Body Dynamics" [19, 32] here to improve its visual presentation. In such a system, the words are treated as "rigid bodies" [32] and the following "mechanical forces" are applied between them :

1. "*Attractive Forces*" [2]: Brings the neighbouring words W_i, W_j closer, to reduce the empty space within a WordCloud. The attractive force between words W_i and W_j is given by

$$f_a(i, j) = k_a \times (1 - S_{ij}) \times \delta l$$

   where S_{ij} is the similarity measure computed previously; δl the minimum distance to be maintained between two words.
2. "*Repulsive Forces*" [2]: Acting as a balancing factor to attractive forces, preventing words from overlapping by bringing them too close. The repulsive force between words W_i and W_j is given by

$$f_r(i, j) = k_r \times \min(\delta x, \delta y)$$

where δx, δy, respectively, are the width and height of the overlapping region.

k_a and k_r are heuristic parameters [2] found to be optimistic for values 15 and 500, respectively.

4 Conclusion

On a conclusive note, the paper intended to provide a detailed description of the architectural framework for generating any visualization based on semantic wordclouds. With an overview of the basic diagram, each stage has been delved into highlighting the various options available. Such a tutorial article can be a guiding material to many of the future researches in semantic wordclouds.

The growing popularity of semantic wordclouds can be attributed to their visual appeal, huge scope for being used in many real-life applications, and above all its design simplicity as observed by even non-technical users. However, wordclouds have proved to be not well efficient as a "data analytics" tool. It is near impossible to obtain the actual values of information encoding or the frequency of words in the text from the wordcloud. Also wordclouds are rarely used for comparing multiple documents due to its lack of ability in giving accurate frequency measures.

Even then wordclouds and its semantic version in particular continues to serve as one of the most admired computer visualization techniques for text (automatic) summarization. The possible future works for semantic wordclouds include developing web-based tools as educational or reading aids wherein the wordclouds can dynamically regenerate as the document is scrolled down through the pages. Each wordcloud can be a representative of the contents in the current context space of a page.

References

1. Barth, L., Kobourov, S.G.: Semantic word cloud representations: hardness and approximation algorithms. In: Latin American Theoretical Informatics Symposium (2014). Springer, Berlin. https://doi.org/10.1007/978-3-642-54423-1_45
2. Barth, L., Kobourov, S.G., Pupyrev, S.: Experimental comparison of semantic word clouds. In: International Symposium on Experimental Algorithms. Springer, Berlin (2014). ISBN:978-3-319-07959-2_21
3. Bin Packing Problem. https://en.wikipedia.org/wiki/Bin_packing_problem. Accessed 21 May 2018
4. Binucci, C., Didimo, W., Spataro, E.: Fully dynamic semantic word clouds. In: IEEE International Conference on Information, Intelligence, Systems and Applications (IISA) (2016). https://doi.org/10.1109/IISA.2016.7785428
5. Feinberg, J.: Wordle-beautiful word clouds. http://www.wordle.net/. Accessed 16 Jan 2018

6. Girvan-Newman algorithm. https://en.wikipedia.org/wiki/Girvan-Newman_algorithm. Accessed 01 Mar 2019
7. Hearst, M., Franconeri, S.: An evaluation of semantically grouped word cloud designs. IEEE Trans. Vis. Comput. Graph. (2019). https://doi.org/10.1109/TVCG.2019.2904683
8. Jo, J., Lee, B., Seo, J.: WordlePlus: expanding Wordle's use through natural interaction and animation. IEEE Comput. Graph. Appl. (2015). https://doi.org/10.1109/MCG.2015.113
9. Kanungo, T., Mount, D.M., Wu, A.Y.: An efficient k-means clustering algorithm: analysis and implementation. IEEE Trans. Pattern Anal. Mach. Intell. (2002). https://doi.org/10.1109/TPAMI.2002.1017616
10. Kaser, O., Lemire, D.: TagCloud drawing: algorithms for cloud visualization. In: Proceedings of Tagging and Metadata for Social Information Organization (WWW'07) (2007)
11. Koh, K., Lee, B., Kim, B., Seo, J.: ManiWordle: providing flexible control over wordle. IEEE Trans. Vis. Comput. Graph. **16**, 1190–1197 (2010). https://doi.org/10.1109/TVCG.2010.175
12. Kucher, K., Kerren, A.: Text visualization techniques: taxonomy, visual survey, and community insights. In: IEEE Pacific Visualization Symposium (PacificVis) (2015). https://doi.org/10.1109/PACIFICVIS.2015.7156366
13. Li, C., Dong, X., Yuan, X.: Metro-wordle: an interactive visualization for urban text distributions based on wordle. Vis. Informat. **2**, 50–59 (2018). https://doi.org/10.1016/j.visinf.2018.04.006
14. Madhulatha, T.S.: Comparison between K-means and K-medoids clustering algorithms. In: International Conference on Advances in Computing and Information Technology, ACITY 2011, vol.198, pp. 472–481 (2011). https://doi.org/10.1007/978-3-642-22555-0_48
15. Mikolov, T., Chen, K., Corrado, G.S., Dean, J.: Efficient estimation of word representations in vector space. In: International Conference on Learning Representations (2013). arXiv:1301.3781
16. Nan, C., Cui, W.: Overview of text visualization techniques. Introduction to text visualization, Chap. 2. In: Atlantis Briefs in Artificial Intelligence, vol. 1 Springer, Berlin (2016). https://doi.org/10.2991/978-94-6239-186-4_2
17. Natural Language processing. https://en.wikipedia.org/wiki/Natural-language_processing. Accessed 02 May 2018
18. Natural language toolkit. https://www.nltk.org/. Accessed 10 Aug 2018
19. Pedersen, S.W.: Simulation of rigid body dynamics. University of Oslo, Department of Informatics (2003)
20. Pennington, J., Socher, R., Manning, C.D.: Glove: global vectors for word representation. In: *Proceedings of the 2014 Conference on Empirical Methods in Natural Language Processing (EMNLP)* (2014). https://doi.org/10.3115/v1/D14-1162
21. Ransom note effect. https://en.wikipedia.org/wiki/Ransom_note_effect. Accessed 12 Feb 2019
22. Rinaldi, A.M.: Document summarization using semantic clouds. In: IEEE Seventh International Conference on Semantic Computing (2013). https://doi.org/10.1109/ICSC.2013.26
23. Rivadeneira, A.W., Millen, D.R.: Getting our head in the clouds: toward evaluation studies of tagclouds. In: Proceedings of the SIGCHI conference on Human Factors in Computing Systems (2007). https://doi.org/10.1145/1240624.1240775
24. Rose, S., Cowley, W.: Automatic keyword extraction from individual documents. In: Text Mining: Applications and Theory (2010). https://doi.org/10.1002/9780470689646.ch1
25. Schubert, E., Spitz, A., Weiler, M., Gei, J., Gertz, M.: Semantic word clouds with background corpus normalization and t-distributed stochastic neighbor embedding (2017). ACM, arXiv:1708.03569v1
26. Seifert, C., Kump, B., Kienreich, W., Granitzer, G., Granitzer, M.: On the beauty and usability of tag clouds. In: 12th International Conference Information Visualisation (2008). https://doi.org/10.1109/IV.2008.89
27. Semantic WordCloud visualization. http://wordcloud.cs.arizona.edu/. Accessed 15 Jan 2018
28. Tag cloud. https://en.wikipedia.org/wiki/Tag_cloud. 01 Jan. 2018
29. Viegas, F.B., Wattenberg, M.: TIMELINES: tag clouds and the case for vernacular visualization. ACM Interact. **15** (2008). https://doi.org/10.1145/1374489.1374501

30. Viegas, F.B., Wattenberg, M., Feinberg, J.: Participatory visualization with wordle.IEEE Trans. Vis. Comput. Graph. **15**, 1137–1144 (2009). https://doi.org/10.1109/TVCG.2009.171
31. Visualization (graphics).https://en.wikipedia.org/wiki/Visualization_(graphics). Accessed 05 June 2018
32. Wang, Y., Chu, X., Bao, C., Zhu, L., Deussen, O., Chen, B., Sedlmair, M.: EdWordle: consistency-preserving word cloud editing. IEEE Trans. Vis. Comput. Graph. (2018). https://doi.org/10.1109/TVCG.2017.2745859
33. Wu, Y., Provan, T., Wei, F., Liu, S., Ma, K.L.: Semantic preserving word clouds by seam carving. In: IEEE Symposium on Visualization 2011 (EuroVis'11) (2011). https://doi.org/10.1111/j.1467-8659.2011.01923.x
34. Xu, J., Tao, Y., Lin, H.: Semantic word cloud generation based on word embeddings. IEEE Pac. Vis. Symp. (2016). https://doi.org/10.1109/PACIFICVIS.2016.7465278

THIRD EYE: A System to Help the Visually Impaired Students in Academics

J. Midhun Chandran, V. Vinayakrishnan, Salam Vaheetha, and S. Nadera Beevi

1 Introduction

The pedagogy used in the cultivation of proper education among the sightless is always a matter of challenge. Beyond any doubt, Braille's language makes the situation less difficult for both the students and the teachers. To a great extent, Braille's system of raised dots remains the foundation of many of today's advanced technology communication devices for the blind and visually impaired. Unfortunately, the accessibility of Braille materials is very much limited and also quite expensive. In a country like India, education for a blind child will definitely put an extra burden on the parents, due to the unavailability of the resources for their child. Currently they are compelled to depend either on Braille texts or on a secondary person for their child's studies. During the time of examinations, the students communicate the answers to the scribe and the scribe writes the examination for them. As per the government norms, the scribe should be studying in lower classes when compared to the blind student. This constraint is a disadvantage to the blind student, as the conveyed idea is not correctly presented in the answer sheets. The system 'THIRD EYE', has two main modules: text-to-speech and speech-to-text. The text-to-speech is used for converting textbooks, and speech-to-text can be used during the time of examination. In the text-to-speech module, the given input that is the textbook is converted to the speech and the corresponding audio is produced as output. Similarly, in speech-to-text module, the given audio (i.e. answers) is converted and the corresponding text is given as output.

J. Midhun Chandran (✉) · V. Vinayakrishnan · S. Vaheetha · S. Nadera Beevi
M C A Department, T.K.M College of Engineering, Kollam, India

M. Palesi et al. (eds.), *Second International Conference on Networks and Advances in Computational Technologies*, Transactions on Computational Science and Computational Intelligence, https://doi.org/10.1007/978-3-030-49500-8_10

2 Related Works

In India, the number of visually impaired people has been increasing steadily over the past few years. The increasing number of blind people in India is a great source of concern. It is estimated to be 31.6 million by 2020. As per a survey conducted in 2004, 71% out of 72,044 visually impaired individuals were found to be illiterate, and 84.6% reside in rural areas [1].

In the recent past, a number of researches and studies have been going on in the field, of how to educate the visually impaired. Most of the works are based on Braille system. In India, a system called Bharati Braille is being used. It is the conversion of the six-dot system for the languages of India. There are different types of Braille devices available for the visually impaired people who are well-versed with Braille. These include Braille Printers [2, 3], Braille Translation [4] and Braille Displays [5–8].

All the above-mentioned systems are dependent on Braille system, and they are also quite expensive. The paper proposes a system that will liberate the visually impaired from Braille and is cost effective and efficient than other systems.

In a smartphone-based hearing assistive system called SmartHear, a system facilitating speech recognition for various target users in classroom is proposed [9]. Transmitter and receiver devices (e.g. smartphone and Bluetooth headset) are used in this system for voice transmission. An android mobile application is used to control and connect the different devices via Bluetooth or Wi-Fi technology.

In text-to-speech conversion, WaveNet technology is used for producing audio files [10–13]. In speech-to-text, a recorded speech may contain noise that is to be eliminated before processing. The noise can be eliminated using spectral subtraction method. The crucial part of the speech recognition is the feature extraction. Many feature extraction techniques are available. Mel-frequency cepstral coefficient (MFCC) is widely used. MFCC is popular because it is considered quite good in representing signal and it considers frequencies with the human perception sensitivity. Gaussian mixture model (GMM) is one of the best techniques, which is used to train the extracted features from MFCC [14–17].

3 Proposed System

The proposed system focuses on developing a user-friendly module to assist blind students in writing exams more efficiently. The text-to-speech (TTS) module captures the text, validates it and converts it into audio (MP3) format. The module has been tested using a set of sample data. The speech-to-text (STT) module will capture audio from the microphone, remove noise and convert it to a text file (.txt). The text file can be reconverted back to audio using TTS and can be verified for its correctness. If required, audio can be recorded again and converted to text format. The module has been tested using a set of sample data. The schematic diagram of the proposed system 'THIRD EYE', is shown in Fig. 1.

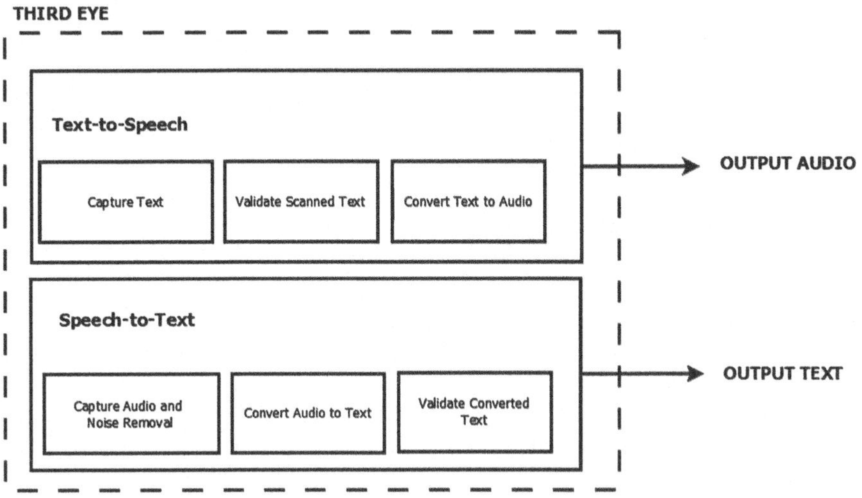

Fig. 1 The schematic representation of THIRD EYE

3.1 Text-to-Speech

The artificial production of human speech is termed as speech synthesis. A text-to-speech (TTS) is a system that converts normal language text into speech. The quality of a speech synthesizer depends on its resemblance to the human voice and its ability to understand clearly. A good text-to-speech program allows those with visual impairments or reading disabilities to listen to written words on the computer.

Capture Text The textbooks that are to be converted are fed as input to the text-to-speech module. The textbooks that are to be converted can be in English or Malayalam.

Validate Scanned Text The input text that is to be fed is accepted only if it is in text format (.txt). The first stage of speech synthesis is based on pre-processing, which is usually a very complex task, depending on the language. The collected sequence of words and symbols from Malayalam and English language will be pre-processed and validated in this stage.

Convert Text-to-Speech The text-to-speech is done using the technology called WaveNet. WaveNet is a deep neural network for generating raw audio (speech), which mimics human voice and sounds more natural. This model is fully probabilistic and autoregressive. This system is able to generate relatively realistic sounding human-like voices by directly modelling wave forms using a neural network method trained with recordings of real speech. WaveNet has the ability to accurately model different voices. WaveNet uses a deep convolutional neural network (CNN). It takes

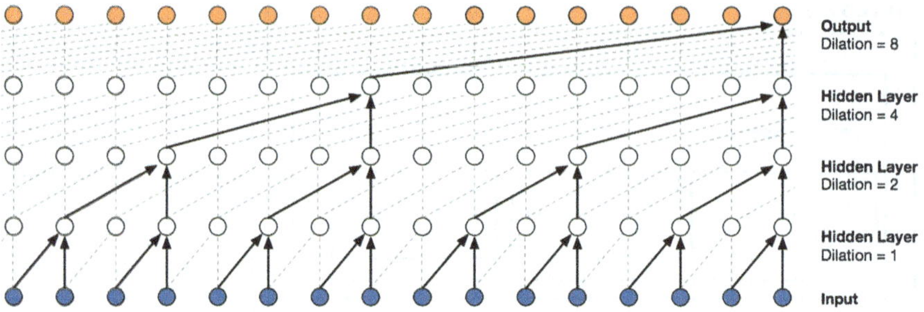

Fig. 2 A representation of dilated casual convolutions with 1, 2, 4, 8 dilations

raw signal as an input and synthesizes an output one sample at a time. The joint probability of a waveform $x = x1, x2, \ldots xT$ is factorized as a product of conditional probabilities as follows:

$$p(x) = \prod_{t=1}^{T} p(x_t | x_1, \ldots, x_{t-1}).$$

WaveNets use casual convolutions, where it is made sure that the model does not violate the order in which the data is modelled. The major disadvantage with this system is that it requires many layers. The drawbacks can be rectified by using dilated convolutions. A filter is applied over an area larger than its length by skipping the input values with a certain step, which is equivalent to a convolution with larger filter derived by the original filter with zeros. In Fig. 2, a representation of dilated casual convolutions with 1, 2, 4, 8 dilations is shown [11, 12].

To model the continual distribution, softmax distribution is used, which is a function that takes as input a vector of K real numbers and normalizes it into a probability distribution consisting of K probabilities. The softmax distribution works better because categorical distribution can more easily model arbitrary distributions as it makes no assumptions about shape.

The activation function used is gated activation function:

$$z = tanh(W_{f,k} * x) \odot \sigma(W_{g,k} * x),$$

where W is a learnable convolution filter, $\sigma(.)$ is a sigmoid function, * denotes a convolution operator, $\odot$ denotes an element-wise multiplication operator, k is the layer index and f and g denote filter and gate, respectively [13].

Output Audio The converted audio files are saved in a secondary storage, as a digital library, which can be used in the future.

3.2 Speech-to-Text

The recorded audio, which is subjected to noise reduction, is then fed as input to the speech-to-text module. Speech-to-text conversion always has many applications. It is one of the most tedious tasks in the field of speech processing. Speech-to-text conversion system is a boon for visually impaired students for their education. If a system can understand what a human speaks, then it is the best method of interaction between a human and a computer. Speech-to-text systems can take speech as input, recognize it and convert it into text. The uttered word, which is the input for the system, is displayed as the output.

Capture Audio and Noise Removal During the time of examinations, the students dictate the answer, and the system records the audio and it is stored. A set of samples are recorded and stored as separate files. Noise reduction on the recorded audio is done using spectral subtraction method.

Convert Audio-to-Text After the recording phase, features are extracted using mel-frequency cepstral coefficient (MFCC). MFCC algorithm includes the following steps: Pre-Emphasizing, Framing, Windowing, Fast Fourier Transform, Mel Filter, Frequency Wrapping and Discrete Cosine Transform.

In order to increase the signal-to-noise ratio, Pre-Emphasizing is used. The audio signal is passed through a high-pass filter, so that the high frequency gets amplified. It will boost the higher-frequency uttered words.

Speech data is to be processed in tiny chunks called frames. Typically, input is divided into frames of 20–30 ms with an optional overlap. To make the signal possible for FFT, frame size is typically taken as power of 2. Otherwise zero padding is to be done to the nearest length of power of 2.

Hamming window is multiplied with each frame in order to keep the last and first points and its continuity in the frame. Let the signal in a frame be denoted by $s(n), n = 0, \ldots, N - 1$, and then after Hamming windowing, the signal can be given as $s(n) * w(n)$, where $w(n)$ is the Hamming window.

Fast Fourier Transform is employed to seek out the magnitude frequency response from every frame. The system uses a set of 20 triangular bandpass filters. Here the magnitude of frequency response is multiplied with these 20 triangular bandpass filters to find the log energy on mel scale. These filters are arranged as equally spaced along the mel frequency. Mel frequency is directly proportional to the logarithm of linear frequency. The spectral envelope is extracted using these filters. The Mel Frequency Wrapping helps to keep only the useful information.

Discrete Cosine Transform (DCT) makes a transformation from the frequency domain into a time-like domain called quefrency domain. The DCT formula is

$$c_m = S_{k=1}^{N} cos\, [m * (k - 0.5) * p/N] * E_k, m = 1, 2, \ldots, L,$$

where N refers to the number of triangular bandpass filters and L is the number of mel-scale cepstral coefficients. The features obtained are similar to a cepstrum, hence it is called as the mel-scale cepstral coefficients.

A Gaussian mixture model is a probabilistic model. It is used as a classifier to compare the extracted features from MFCC with stored templates. It assumes that all the data points are found from a mixture of a finite amount of Gaussian distributions with unknown parameters. A weighted sum of M component Gaussian densities, bi the gaussian mixture density and is expressed as

$$p(\bar{x}/\lambda) = \sum_{i=1}^{M} p_i b_i(\bar{x}).$$

The mean, vectors, co-variance matrices and mixture weights from all component densities are used for describing GMM. Euclidean distance between various recordings is found for the matching purpose, and hence a correct match is found.

Validate Converted Text The converted text is validated by replaying the audio (converted text's audio) and correcting it. The text file can be reconverted back to audio using TTS and can be verified for its correctness. If required, audio can be recorded again and converted to text format.

4 Experimental Results

To measure the performance of the system, the text-to-speech and speech-to-text modules are evaluated separately. For text-to-speech, samples related to various contexts are used in both languages. For speech-to-text, speech signals of different persons with different slangs are collected for the same sample for both the languages. The samples used are as follows:

Sample 1: **The pain you feel today, will be the strength you feel tomorrow.**
Sample 2: **When you feel like quitting, remember why you started.**
Sample 3: **That which we call civilization is merely the accumulated debris of a chilling number of bad nights.**
Sample 4: **This is very real, but safe escapade with resplendent scenery, wonderful camaraderie and truly ineffable ancient art.**
Sample 5: **Exasperating farrago of distortions, misrepresentations and outright lies being broadcast by an unprincipled showman masquerading as a journalist.**
Sample 6: മാത്ര അന്ജക്ഷരം മൂന്നിൽ വരുന്നൊരു ഗണങ്ങളെ
എട്ടു ചേർത്ത് ഉള്ളോരിടക്ക് ചൊല്ലാം കാകളി എന്ന് പേർ
Sample 7: ശ്ലഥ കാകളി വൃത്തത്തിൽ രണ്ടാം പാതത്തിൽ അന്ത്യമാം
രണ്ടക്ഷരം കുറന്ന്ജിടിൽ അത് മഞ്ജരി ആയിടും
Sample 8: അങ്കണത്തൈമാവിൽനിന്നാദ്യത്തെ പഴം വീഴ്കെ
അമ്മതൻ നേത്രത്തിൽ നിന്നുതിർന്നൂ ചുടുകണ്ണീർ
നാലുമാസത്തിൻ മുൻപിലേറെനാൾ കൊതിച്ചിട്ടി
ബാലമാകന്ദം പൂവിട്ടുണ്ണികൾ വിരിയവേ
അമ്മതൻ മണിക്കുട്ടൻ പൂത്തിരികത്തിച്ചപോൽ
Sample 9: മറ്റൊന്നിൽ ധർമ യോഗത്താൽ അത് താനല്ലയോ ഇത് എന്ന്
വർണ്യത്തിൽ ആശംഗ ഉൽപ്രേക്ഷക അലംകൃതി
Sample 10: ഇട്ടിരിക്കാൻ പൊൻതടുക്ക ഇട്ടുണ്ണാൻ പൊൻതിളക
കൈ കഴുകാൻ വെള്ളിക്കിണ്ടി കൈ തോർത്താൻ പുള്ളിപ്പട്ട്
കളിപ്പാനോ കളം തരുവേൻ കുളിപ്പാനോ കുളം തരുവേൻ

Train Data	Total No: of words	No: of correctly pronounced words
Sample 1	12	12
Sample 2	9	8
Sample 3	17	14
Sample 4	17	14
Sample 5	18	14
Sample 6	12	9
Sample 7	11	8
Sample 8	16	12
Sample 9	11	8
Sample 10	16	12

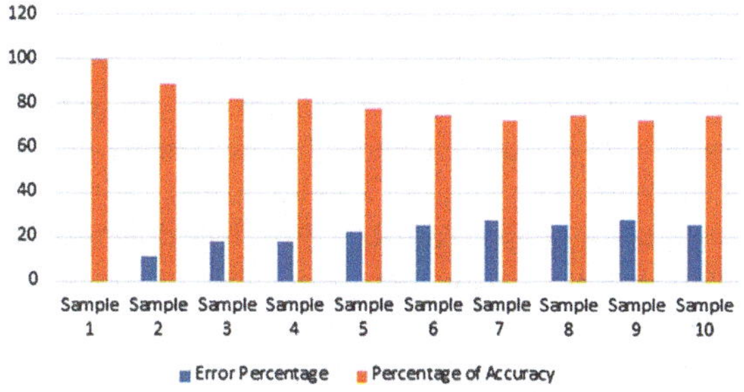

Fig. 3 Performance analysis of text-to-speech module

Train Data	No: of Times the Test is done	No: of correct results
Sample 1	30	27
Sample 2	30	26
Sample 3	30	26
Sample 4	30	22
Sample 5	30	22
Sample 6	30	22
Sample 7	30	24
Sample 8	30	22
Sample 9	30	23
Sample 10	30	22

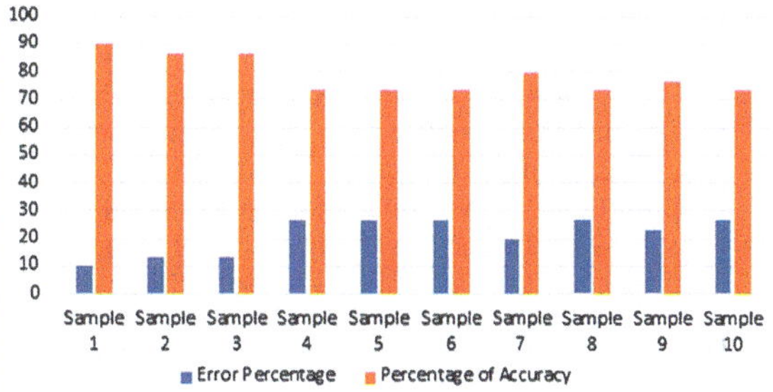

Fig. 4 Performance analysis of speech-to-text module

4.1 Text-to-Speech

For the first module, TTS, the system was trained using 10 different sentences.

The sentences include both English and Malayalam and is shown in Fig. 3.

The overall accuracy of the system is approximately 80.1%.

4.2 Speech-to-Text

For the other module, STT, the system was trained using 10 different samples, each stored as separate audio files. Each sample is tested 30 times. The samples include both English and local languages like Malayalam. The overall accuracy of the system is approximately 78.66% and is shown in Fig. 4.

The overall accuracy of the system is approximately 78.66%.

5 Conclusion

The system is designed as a helping aid for the visually impaired students. The system 'THIRD EYE' can be accessed via web or android app. This system will undoubtedly be advantageous for the visually impaired students. The text-to-speech module has an accuracy of approximately 85%, and the speech-to-text module has an accuracy of approximately 80%. The drawback of this system is with Malayalam text-to-speech, where the naturalness of the audio is comparatively low, and in speech to text, spontaneous speech and dialect cause error in the converted text. We would like to add translation as a future enhancement, whereby the notes in English will be converted to Malayalam, which will help the students in their studies.

References

1. Dandona, L., Dandona, R., John, R.K.: Estimation of blindness in India from 2000 through 2020: implications for the blindness control policy. Natl Med. J. India **14**(6), 327–334 (2000)
2. Bo, L., Junbiao, L., Zhiping, W., Guangrong, F.: Graphic printing method for new braille printer. In: Youth Conference on Information, Computing and Telecommunications, pp. 150–153. IEEE, Beijing (2010)
3. Ioan, L., Daniel, A.V., Alin, G.M.: Experimental module for assistive technologies applications. In: 22nd International Symposium for Design and Technology in Electronic Packaging (SIITME), pp. 304–307. IEEE, Romania (2016)
4. Paul, B., Gareth, E.: Automated braille production from word-processed documents. IEEE Tran. Neural Syst. Rehabil. Eng. **9**(1), 81–85 (2001)
5. Sariat, S., Aaphsaarah, R., Fyaz, H.C., Hasan, U.Z.: A novel braille pad with dual text-to-braille and braille-to-text capabilities with an integrated LCD display. In: International Conference on Intelligent Computing, Instrumentation and Control Technologies (ICICICT), pp. 195–200. IEEE, Kannur (2017)
6. Raj, D.S., Himanshu, S.S., Sudhir, R.B., Sajid, A.K., Darshan, H.M.: Refreshable braille display for the visually impaired. In: 14th IEEE India Council International Conference (INDICON). IEEE, Roorke (2017)
7. Alexander, R., Brent, G.: Refreshing refreshable braille displays. IEEE Trans. Haptics **8**(3), 287–297 (2015)
8. Premkumar, T.: Braille display by rotating multi-octagonal segment. In: Innovations in Power and Advanced Computing Technologies (i-PACT), pp. 1–4. IEEE, Vellore (2017)
9. Chern, A., Lai, Y.H, Chang, Y.-P., Tsao, Y., Chang, R.Y., Chang, H.W.: Smartphone-based multi-functional hearing assistive system to facilitate speech recognition in the classroom. IEEE Syst. J. **12**(1), 20–29 (2017)
10. Dhananjaya, M.S, Niranjana, K.B., Sushma, R.: Kannada text to speech conversion- a novel approach. In: International Conference on Electrical, Electronics, Communication, Computer and Optimization Techniques (ICEECCOT), pp. 168–172. IEEE, Mysuru (2016)
11. Arun, G., Shobana, D.P., Sajini, T., Bhadran, V.K.: Implementation of Malayalam text to speech using concatenative based TTS for android platform. In: International Conference on Control Communication and Computing, pp. 184–89. IEEE, Thiruvananthapuram (2013)
12. van den Oord, A., Dieleman, S., Zen, H., Simonyan, K., Vinyals, O., Graves, A., Kalchbrenner, N., Senior, A., Kavukcuoglu, K.: WaveNet - a generative model for raw audio, pp. 1–15. Arxiv, London (2016)

13. WaveNet: a generative model for raw audio. https://deepmind.com/blog/wavenet-generative-model-raw-audio/. Accessed 24 March 2019
14. Virendra, C., Shobhana, D., Pooja, K., Potdar, S.M.: Speech to Text Converter using Gaussian Mixture Model (GMM). Int. Res. J. Eng. Technol. **3**, 160–164 (2016)
15. Rishiraj, M.: Speaker recognition using shifted MFCC. Graduate Theses and Dissertations (2012)
16. Koustav, C., Asmita, T., Savitha, U.: Voice recognition using MFCC algorithm. Int. J. Innov. Res. Adv. Eng. **1**(10), 158–161 (2014)
17. Tahira, M., Memoona, K., Malik, S.H.K., Ruqia, B.: Speaker identification using GMM with MFCC. Int. J. Comput. Sci. Issues **12**(2), 126–135 (2015)

A Survey on Different Search Techniques Over Encrypted Data in Cloud

Amrithasree Haridas and L. Preethi

1 Introduction

Cloud computing provides the facility to the variety of applications operating over thousands of computers and servers to concurrently access the services through Internet. With the evolvement of cloud computing it has become easier for users to store, retrieve, and share their data among themselves. It offers various benefits to users as well as to service providers. It provides flexibility to work from anywhere at any time. The most extensively adopted application of cloud computing is cloud storage. A tremendous amount of information is being stored by users on cloud servers every day. This information needs protection from different kinds of cyber threats. To maintain data confidentiality and secure storage, various types of encryption algorithms are used for protecting information from unauthorized disclosure. However, searching over encrypted data was difficult to attain. Therefore, keyword based searching has been introduced where the desired file is retrieved when searched for a particular keyword. Numerous searchable encryption schemes are existing in the literature. The common factor in these existing method is that the user data must be encrypted (generating trapdoor) before sending to the cloud server. Upon receiving the search query the cloud server searches in the encrypted document (represented by an encrypted index) and returns the search result to the users. However it must be ensured that while performing the search it should not cause any information leakage. This survey makes the comparative study of some recent single/multi-keyword search techniques on large-scale encrypted data in cloud.

A. Haridas (✉) · L. Preethi
College of Engineering Trivandrum, Thiruvananthapuram, India
e-mail: amrithasreeharidas@gmail.com; preethi@cet.ac.in

M. Palesi et al. (eds.), *Second International Conference on Networks and Advances in Computational Technologies*, Transactions on Computational Science and Computational Intelligence, https://doi.org/10.1007/978-3-030-49500-8_11

1.1 Cloud Security Requirements

The infrastructure of cloud must be capable enough to implement the appropriate security measures at its premises. Although the services provided by the cloud are regularly being improved, still there is a great need for protection of data stored in the cloud. For this, the most important requirement is to build up trust between the user and service provider.

To protect cloud data, following security measures need to be implemented:

- **Authentication**: This technique helps the communicating entities to prove its identity and assures authentic communication This service also guarantees that no other unauthorized entity can masquerade itself as authorized entity to take undue advantage of ongoing communication.
- **Access control**: Authentication and identification of entity must be carried out to give access rights to the entity. It is the process of imposing the restriction to access systems and applications according to the level of security requirements.
- **Confidentiality**: The attacker is not allowed to look at frequency, length, and other attributes of traffic flowing through the network. Unauthorized exposure of information must be protected to maintain the confidentiality of sensitive cloud data.
- **Integrity**: The received data must be free from duplication, modification, and reordering. Only authorized users can make changes to it. This service assures the correctness and validity of data being transmitted through the network.
- **Availability**: To maintain it offsite backup should be done regularly, and the systems must be prevented by Denial of Service attacks. This service assures that information is available to authorized users whenever required.
- **Non-Repudiation**: This service provides the proof that the authorized sender and receiver have sent and received the information, respectively. For this, accurate and traceable records must be maintained.

1.2 Architecture of Search Over Cloud Data

Figure 1 demonstrates the architecture of the system. The different modules of the system are data user, data owner, and the cloud server. Data stored in the cloud is (Fig. 2) in encrypted format for the purpose of security. Only authenticated users can use the data in the cloud and hence security is improved. Data owner is the person who uses the cloud for storage. Also he can share the data with other peoples. Data users are the person who can access the data uploaded by data owners.

Searchable encryption permits clients to correctly retrieve the encrypted information. This method has two major disadvantages; First one is, that users who do not essentially have pre-knowledge of the encrypted cloud information have to post-process each file obtained after search, in order, to realize the one most matching

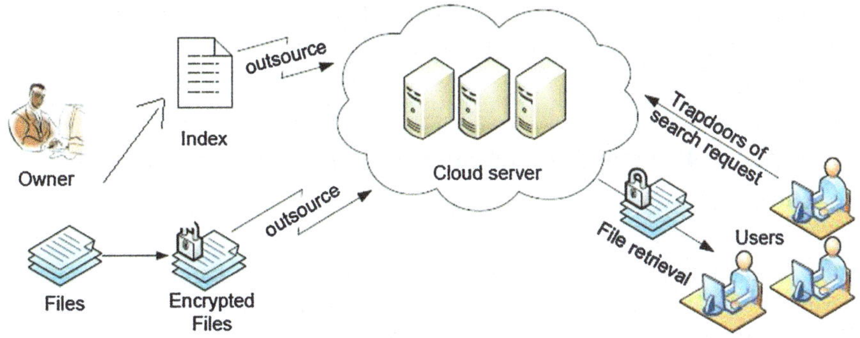

Fig. 1 Architecture of the system [1]

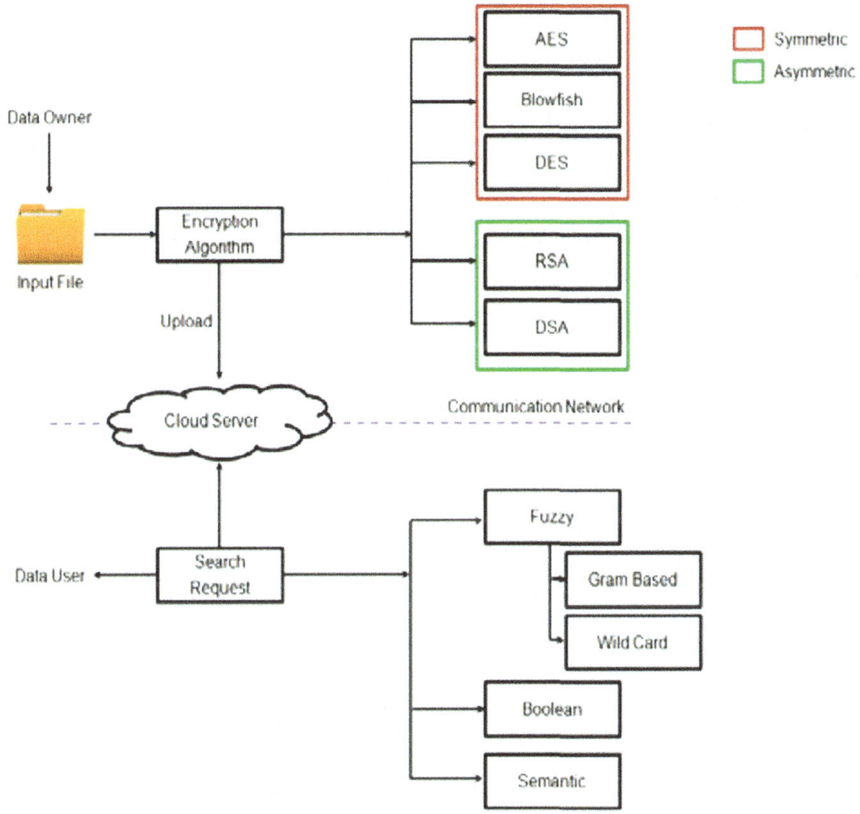

Fig. 2 Search architecture in cloud data [2]

their interest; second disadvantage is when multiple records containing the queried keyword are retrieved in the search this causes unnecessary network traffic.

2 Types of Keyword Based Searching in Encrypted Cloud Data

Song et al. introduced the concept where each word of the file is encrypted separately. But this technique resulted in higher cost as the word by word scanning of the documents is required. They suggested a sequential scan which could be executed with or without an index. When the documents in the dataset are large, then the index based scheme is preferred since it gives faster search results. But this system causes trouble in the situation where storage and updating of records are needed.

3 Searchable Encryption

With searchable encryption technique user can store and share their data in encrypted format to improve the security of data, and other users can search in this encrypted secure data. Figure 3 shows different types of searchable encryption.

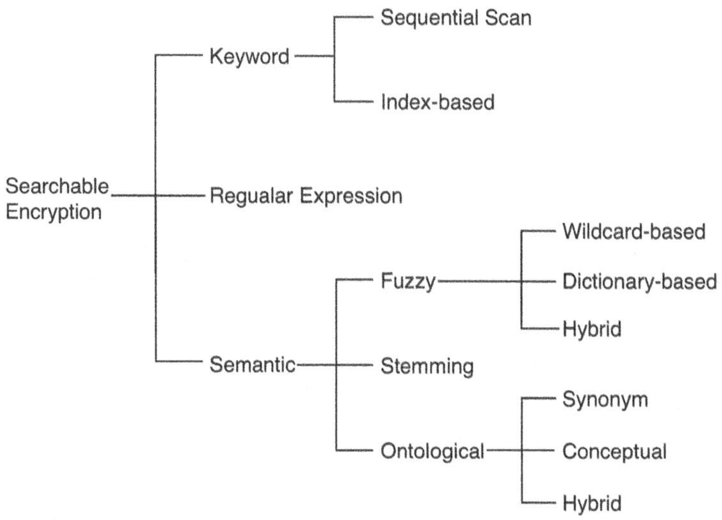

Fig. 3 Classification of the current searchable encryption system

Various encryption techniques can be used to achieve this, such as AES, ECC, etc. This can work on large-scale data.

3.1 Keyword Based Approaches

Ranked Single Keyword Search

C. Wang [3] proposed an efficient solution for supporting ranked keyword search problems. In this technique, single random keyword is the input to the cloud server and the cloud server generates the most related file that matches with the input keyword. Ranked keyword search generates the results rank wise instead of just providing matched results. It will reduce the cost of searching and also provide the most related results to improve the user experience.

Data owner outsources their n data files to the cloud. Before outsourcing the files to the cloud server, they encrypt their file for the security purpose. Even though they are encrypted they can also retain their ability to search for effective data usage. Before outsourcing their information, create an index i by using a set of k distinct keywords extracted from the file collection D, and store both the index I and the encrypted file collection D on the cloud server. After this, on receiving search query T_w from data user, the server is actually responsible for finding the index i and provides results without revealing the actual content of sensitive data. For data user, to search for a keyword w authorization is performed with trapdoor generation T_w and submitted to the cloud server.

Multi-Keyword Search

To make searching system more practical, system can support multi keyword search in place of single keyword search. Zhihua Xia et al. [4] proposed a scheme, where input search query may contain more than one keyword, this improves the accuracy of search query. Multiple keywords have the capability to explain the search query accurately.

It also supports dynamic functions such as inclusion and removal. In particular, the widely used TF X IDF model and the vector space model have been integrated into index building and generation of queries. It specifies an index structure based on tree structure and suggests "Greedy Depth-First Search" algorithm for ranked multi-keyword search. And for encrypting query vectors it uses a secure kNN algorithm, then accurately calculates the correct score between the encrypted query vectors and the index.

Ranked Multi-Keyword Search

Ranked keyword search generated the results rank wise instead of just providing matched results. Cloud server then generates the most related file that matched with input keywords. In this technique, multiple random keywords are input to cloud server. N. Cao [5] tried to improve the previous works in this field by adding the ranking functionality together with search over encrypted data in cloud having multi-keywords and thus bettering the user search experience.

Here is used a privacy-preserving, similarity based text retrieval scheme where the search results are hidden from the unauthorized entities. Also, the server is unable to reconstruct the term composition of documents and queries performed. They employed similarity measure of "coordinate matching" organized as multi-keyword semantics. It uses "inner product similarity" to quantitatively evaluate the similarity measure. But there were two shortcomings of this scheme. First of all, it requires the reorganization of static dictionary each time with the entry of a new keyword. As the size of the collection of records grows exponentially the time for search also increases exponentially. The system is MRSE based on secure inner product computation. But in this technique the synonym searching was not taken into consideration. And therefore the searching time was increased intensely.

Fuzzy Keyword Search

To make the search process more user interactive, this were introduced over encrypted data in cloud. Initially, the concept of search using fuzzy keyword in encrypted cloud data was given by J. Li [6]. It states that the search system used in this method can give the accurate result even if the keyword is slightly misspelled by the user. This technique attempted to make the search procedure user interactive. On the other hand, in traditional techniques, no result is found when there are minor errors in spelling of keywords entered, and hence it makes the user's task very complicated. To handle this problem, J. Li [6] implemented fuzzy keyword searching. It also focused on preserving the privacy of keywords. If user spell incorrectly then integrate edit distance with wildcard-based technique to build fuzzy keyword sets, to address minor misspelling issues and format inconsistency and by using this method it calculate the closest matching keyword. To diminish the difficulty in storage and to handle the issues in representation, they developed keyword dictionary. They demonstrated that their work was proficient in maintaining the privacy and security employing detailed security analysis. It also showed the utility of this technique.

A Conjunctive Keyword Search

C. Wang [7] proposed a method where, a query request has multiple keywords and for each keyword trapdoor is generated. The final result is the intersection of results of each keyword in search request.

To be precise, it is the statistical measure approach, that is, it calculates the relevant scores. It generates secure search results from information and builds up a one-to-many order-preserving mapping techniques to appropriately protect these sensitive score information. Order-preserving encryption (OPE) is a practical method to support fast ranked search. OPE is used to encrypt relevance scores in the inverted index, and so often the data privacy cannot be assured in applications. The OPE has improved by Wang et al. [7], in their secure keyword search scheme to "One-to-many OPE," where they tried to build a probability encryption scheme and hide the distribution of plaintext. The server side rankings will have an effective design without losing data. These methods, however, cause substantial overhead in communicating due to sharing the secret and increasing computing costs due to the bilinear mapping.

3.2 *Semantic Based Approaches*

MRSE System Supporting Synonym Query

The search results are purely based on user authentication and only authorized user can input the search request. It increases the flexibility of search when the user forgets the exact keyword, and the user can search by using some of its synonyms. Zhangjie Fu [8] suggested a synonym based multi-keyword search system in an encrypted data in cloud. This is the first method suggested based on semantics.

Apart from other methods the main difference in this method is that, here the keyword set is extended by adding some of its synonym also, for that keywords need to be extracted from the file collection before outsourcing it to the cloud. Here uses a better text feature weighting method, which adds a new component module to indicate the distinguishability of the term on the basis of the original TFIDF (term frequency-inverse document frequency) method (term frequency–inverse document frequency) method. The new element C_d has been added to the equation of TFIDF,

$$\text{Weightingfactor} = \text{TF}X\text{IDF}XC_d \tag{1}$$

To accomplish an efficient meaningful search for outsourced data, the keyword set should be extended by adding synonyms. If a keyword has more than 2 synonyms, then all synonyms are added into the keyword set. So using this improved method will extract keywords from outsourced text files. All keywords separated from a single text form a keyword set at the end. The redundant keywords are deleted to reduce the burden of storage.

Semantic Search Using Stemming Algorithm

T. Moataz [9] incorporated a stemming algorithm with an efficient searchable encryption technique. In the case of keyword based approach each keyword is mapped to all the documents having this word. In a semantic search not only the

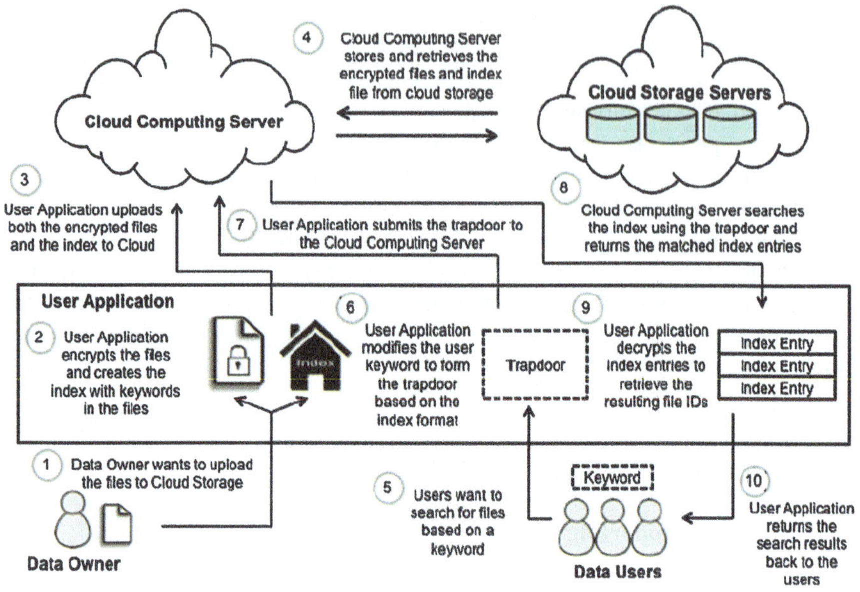

Fig. 4 An overview of semantic searches using encrypted data in cloud computing

documents containing the keyword itself but also the documents with words related to the keyword have to be mapped. In the search request the user inputs the query and the keywords are extracted from the query. Then the root of this keyword is found and these roots are sent instead of the keywords in the search request. Related keywords have the same root keyword, so that it only stores the root of these keywords in the index. By combining a searchable encryption algorithm with a stemming algorithm an efficient semantic search technique over encrypted cloud data can be achieved. As a result, the user will retrieve all the encrypted documents that contain all related terms. Figure 4 shows this method.

Semantic Search Using Online Ontological Network

Z. Jason Woodworth [10] proposed a semantic search using online ontological network. The basic idea of this methodology is to consider the frequency of occurrence of keywords in document. It ignores the importance of the terms in the text or the request. For that reason, change every instance of a particular word in each document into a similar token, and subsequently apply a similar change when that word shows up in the search query. For normal text recovery, Okapi BM25 algorithm will be used. This is a TF and IDF model. Algorithm is never considering the true meaning of each term in the document. This feature makes this algorithm very applicable in this method. This technique has 2 phases: upload and search.

Upload Process
The purpose of this process is to parse the file to the indexable format and encrypt it before sending it to the cloud. Term frequency of each keywords for the text is the product of TF and IDF collected. For each word it finds its hashed value and writes them to a temporary index file and sends it to the cloud with encrypted documents. A subset of words (usually called keywords) is usually taken from the document to represent the semantics of that document.

Search Process
This process has mainly two components: One is Query modification and the other is index searching and ranking. Query modification is performed on the client side and searching is performed on the cloud processing server. The process of query modification involves splitting, semantic expansion, and weighting. The clients query cannot be semantically identified by this methodology alone. To achieve this it uses more advanced ontological networks. For example, it utilizes the contents from Wikipedia and performs keyword extraction on them to get related words and expressions. These related terms are merged into modified query set Q. As a result, users can retrieve the documents containing ideas related to the query.

3.3 Comparison of Different Searchable Encryption Techniques

There are several methods for performing a search over encrypted data in cloud, using keyword as a factor for searching. Single keyword and multi-keyword searches are possible. To enhance the efficiency and fastness of searching multi-keyword technique can be used. A comparative study on different search algorithms in the cloud is described in Tables 1 and 2.

Table 1 Comparative analysis of various semantic based searching techniques

Methods	Keyword search (Single/Multiple)	Process used
Multi-keyword ranked search supporting synonym query	Multi-keyword	Input the synonyms of the extracted keyword
Semantic search using stemming algorithm	Multi-keyword	Stemming algorithm finds the root of queried keyword in the search request. And the e root of queried keyword is used for search instead of keywords in the request
Semantic search using online ontological network	Multi-keyword	Use online ontological network

Table 2 Comparative analysis of various semantic based searching techniques

Sr.No	Method	Process used	Advantage	Disadvantage
1	Searchable Encryption Scheme	Symmetric Public Key Encryption	Secure search employed over encrypted data on cloud	Costly in terms of computation
2	Boolean Keyword Searchable Encryption Scheme	Structural and Boolean keyword search using Boolean operators AND, OR, and NOT	Comfortable enough to express small, easy information needs	Excess network traffic. Efficient document ranking is not supported
3	Single keyword searchable encryption scheme	Encrypted searchable index	5 keyword frequency utilization to rank results	Not comfortable enough to precise complex information needs
4	Ranked keyword searchable encryption scheme	Relevance score is employed to make a secure searchable index. Order-preserving mapping function	Enhances system usability by returning the matching files in a ranked order concerning to certain relevance criteria. Eliminate excess network traffic	Compromise the privacy
5	Fuzzy keyword searchable encryption scheme	Wildcard-based technique	Eliminates the requirement for enumerating all the fuzzy keywords	Supports only Boolean keyword search. Huge storage complexity
6	Plaintext Fuzzy keyword searchable encryption scheme	Plaintext searching, string matching algorithm	To find relevant information it allows user to search using try and- see approach	Statistics and dictionary attacks and fails to attain the search privacy
7	Conjunctive Keyword Searchable Encryption Scheme	Decisional Diffie–Hellman (DDH) and hardness assumption	Solution to Boolean keyword search problem	Privacy overhead
8	Multi-keyword Searchable Encryption Scheme	Provides secure index structure, generates secret trapdoors	Documents confidentiality and privacy of index, trapdoor, trapdoor unlinkability	CSPs that keep the data for users may access users sensitive information without authorization

4 Performance Analysis

Here compare the time used for searching with different size of dataset in different methods. The first figure (Fig. 5) shows the search time in case of synonym query and the second figure (Fig. 6) shows search in case of synonym query by using ontological network. The final figure (Fig. 7) shows search in case of by using stemming algorithm.

5 Conclusion

This survey makes the comparative study of some recent single/multi-keyword search techniques on large-scale encrypted data in cloud. The traditional methods of keyword searching were limited to the exact keyword search. But recently, many researchers have implemented fuzzy keyword searching in which the encrypted file is retrieved when the keyword matches exactly or when it is slightly misspelled and preserving the privacy of keywords at the same time. Further suggested a synonym based multi-keyword search system in an encrypted cloud data. Then the user can do searching by similar meaning words. Such techniques maintain the privacy and security of data during search. It might be possible that user forgets the exact

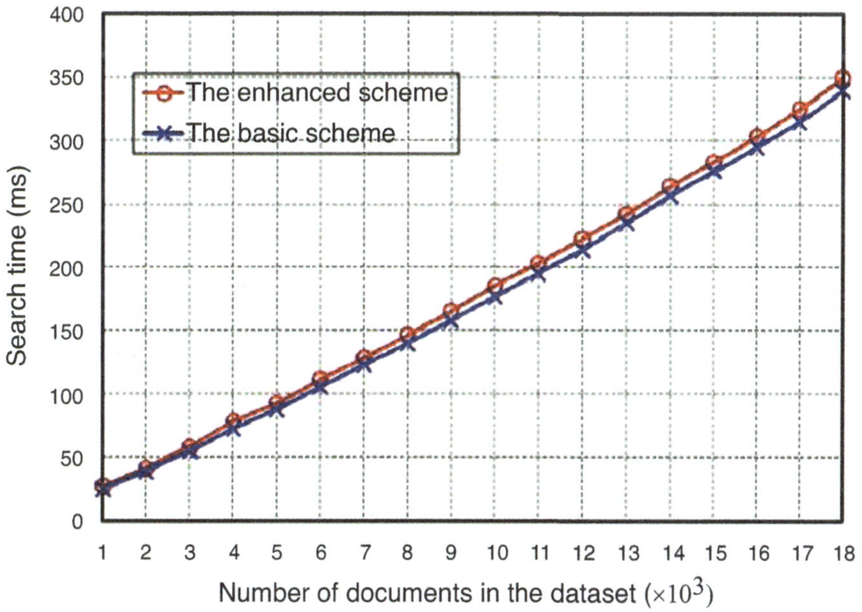

Fig. 5 Time taken to search for MRSE system supporting synonym query [5]

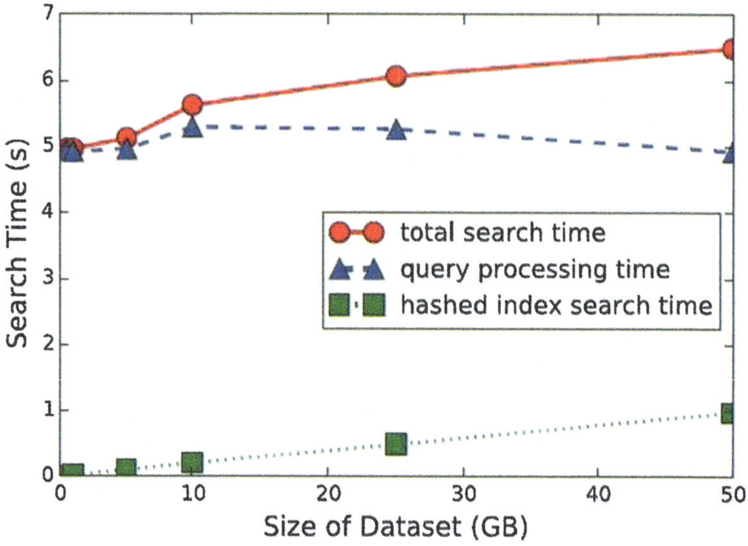

Fig. 6 Time taken to search for semantic search using online ontological network [10]

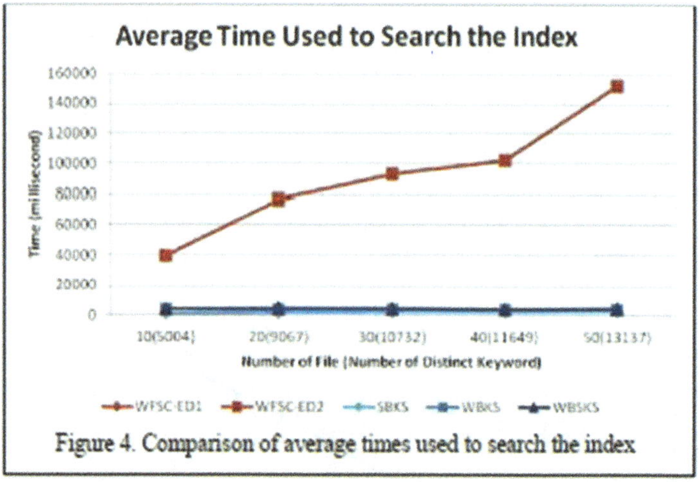

Fig. 7 Time taken to search for semantic search using stemming algorithm [9]

keyword. In this comparative analysis, compare them on the basis of various criteria such as, key idea of approach, their advantages and disadvantage. And this survey also identifies the limitations of existing system.

References

1. Mistry, S., Tandel, P.: A survey on context based search over encrypted cloud data techniques. In: Proceedings of the International Journal of Modern Trends in Engineering and Research (2015)
2. Ingale, S., Phulpagar, B.D.: A survey on different keyword-based search techniques over encrypted data. In: Proceedings of the (IJCSIT) International Journal of Computer Science and Information Technologies (2016)
3. Wang, C., Cao, N., Li, J., Ren, K., Lou, W.: Secure ranked keyword search over encrypted cloud data. In: Proceedings of the IEEE 30th International Conference on Distributed Computing Systems (ICDCS 10) (2010). https://doi.org/10.1109/ICDCS.2010.34
4. Xia, Z., Wang, X., Sun, X., Wang, Q.: A secure and dynamic multi-keyword ranked search scheme over encrypted cloud data. IEEE Trans. Paral. Distri. Syst. **27**, 340–352 (2015). https://doi.org/10.1109/TPDS.2015.2401003
5. Cao, N., Wang, C., Li, M., Ren, K., Lou, W.: Privacy-preserving multi-keyword ranked search over encrypted cloud data. IEEE Trans. Paral. Distri. Syst. **25**, 222–233 (2014). https://doi.org/10.1109/TPDS.2013.45
6. Li, J., Wang, Q., Wang, C., Cao, N., Ren, K., Lou, W.: Fuzzy keyword search over encrypted data in cloud computing. In: Proceedings of the IEEE INFOCOM, San Diego, CA (2010). https://doi.org/10.1109/INFCOM.2010.5462196
7. Wang, C., Cao, N., Ren, K., Lou, W.: Enabling secure and efficient ranked keyword search over outsourced cloud data. IEEE Trans. Paral. Distri. Syst. **23**, 1467–1479 (2012). https://doi.org/10.1109/TPDS.2011.282
8. Fu, Z., Sun, X., Xia, Z., Zhou, L., Shu, J.: Multi keyword ranked search supporting synonym query over encrypted data in cloud computing. In: Proceedings of the IEEE 32nd International Performance Computing and Communications Conference (IPCCC2013), San Diego, CA (2013). https://doi.org/10.1109/PCCC.2013.6742783
9. Moataz, T., Shikfa, A., Cuppens-Boulahia, N., Cuppens, F.: Semantic search over encrypted data. In: Proceedings of the 20th International Conference on Telecommunications (ICT) (2013). https://doi.org/10.1109/ICTEL.2013.6632121
10. Woodworth, J.Z., Salehi, M.A., Raghavan, V.: S3C-An Architecture for space-efficient semantic search over encrypted data in the cloud. In: Proceedings of the IEEE International Conference on Big Data (Big Data) (2016). https://doi.org/10.1109/BigData.2016.7841040

Modeling and Verification of Launch Vehicle Onboard Software Using SPIN Model Checker

Ranjani Krishnan and Lalithambika V. R.

1 Introduction

In a launch vehicle, the avionics system plays a pivotal role. It consists of the onboard computer that serves as the brain of the rocket, control electronics systems to maneuver the vehicle along the required trajectory, stage sequencing systems, communication systems, and other critical hardware. Equally important is the onboard software that consists of distinct components like navigation, guidance, digital autopilot, sequencing, and a scheduler. The fault-free and synergistic operation of these different hardware and software systems ensures that the objectives of a launch vehicle mission are successfully accomplished.

The testing of onboard software is as significant as the design and development. Over the years, several techniques have been employed to test such mission-critical software and ensure its reliability. But, traditional methods cannot guarantee the complete absence of errors. In such scenarios, the application of formal verification is gaining prominence. Formal methods are based on Mathematical techniques and can ensure that the software satisfies properties specified by the user. Many tools are available for formal verification of embedded software, some based on theorem proving, some based on model checking, and some for static analysis.

Our aim is to model a part of the onboard software in a launch vehicle using a suitable tool and check whether it satisfies specified properties. The SPIN model checker [1] has been adopted for our work, considering its heritage and usage history in the space and aviation sectors. It has been used for verification of spacecraft

R. Krishnan (✉)
Vikram Sarabhai Space Centre, Trivandrum, Kerala, India

. Lalithambika V. R.
ISRO Headquarters, Bengaluru, Karnataka, India

M. Palesi et al. (eds.), *Second International Conference on Networks and Advances in Computational Technologies*, Transactions on Computational Science and Computational Intelligence, https://doi.org/10.1007/978-3-030-49500-8_12

software for different missions by NASA [2–6]. In [2], the application of model checking to legacy flight software from NASA's Deep Space 1 (DS1) mission and the results are explained. [3] describes two separate formal verification efforts of the Remote Agent (RA) autonomous spacecraft controller developed at NASA. A part of RA executive, while it was under development, was modeled in SPIN and analyzed that led to the discovery of five errors in the LISP code. In [4], validation of a dually redundant system (DRS) for a spacecraft controller with a checkpoint and rollback scheme for fault tolerance, through model checking is described. [5] elucidates the application of SPIN to formally verify a multi-threaded plan execution programming language named Executive Support Language (ESL). In [6], the authors explain the verification of a complex fault protection (FP) scheme in Dawn spacecraft using SPIN. [7–11] present some recent, related work on formal verification of software in spacecraft, missile, aircraft, and automotive systems. Different techniques and tools have been used for modeling and verification in these. Literature on modeling and verification of launch vehicle systems is not common and has been attempted in this work.

The rest of this paper is organized as follows: Section 2 gives the necessary background, viz. a short introduction to the software configuration and scheduling scheme in the Onboard Computer (OBC) of a launch vehicle. Section 3 describes the overall scheme of the modeling effort in SPIN and details on each part of the model. The results of the simulation and verification are discussed in Sect. 4. Finally, Sect. 5 concludes the paper and also sketches the scope and course of future work.

2 OBC Software

The onboard software in the launch vehicle is responsible for different functions, like acquiring and processing data from position sensors, determining the optimum trajectory, computing the control commands based on appropriate algorithms, generating ignition and separation commands based on launch events, etc. There is also an OS-like component which establishes the necessary software and hardware interfaces and schedules the different tasks in required periodicities. This critical component carries out various functions like managing the processor state during booting, Input/Output operations, synchronization with other units, fault handling, and so on.

The onboard software of a typical rocket is considered for our case study. The scheduling scheme used in this software is depicted in Fig. 1. In this system, the tasks are organized such that they execute in two periodicities. The shorter period is referred to as minor frame and the longer period as major frame. The tasks running in the higher frequency (minor frame tasks) are executed in sequence, without any interruption, till they finish. A fixed priority, deterministic scheduling scheme is followed. In every minor frame, after the minor tasks have completed execution,

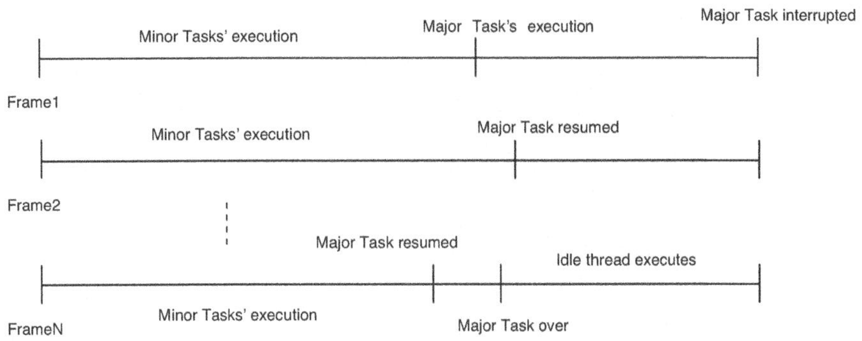

Fig. 1 Scheduling scheme of onboard software

the execution of major frame tasks is commenced. The minor frame is implemented through a hardware timer that raises an interrupt at the end of the time period. If this interrupt occurs when the major task is executing, the execution is suspended and the required context is saved. It will resume from this point in the next minor frame after completion of minor tasks. The interrupt service routine carries out initializations for the next frame on the occurrence of every timer interrupt and the cycle repeats. An indefinite loop, named idle task executes once all major tasks have finished execution.

3 Modeling in SPIN

In SPIN model checker, the system to be modeled and verified is represented in the language Promela (Process Meta Language), which has a syntax similar to C language. Distributed systems are modeled using concurrent processes in SPIN. The communication and interfacing between the processes is facilitated through channels, which may be synchronous or buffered.

3.1 Scheduler Model

A model of the scheduling scheme, elaborated in the previous section was developed in the SPIN model checker (~500 LOC). The scheduler and timer are represented by two processes running simultaneously. The variables, constants, channels, and macros are declared as necessary (Listing 1). A user-defined data structure is used to model the tasks, with task number, state, and completion status bit as parameters. An enumerated data type consisting of three values (Ready, Executing, Suspended) represents the execution state of the task. To interface between the timer and scheduler processes, a channel named "FromTimer" is declared. It is assumed that there are three minor and major frame tasks.

Listing 1 Variable declarations

```
/* Discrete time macros */
#define timer int
#define State mtype
/* Minor frame time parameters */
#define MinorFrame 20
/* Constants */
#define OK 55
#define NotOK 1010
/*timers*/
timer MinFrame;
/* Global constants/ variables */
short MinorID = 0, MajorCount = 1, MinorTaskNum = 0,
  MajorTaskNum = 1, CurrentTask = 1;
bit TimerInterruptMn = 0, AllMinorComplete = 0,
  AllMajorComplete = 0, RandomError = 0;
short MinCount = 0, ErrorFlag = 0;
short Min1toMin2,Min2toMaj1,Maj1toMaj2,Maj2toMin1;
short Min1toMin2C,Min2toMaj1C,Maj1toMaj2C,Maj2toMin1C;
short var3,var3C,var4,var4C;
short SavedContext[3],Reg[3];
byte BC_Health;
State = { READY, EXECUTING, SUSPENDED }
typedef Task{
byte TaskNum;
State state;
bit Complete;
};
Task MinTask[3];
Task MajTask[3];
/* channels */
chan FromTimer = [0] of {bit};
```

In the timer process, a fixed value counter that down counts till zero is modeled (Listing 2). On completion, it transmits an interrupt signal to the scheduler process through channel FromTimer. The counter is then reloaded with the initial value and it starts down counting again, indicating a new minor frame. Through this, the periodic behavior of the onboard software in real time is represented.

Listing 2 Timer process

```
proctype Timers()
{
   do
   ::MinFrame> 0 ->
         MinFrame = MinFrame - 1;
   ::MinFrame<= 0 ->
         TimerInterruptMn = 1;
         FromTimer!TimerInterruptMn;
         MinFrame = MinorFrame;
   od;
}
```

The actions to be carried out on the occurrence of a timer interrupt are implemented in the Interrupt Service Routine (ISR). It is modeled as an inline function that is invoked by the scheduler whenever the interrupt is raised (Listing 3). In this ISR, the context of the interrupted major task is saved, the count of minor frames is incremented, the error flag and completion status bits of tasks are cleared, and the execution state of all tasks is set as "Ready." A similar ISR is modeled for completion of a major frame also.

Listing 3 Inline function for minor ISR

```
inline MinorISR()
{
   d_step{
      int i;
      CurrentTask = MajorTaskNum;
      RealTimeCount++;
      if
      :: (MinTask[0].Complete == 0) || (MinTask[1].
     Complete == 0) || (MinTask[2].Complete == 0) ->
         BC_Health = NotOK;
      :: else ->
         BC_Health = OK;
      fi;
      for (i : 1 .. 3){
         MinTask[i-1].state = READY
      }
      MinTask[0].Complete = 0;
      MinTask[1].Complete = 0;
      MinTask[2].Complete = 0;
      AllMinorComplete = 0;
      ErrorFlag = 0;
   }
}
```

The minor and major tasks are represented as inline functions (Listing 4). A few variables and minimal computations are included. The tasks in different periodicities interact through exchange of data. The validity and correctness of the data are ensured by passing the data along with its logical negation. Before using the data, it is checked whether the negation of the variable matches its complement, thus ensuring its integrity. The error flag is set in case the check does not pass. The three minor frame tasks are named MinTask1–3; the major frame tasks are MajTask1–2 and idle task.

Listing 4 Inline function for a minor task

```
inline MinTask1()
{
   MinTask[0].state = EXECUTING;
   short var1,var1C;
   var1 = 11;
   var1C = -11;
   Min1toMin2 = var1;
   Min1toMin2C = var1C;
```

```
        if
        :: (Maj2toMin1 + Maj2toMin1C) != 0 ->
           ErrorFlag = ErrorFlag | 8;
        :: else ->
           ErrorFlag = ErrorFlag | 0;
        fi;
        MinTask[0].Complete = 1;
}
```

For our model, five minor frames per major frame are considered. At the end of execution of each task, its completion bit is set to 1. The context saving of interrupted major tasks is modeled as saving essential register values by calling an inline function. Another function is used to model the procedure of restoring the saved context when the major frame task resumes execution. The periodicity of the scheduler is simulated using the "unless" construct (Listing 5). The overall structure of the scheduler process is as follows.

Listing 5 Scheduler process

```
proctype Scheduler()
{
   do
   ::  FromTimer?TimerInterruptMn;
   TimerInterruptMn = 0;
       {
         MinorISR();
         if
         :: MinorID< 5 ->
            MinorID++;
         :: MinorID == 5 ->
            MajorISR();
         fi;
         MinorTaskNum = 1;
         MinTask1();
         MinorTaskNum = 2;
         MinTask2();
         MinorTaskNum = 3;
         MinTask3();
         AllMinorComplete = 1;
         if
         :: CurrentTask == 1 ->
            if
            ::  MajTask[0].state == SUSPENDED ->
               Restore_Context();
            :: else -> skip;
            fi;
            MajTask1();
            d_step{
            MajTask[0].state = READY;
            MajorTaskNum++;
            CurrentTask = MajorTaskNum;
            }
         :: else -> skip;
         fi;
```

```
                    if
                    :: CurrentTask == 2 ->
                       -------------
                    :: else -> skip;
                    fi;
                    AllMajorComplete = 1;
                    if
                    :: CurrentTask == 3 ->
                       if
                       :: MajTask[2].state == SUSPENDED ->
                          Restore_Context();
                       :: else -> skip;
                       fi;
                    IdleTask();
                    :: else -> skip;
                    fi;
                    } unless {TimerInterruptMn == 1;
                     Interrupt_Task() };
              od;
          }
```

4 Simulation and Verification

The OBC software model coded in Promela was simulated in SPIN. It was ensured that in the nominal case, the scheduling and execution of the minor and major frame tasks was taking place as expected in the simulation. The execution sequence is visible in the tool, with each frame and the events happening in each frame displayed in one window. In order to check if the execution enters an erroneous path in any scenario, assertion statements were introduced in the code. For instance, the assertion statement assert(ErrorFlag == 0) was inserted in the listing for MinTask1. The model checker was running on an x86-based PC with 4GB RAM. The following Table 1 gives a summary of the memory, state space, and time taken for the verification run.

It was found that the assertion was violated in one particular execution sequence. This sequence is depicted in Fig. 2. On analyzing the error trail, it was found that this was due to failure of the complement check in a minor task. The series of events that led to setting of the error flag is as follows:

1. In a major task executing in a particular minor frame, one variable that is passed to a minor task is updated.

Table 1 State space, time, and memory for verification

Parameter	Value
Number of reachable states during verification	190,468
Elapsed Time (in seconds)	45
Memory (MB)	51.504

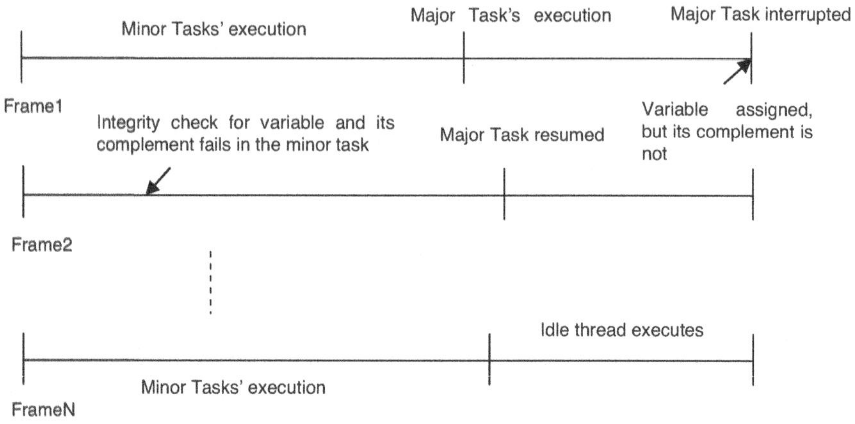

Fig. 2 Error sequence

2. A timer interrupt is generated, indicating the end of that minor frame before the complement of the variable is passed to the minor task.
3. In the next minor frame, data integrity check is carried out on this variable in a minor frame task. Since the complement is not updated yet, the check fails.
4. Error flag is set, leading to violation of the assertion.

5 Discussion and Conclusion

In our work, we have modeled a portion of the flight software, namely the scheduler. We have also represented some minimal tasks and inserted assertion statements. One possible problematic execution sequence, which leads to setting of the error flag was detected during simulation. It is a realistic execution scenario which could manifest during simulations or flight. The missing atomicity of two statements was leading to this error trail.

For safety-critical systems like aircrafts, launch vehicles, and spacecrafts, the onboard software should be highly reliable. Even a minor bug can lead to catastrophes. Hence, advanced techniques like formal methods are essential so that errors can be detected as early as the requirements stage or design phase. In future, it is planned to model the entire onboard software and verify it using model checking. It is anticipated that several errors can be found out by means of verification with the SPIN model checker.

References

1. Holtzman, G.J.: The SPIN Model Checker, primer and reference manual. Addison-Wesley, Boston (2003)
2. Gluck, P.R., Holzmann, G.J.: Using SPIN model checking for flight software verification. In: Proceedings of the 2002 Aerospace Conference, IEEE, New York (2002)
3. Schneider, F., Easterbrook, S.M., Callahan, J.R, Holzmann, G.J.: Validating requirements for fault tolerant systems using model checking. In: Proceedings of the Third IEEE International Symposium on Requirements Engineering, Colorado Springs (1998)
4. Havelund, K., Lowry, M., Park, S.J., Pecheur, C., Penix, J., Visser, J., White, J.L.: Formal analysis of the remote agent before and after flight. In: Proceedings of Fifth NASA Langley Formal Methods Workshop, Williamsburg, VA (2000)
5. Havelund, K., Michael, R.: Lowry and John Penix: formal analysis of a space-craft controller using SPIN. IEEE Trans. Softw. Eng. **27**, 749–765 (2001)
6. Horvath, G., Jones, G., Joshi, R.: A model-based approach to verification of Spacecraft Software using the SPIN Model Checker. In: AIAA SPACE 2009 Conference & Exposition, Pasadena, CA (2009)
7. Kaslow, D.C., Anderson, L.V., Asundi, S., Ayres, B.J., Iwata, C., Shiotani, B., Thompson, R.E.: Developing a CubeSat Model-Based System Engineering (MBSE) reference model—interim status. In: Proc. 2015 IEEE Aerospace Conference (2015)
8. Kolcio, K., Fesq, L.M.: Model-based off-nominal state isolation and detection system for autonomous fault management. In: Proc. 2016 IEEE Aerospace Conference, pp. 1–13 (2016)
9. Gross, K.H.: Formal specification and analysis approaches for spacecraft attitude control requirements. In: Proc. 2017 IEEE Aerospace Conference, pp. 1–11 (2017)
10. Albiol, L., Batlle, J., Cebrian, J., Gutiérrez, G., Pita, F., Vega, I., Acar, G., Cioni, S., Rio, J.D.: Validation of a new satellite communications protocol for long-term ATM needs. In: Proc. 2015 IEEE/AIAA 34th Digital Avionics Systems Conference (DASC), pp. 2B5-1–2B5-10 (2015)
11. Schrammel, P., Kroening, D., Brain, M., Martins, R., Teige, T., Bienmüller, T.: Successful use of incremental BMC in the automotive industry. In: IEEE Workshop on Industrial Strength Formal Specification Techniques (2015)

Multichannel Probabilistic Framework for Prenatal Diagnosis of Fetal Arrhythmia Using ECG

K. Surya and K. K. Abdul Majeed

1 Introduction

The infant mortality rate is the number of newborns dying under 1-year-old per 1000 live births [1]. The infant mortality rate of the world is 49.3 according to the United Nations and 34.2 according to the CIA World Fact book. So from this recording, we could see it is almost nearest to 50% i.e. the survival rate of the newborns is only 50%. The fetal electrocardiogram can provide valuable information about the fetal wellbeing [2]. During pregnancy period, fetal heart rate (FHR) variations have commonly been beheld as indirect indications of fetal health conditions. Peaks present within the signals show the important information about the fetal health conditions. Due to unwanted noise and movement of the fetus, R-peak detection is challenging due to the low signal to noise ratio [3]. Biomedical signals are electrical signals acquired from the maternal abdomen that represents a physical health information of our body i.e., that signals constantly communicate health variations. Fetal ECG (fECG) is a biomedical signal that gives an electrical representation of FHR conditions. Sometimes the fECG is the only information provider in early-stage diagnostic of fetal health conditions. Infant mortality rate, a measure of human infant deaths in a group younger than 1 year of age [4], is an important factor and indicator of the overall physical health of a community.

Most of the birth defects happen during the first stage of pregnancy; if the defect occurred on the heart it will lead to death. During pregnancy, fetal health must be carefully monitored. In general, the fetal ECG can be monitored from the maternal

K. Surya (✉)
MEA Engineering College, Affiliated to A.P.J Abdul Kalam University, Malappuram, India

K. K. Abdul Majeed
School of Electronics Engineering, Vellore Institute of Technology, Vellore, India

M. Palesi et al. (eds.), *Second International Conference on Networks and Advances in Computational Technologies*, Transactions on Computational Science and Computational Intelligence, https://doi.org/10.1007/978-3-030-49500-8_13

abdomen by placing electrodes on the abdomen of the pregnant women. But the extracted signal contains the artifacts due to noise. So the recorded signal has low signal to noise ratio. This work is based on detecting the normal sinus rhythms, and it also detects arrhythmia of the fetus [5]. The fECG signal is interfered by various noises with unknown spectral and temporal attributes; these noises are called power line noise and baseline wondering noise. The fECG signals are often unhinged by electrical noise from other sources, which will corrupt fECG signals significantly [6]. In this work a hierarchical probabilistic framework is introduced to estimate the R-peak detection of fetal ECG. This work is based on detecting the normal sinus rhythms and the detection of diseases called arrhythmia of the fetus [7].

The abdominal fetal electrocardiogram can provide worthwhile information about fetus. As per the infant mortality statistics every year many babies are born with a congenital heart disorder. If the defect is not crucial then the baby may appear healthy and the chances of detecting the defect may be low [8]. Inherited heart defect originates in the early stage of pregnancy when the heart is forming and the defect may affect any part or function of the body. Severe cardiac anomalies may be caused due to a genetic syndrome inherited disorder, infection, and drug misuse; it may also affect the child [9]. Monitoring fetal heart rate and interpretation of electrocardiogram during pregnancy is an important aspect and it may support medical decision-making [10].

In this proposed study an adaptive multichannel framework is used to find the R-peaks within the selected signals. Maternal ECG suppression is proposed as the initial work and a preprocessing scheme is introduced to remove baseline wandering, high-frequency noise, and power line interference. Independent component analysis is used to further enhance the fECG. Initially, the proposed method is evaluated by using a single channel approach and then extended to multichannel model. In abdominal fECG recordings, the amplitude of the QRS complex is generally large compared to other segments. Therefore, this model is limited to describing the fetal QRS complex. In this approach, the QRS complex is modeled by using state space machine logic thresholding method. By this method initially a threshold value is adaptively set for detecting the highest peak within the signal. The adaptive threshold is set based on the sample signal. After that the signal amplitude, position, peaks, intervals are detected based on this threshold. The values are stored as a feature vector. A classifier is used here to classify the signal based on the feature vector. Naive Bayes classifier is generally used for the classification and the sample data and the feature vector are compared with the signal. So from that comparison we can separate the signal as normal and abnormal.

2 Methodology

The electrocardiogram signal of fetus, i.e. fECG express very clear information which helps doctors in making appropriate and timed decision during labor. The profound interest of fECG analysis is in the field of biomedical applications

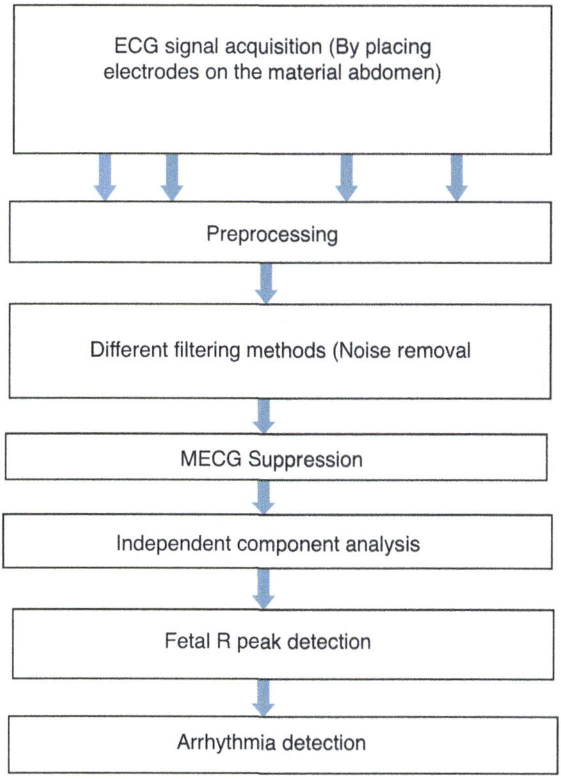

Fig. 1 Block diagram of fetal arrhythmia detection

and clinical analysis [11]. Fetal ECG is extracted from composite abdominal signals using advanced methodologies, and plays a pivotal role in automated fetal monitoring systems. This work proposes various strategies and existing algorithms for fECG detection and analysis to facilitate proficient and detailed understanding of fECG and its role in monitoring of fetal health [12]. A comparison has been drawn to show the accuracy and performance of methods used for fECG signal analysis. This approach clearly opens up a section for analysts, doctors, and end clients to promoter a superb comprehension of fECG sign and its investigation systems for observing framework for fetal heart rate. The R-peak detection is the challenging task of fetal ECG due to the unwanted noise effect. Multichannel probabilistic approach is introduced to detect the R peak of the fetal ECG (Fig. 1).

2.1 Preprocessing

The ECG signals can be acquired by placing the electrodes on the maternal abdomen [13]. Electrodes can be placed by using conductive gels. The noise component present in the signal can be eliminated by different filtering methods like Notch filtering, Butterworth filtering, Discrete Wavelet Transform (DWT), FIR filtering, and normal denoising method. From the MSE and PSNR values of the signal the discrete wavelet transform shows better result.

2.2 Signal Extraction

The signal captured from the maternal abdomen contains many noises like power line noise, baseline wondering, etc. The signal is contaminated with respiratory as well as muscle contraction. So from that collection of mixed noise fetal ECG is needed to be separated. This can be done by using independent component analysis (ICA). As the name suggest the source is unknown and from the mixture of signal independent source may separate [14]. The extracted signal amplitude is measured to know which one is the fetal component. The simple method to know more about IAC is cocktail party problem. In this study, Extraction of fetal ECG has very important role during pregnancy. From the extracted signal we can detect any variations within the heart rate. The detection of fetal heart conditions gives the information about the fetal health conditions during pregnancy. Fetal ECG extraction from the maternal abdomen is difficult due to the unwanted noise effect within the waveform. There are different noise components that affect the ECG signals, which are Electromyogram noise, Additive white Gaussian noise, and power line interference. Initially we wanted to remove the noise components within the signal [15]. So we have to perform different filtering methods. Notch filtering, Butterworth filtering, Fir filtering, Normal denoising method, and wavelet decomposition methods are used in this work. From the result it shows wavelet decomposition method shows the better result. The main component of noise within the ECG waveform is maternal ECG; it is a high component noise with higher amplitude. By using independent component analysis (ICA) maternal ECG can be effectively separated. It is very useful to remove the maternal ECG component. ICA can be used in many fields to separate the mixed signals.

2.3 Peak Detection

After signal separation fetal waveform must be evaluated. Fetal peaks are detected using state machine logic. Waveform is first processed to produce the set of weighted unit samples at the location of maxima. In this algorithm a normal beat is used for experimentation; for this beat the autocorrelation is zero. Here a 1 second window has been determined on the ECG signal and average of the window is calculated (m). At zero state, the product of m and an initial weight is determined. If the product is less than the mean value calculated in a 15 sample interval from the one second interval. And if the amplitude of the studied sample is greater than the rest of the samples. The amplitude and the location of the first sample are stored as the R peak.

Feature extraction can be done by using the peak detection algorithm. It finds the location and the amplitude of the fetal ECG. In this algorithm, peak detection is done by using state machine logic by finding threshold for each segments within the signal. If the observed signal sample satisfies the condition it is stored as the corresponding peak. After identifying each segments within the signal the algorithm measures all the intervals, i.e., R-R interval, S-T interval, QRS width, and amplitude of Q, R, S, T segments. Normal ECG has standard value for all the component present in the signal. If the input is an abnormal signal, then the abovementioned conditions are different. So from the variations we can detect the abnormality within the signal.

2.4 Signal Classification

If the extraction and feature selections are successfully done the next step is the classification of the inputs. There are many classifiers available for classification, ANN, Support Vector Machine (SVM), Quadratic Discriminant Analyzer (QDA), Linear Discriminant Analyzer (LDA), and Naive Bayesian Classifier [16]. Here Naive Bayesian Classifier is used for classification. The main advantage of this classifier is, it requires less features for classification. It is suitable because the classifier only requires less number of data. In this classifier two steps are performed first one is training and second is testing. In training feature vector is used to train the signal and in testing a new sample is given as an input, it measure posterior probability based on the features. When a new signal is received as an input, it will compare the feature values of new input with the existing trained set. And the classifier evaluates the probabilities of the signal having arrhythmia or not. If the signal is an abnormal signal i.e., if the fetus having arrhythmia the signal from that fetus is considered as abnormal it moves to the most probable set and concludes it as an abnormal arrhythmia signal. The disease detection can also be processed using the beats per minute of the signal. The beats per minute can be calculated from the R-R interval of the signal. From the obtained RR interval calculate the beats per minute value using equation. From that we can classify the signals.

3 Result and Discussion

The fetal extraction is difficult mainly due to the collective noise present within the extracted signal. To evaluate the performance of the designed algorithm, collection of experiments are performed based on the dataset available from the physionet library. For performing experiments Abdominal and direct fetal ECG database is used from the physionet library. For evaluation, Non-Invasive fetal ECG Arrhythmia database is used. All the recorded data are in millivolt. The signal is acquired from the maternal by placing the electrodes on the maternal abdomen. Conductive gel

Table 1 Preprocessing using different filtering methods

	Filtering methods	MSE	PSNR
Signal 1	Notch filtering	0.012016	19.202417
	Normal denoising methods	0.000718	31.440944
	Discrete wavelet transform	0.000001	75.322259
	FIR filtering	0.000386	34.137885
	Butterworth filtering	0.000292	35.342858
Signal 2	Notch filtering	0.006502	21.869372
	Normal denoising methods	0.000192	37.163700
	Discrete wavelet transform	0.000001	74.598686
	FIR filtering	0.000351	34.542936
	Butterworth filtering	0.000216	36.653803

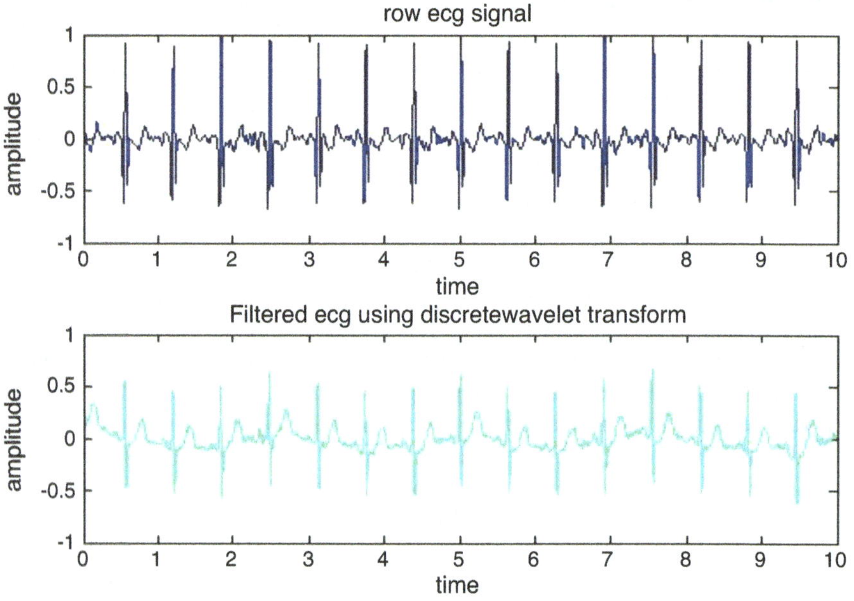

Fig. 2 Filtered ECG using discrete wavelet transform

is used to fix the electrodes on the abdomen. AG-AGcl conductive gel is generally used for fixing the electrodes. Different filtering methods are used for preprocessing to remove the unwanted noise components from the signal, such as FIR filtering, Butterworth filtering, discrete wavelet transform, conventional denoising method, etc. Discrete wavelet transform shows the better result based on the MSE and PSNR values of the signal (Table 1, Fig. 2).

Independent component analysis separates the mixed signal into individual source signal. Joint Approximation Diagonal Eigenvector is used for comparison. JADE uses fourth order cumulant approximation for separating the mixed signals.

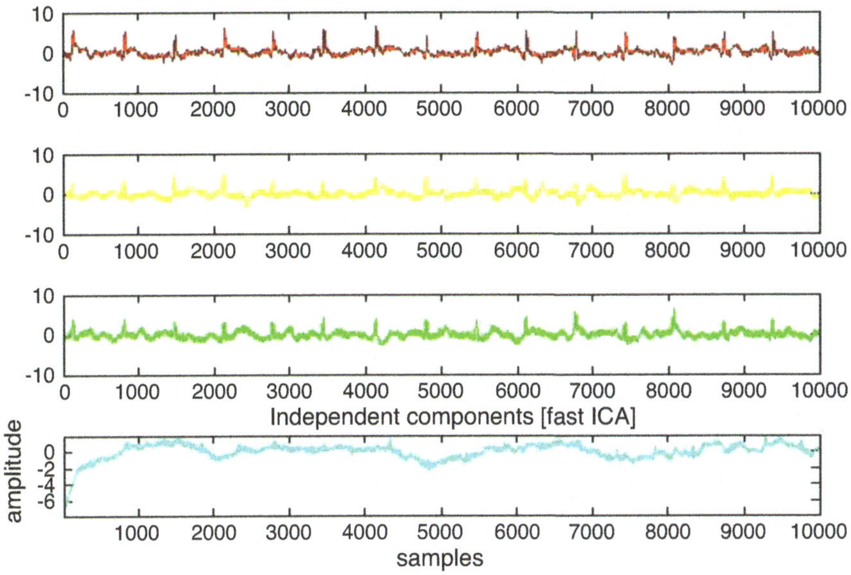

Fig. 3 Independent component analysis of the selected sample signal

Fourth order cumulant is the non-Gaussianity of the signal. Independent component analysis separates the mixed signals into different source component. This can be done by performing the fourth order cumulant of the signal i.e. in general for the probability density function of the signal, the zero order cumulant is the total probability (Fig. 3).

First order moment is mean. Second order moment is variance, third order moment is skewness, and fourth order moment is kurtosis. To find the kurtosis of a signal, the non-Gaussianity of the signal is evaluated. Here we assume that the signal is in unit variance that is zero mean. For a gaussian random variable 4th order moment is zero i.e., kurtosis is zero. So using kurtosis function we can estimate the independent components present in the signal. After separating each fetal ECG component peak detection is performed. R-peak detection is used to find out the variations within the RR interval. From the variations the disease detection can be possible. From the detected peaks each RR interval variation is evaluated and heart beat per minute is evaluated from the estimated interval. Heart rate can be evaluated from the equation 60/RR-Interval. RR interval variations show variations in the heart rate. In normal condition the heart rate variation is in between 120 and 180 beats per minute. So if it moves below 120 it is considered as a disease called bradycardia and if it moves above 180 it is considered as a disease tachycardia. It indicates the low and high heart rate of the fetus (Fig. 4).

For Arrhythmia detection each peaks within the signal is evaluated. Threshold level comparison is done for finding feature vector of the signal. The threshold value is set based on the selected signal; the mean value is evaluated for finding threshold.

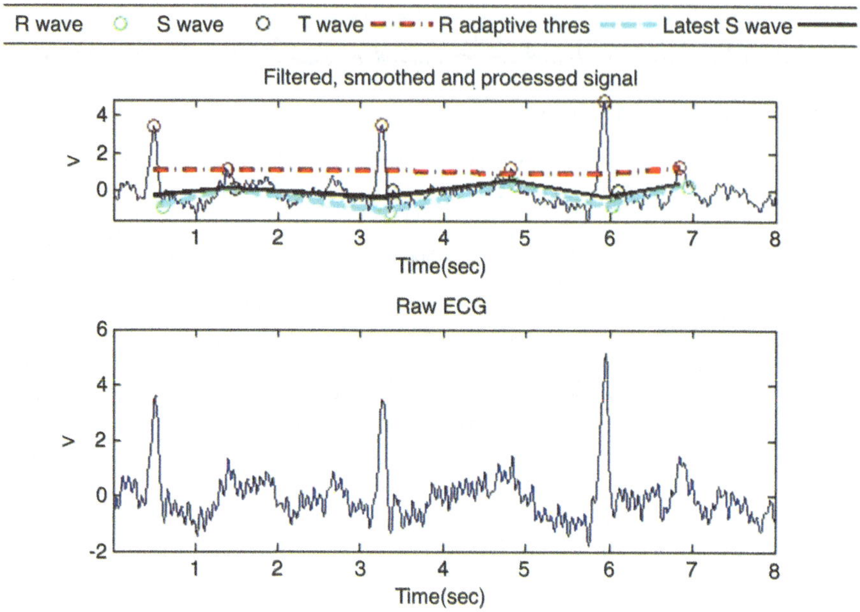

Fig. 4 Detected QRS complex of fetal ECG

A state machine logic algorithm is used for finding the peaks within the signal. Initially zero state is set based on the threshold and the window is also set for finding the peaks; for a window at least two peaks are needed for peak detection. After finding the R peak i.e. the highest peak within the ECG signal it moves to the next state. The next state is for the T peak prediction and each peaks are evaluated based on the threshold. After peak detection, features of each peak are evaluated and stored as a feature vector. From the feature vector the signal classification is performed.

Feature vector is crated based on all information obtained from the selected signal i.e. Amplitude component of all peaks, Interval between each peak and each position of the peak. Naïve Bayes classifier is used for classification. It is generally used in machine learning and for classification. Naive bayes classifier uses features selected from the signal to make decisions, and it also uses posterior probability density function to determine signal falls in which category i.e., weather the signal falls in normal or abnormal class. Based on the values of heart rate we can classify the signals with normal heart rate, Tachycardia, Bradycardia, and abnormal condition. For detecting accuracy and sensitivity, a classification is also used. Naive Bayes classifier is used for classification. There are different classification methods available such as K nearest neighbor, state vector machine algorithm, etc. Naïve Bayes algorithm shows better result. In Naive Bayes algorithm both testing and training can be performed using the feature of the signal. Initially we have to train the data training is used to determine weather the data is normal or abnormal, after finding decisions from already exist data a new signal is given as the input and

Table 2 Abnormal and normal signal classification based on Naïve Bayes classifier

Signal	Output
NR_04m	Normal
NR_05m	Normal
ARR04	Abnormal
ARR05	Abnormal
ARR06	Abnormal
NR_06m	Normal

then tested. In testing posterior probability of the signal is evaluated, from that we can estimate the given signal is normal or abnormal. The normal signal is normal original signal and abnormal signal indicates the arrhythmic signal. For finding accuracy and sensitivity of the work number of true positive, true negative, false positive, false negatives are evaluated. True positive: Is the outcome of the model that correctly predict the positive class. True negative: Is the outcome of the model that correctly predict the negative class. False positive: Is the outcome of the model that incorrectly predict the positive class. False negative: Is the outcome of the model that incorrectly predict the negative class.

$$\text{Accuracy} = \text{TP} + \text{TN}/N \tag{1}$$

$$\text{Sensitivity} = \text{TP}/\text{TP} + \text{FN} \tag{2}$$

where TP is the true positive value. TN is the true negative value. FN is the false negative value. The average sensitivity and accuracy of the system is 88.88% and 94.11%, respectively, by choosing 17 data from the physionet library. Samples are normal data and nine samples are arrhythmic data. The classifier trained the data using the feature of normal and abnormal signal so from that the signal can be classified as normal and abnormal (Table 2).

This classifier used for classification is the Naïve Bayes classifier. Classification is the form of data analysis that extracts models using important data classes. In this work the data extracted from detected peaks is classified using the machine learning classifier. The performance of the selected data is measured using the standard familiar matrix like sensitivity and accuracy. This can be estimated from the true positive, true negative, false positive, and false negative values.

4 Conclusion

Most of the birth defects happens during the first stage of pregnancy. If the defect occurred on the heart it leads to death. So during pregnancy fetal health must be carefully monitored. So this work is based on detecting the normal sinus rhythms, heart rate variation, and disease of the fetus. This method proposes an earlier detection of fetal arrhythmia. The performance of the proposed methods is validated by using the arrhythmic data from the physionet library. The performance of the particular method is evaluated by finding the accuracy and sensitivity of the signal. Experimental result shows that the average sensitivity of 88.08% and accuracy of 94.11%, respectively.

References

1. Sahoo, G.K., Ari, S., Patra, S.K.: Proceedings of 2013 IEEE Conference on Information and Communication Technologies (2013)
2. Warmerdamand, G.J.J., Vullings, R., Schmitt, L., Vanler, J.O.E.H., Begmans, J.W.M.: Hierarchical probabilistic framework for fetal R-peak detection using ECG waveform and heartrate information. IEEE Trans. Signal Process. **66**, 16 (2018)
3. Gaikwad, K.M., Chavan, M.S.: Removal of high frequency noise from ECG signal using digital IIR Butterworth filter. In: IEEE Global Conference on Wireless Computing Networking (GCWCN), pp. 121–124 (2014)
4. Lin, C., Yeh, C., Wang, C., Serafico, B.M.F., Wang, C., Juan, C., Young, H.V., Lin, Y., Yeh, H., Lo, M.: Robust fetal heart beat detection via R-peak intervals distribution. IEEE Trans. Biomed. Eng. **66**, 3310–3319 (2019)
5. Peters, C., Vullings, R., Bergmans, J., Oei, G., Wijn, P.: Heart rate detection in low amplitude non-invasive fetal ECG recordings. In: International Conference of the IEEE Engineering in Medicine and Biology Society, pp. 6092–6094 (2006)
6. Apsana, S., Suresh, M.G., Aneesh, R.P.: A novel algorithm for early detection of fetal arrhythmia using ICA In: 2017 International Conference on Intelligent Computing, Instrumentation and Control Technologies (ICICICT), pp. 1277–1283 (2017)
7. Nikam, S.D.: Fast ICA based technique for non-invasive fetal ECG extraction. In: 2016 Conference on Advances in Signal Processing (CASP), pp. 60–65 (2016)
8. Nagarkoti, S.K., Singh, B., Kumar, M.: An algorithm for fetal heart rate detection using wavelet transform. In: 2012 1st International Conference on Recent Advances in Information Technology (RAIT), pp. 838–840 (2012)
9. Kharabian, S., Shamsollahi, M.B., Samen, R.: Fetal R-wave detection from multi- channel abdominal ECG recordings in low SNR. In: Annual International Conference of the IEEE Engineering in Medicine and Biology Society, pp. 344–347 (2009)
10. Liu, C., Li, P.: Systematic methods for fetal electrocardiographic analysis: determining the fetal heart rate, RR interval and QT interval. In: Computing in Cardiology, pp. 309–312 (2013)
11. Patel, P., Mahajani, P.: Fetal ECG separation from abdominal ECG recordings using compressive sensing approach. In: International Conference on Inventive Research in Computing Applications (ICIRCA), pp. 831–834 (2018)
12. Nikam, S., Deosarkar, S.: Fast ICA based technique for non-invasive fetal ECG extraction. In: 2016 Conference on Advances in Signal Processing (CASP), pp. 60–65 (2016)
13. Gaikwad, K.M., Chavan, M.S.: Removal of high frequency noise from ECG signal using digital IIR butterworth filter. In: 2014 IEEE Global Conference on Wireless Computing Networking (GCWCN), pp. 121–124 (2014)
14. Bhogeshwar, S.S., Soni, M.K., Bansal, D.: Design of Simulink Model to denoise ECG signal using various IIR amp; FIR filters. In: 2014 International Conference on Reliability Optimization and Information Technology (ICROIT), pp. 477–483 (2014)
15. Niknazar, M., Rivet, B., Jutten, C.: Fetal QRS complex detection based on three-way tensor decomposition, In: Computing in Cardiology 2013, pp. 185–188 (2013)
16. Dıker, A., Avci, E., Cömert, Z., Avci, D., Kaçar, E., Serhatlioğlu, İ.: Classification of ECG signal by using machine learning method. In: 26th Signal Processing and Communications Applications Conference (SIU), pp. 1–4 (2018)

High-Capacity Reversible Data Hiding in Encrypted Images by Second MSB Replacement

Jeni Francis and Ancy S. Anselam

1 Introduction

Image security has a major role in all told fields, particularly in extremely confidential areas just like medical field and military purposes. Development of cloud computing has been a major issue which led to many serious problems in which integrity, confidentiality, and authentication are perpetually vulnerable, by outlaw methods like repetition, hacking, and unwanted usage of information. Main use of encoding strategies may be is to ensure confidentiality of information by absolutely or partly varying the original data of an input image that has to be transmitted. When the encrypted image is being transmitted, it is always necessary to decrypt the image even by not keeping the actual data of the input image or the secret key used while encrypting the original image. The methodology of "hiding data reversibly in the encryption phase" (RDHEI) is meant for both information quality increase and to maintain confidentiality and during encryption where the encryption method may be done first making a cloud computing situation easier to handle. When the data of the original image and the encryption key that is used for enciphering the image are not known, it has the potential to hide a secret message within the image that has been encrypted. During the decryption part, the hidden message and image should be utterly retrievable losslessly. So, there must be a trade-off between the payload and the quality of the reconstructed image. Several methods using this concept have been designed in the recent years. During the coding part, the space where the secret message has embedded could be vacated before the encryption part (VRBE) or after the encryption part (VRAE). The original image reconstruction and the secret

J. Francis · A. S. Anselam (✉)
MBCET, Trivandrum, India
e-mail: ancy.anselam@mbcet.ac.in

M. Palesi et al. (eds.), *Second International Conference on Networks and Advances in Computational Technologies*, Transactions on Computational Science and Computational Intelligence, https://doi.org/10.1007/978-3-030-49500-8_14

message retrieval during the decryption part can be obtained at identical time or on as individual basis.

None of the methods used or the previously proposed strategies was able to produce sensible image quality of the reconstructed image with a high payload. Many methods were able to produce a high embedding capability nearer to 0.5 bpp but the quality of the image that was reconstructed was highly varied i.e. PSNR of 40 dB comparing it with the original image. Some other methods fmay carry very good payload, but may have potential just to hide 0.1 bpp at a maximum. The data embedding was mostly performed using LSB (least significant bit) substitution in many of the existing technologies. But when the encryption process is going on, it may be difficult to find out whether the encrypted image contains a secret message within it since the encrypted image pixels have pseudo-random values. So it is potential to use MSB rather than LSB values for secret message substitution. It is easier to predict the MSB values than the LSB values since in gray scale images, neighboring pixels values transformation is smooth than the color image and also the confidentiality of the encrypted images remains same. In this method, second MSB prediction method is used to obtain good capability during hiding data reversibly in the encrypted images. Since in gray scale images, there is pixel correlation between the neighboring two pixels in an image making the nearer pixel values terribly equal. Therefore it becomes easier for the prediction of a pixel value by averaging the top and the left pixels which have been already decrypted, which was the available coding technique. Here, the technique used is HCRDH (high-capacity reversible data hiding). This methodology takes data hiding method into another condition where the prediction errors are corrected before encryption stage. The preprocessing stage is done first to remove the errors caused by prediction and then the resultant image of this stage is encoded i.e. the pixel values are pseudo-randomly changed in order to preserve confidentiality. After encrypting the original image, the data hiding part is done, in which second MSB of every offered pixel will be embedded by one bit of hiding message. Then decryption part is done where the inserted message may be retrieved and with aid of prediction of second MSB scheme, the image may be reconstructed.

The content of this paper is as follows: Section 2 explains an overview of the previous works of data hiding. Section 3 describes the proposed method. Results are provided in Sect. 4. Section 5 draws the conclusion.

2 Related Works

The major field that is emerging nowadays is the techniques of hiding data reversibly in the images. These techniques are highly appreciated because of the reason that the hidden message within the images can be absolutely retrieved and the original image can be reconstructed without degradation. Here confidentiality of the original image is preserved. Many of the previous strategies achieve data hiding by vacating area reversibly within the image. But these methods have a drawback that there are higher chances of errors that may arise during the original image reconstruction.

In 2008, Puech and Chaumont [1] proposed that hiding information in images in purposeful manner that does not have an effect on the original image pixels or those methods that can cause a permanent effect during the decoding stage may be considered as reversible data hiding. These methods make sure that the image legitimacy and confidentiality is highly preserved without any distortion on the digital images. These methods ensure the possibility of detection by the content owner when there is any modification done to the encrypted image. Hiding data reversibly in these methods uses a completely unique algorithm that employs an exceedingly divisible manner. In [2], a completely unique technique for information hiding reversibly in images is employed. Here a cryptographic key is employed for data encryption. Within the encrypted image, data is hidden by compressing the LSB of every pixel by using the information hiding key. If the information content is not large, image can be recovered losslessly. Hong et al. proposed a method that improved the above technique by a side matching technique. He also proposed a formula to evaluate the block smoothness [3]. But in many cases, the quality of the image that was recovered remained to be below 35 dB even when the number of embedded bits was still higher [4]. A joint method was proposed by Zhou et al. where the partial method was encryption [5]. Puteaux, and W. Puech [6] in 2018 proposed the most useful and available technique that was used to embed data within an image while the image is being encrypted.

The method in [7] describes a highly efficient technique that retrieves data without any distortion. The algorithm explained matches the pixels values corresponding to the gray scale images with that of the maximized peaks of the histogram [8]. Explains a conventional data hiding method accompanied with vacating spaces before encryption process is employed. These techniques are highly appreciated because of the reason that the hidden message within the images can be absolutely retrieved and the original image can be reconstructed without degradation. X. Zhang [2] uses a method of employing a key for hiding data and the correlation between the pixels in an image, the hidden information can be retrieved, and the image may be utterly reconstructed.

3 Proposed Method of Reversible Data Hiding

Many of the prevailing strategies employed in many applications did not successfully achieved high payload equal to 1 bpp together with the reconstructed image quality with PSNR of 50 dB. LSB values were used to replace with the secret message with supporting strategies like analyzing the prediction errors. But in some cases, most of the existing methods failed due to the reason that it is not possible to identify whether an image contains a secret message or not if the image is encrypted. This is due to the fact that no correlation between the adjacent pixels remains. So the most efficient technique that can be employed is to use second MSB value to replace and embed the secret message without using the LSB values. Using this method, authentication and integrity may be preserved during the process with an aid that the second MSB values may be easily predicted than any of the existing techniques used in many of the confidential areas.

The method of reversible data hiding with the most effective technology in the encryption phase is employed here. Second MSB substitution is the methodology used to hide the message. When the process of data hiding is done, the second MSB values of replaced pixels will be lost. These values should be correctly predicted when the image is decrypted so as to maintain the reconstructed image quality. This high capability approach may obtain absolute reversibility with PSNR equal to infinity and efficient capacity equal to 1 bpp that can be bestowed intimately taking under consideration the foremost necessary constraint. The method that is used to embed a single bit per pixel can be considered as high-capacity reversible data hiding approach.

3.1 Overview

The goal of this method is to consider an image I with pixels $m \times n$. This image is encrypted using a secret key and from this we get an encrypted image. A data hider can embed a message within this image without even knowing the original content of the image. By embedding this message, we get a marked encrypted image that precisely has the original image size. Three main steps can be included during this process while encryption is done:

- Image encryption
- Secret message encryption
- Second MSB replacement for data hiding

3.2 Image Encryption Phase

An encryption key K_e is used to encrypt the image to make the image I indecipherable. The seed generator is fed with this key as its primary parameters. This generates a set of pseudo-random numbers $s(i, j)$ and then the pixel values after encryption are calculated by XORing the original pixel values with that of the pseudo-random byte sequence.

$$p_e(i, j) = (s(i, j) \bigoplus p(i, j))\ mod\ 128 \tag{1}$$

The clear image may be perfectly reconstructed without any errors. This means that the encryption part may be absolutely reversible. The image may be reconstructed without overflow.

3.3 Embedding of Information

Information can be embedded by a secret message owner into the original image which has been encrypted during the data embedding part while not knowing the key used during encryption or even not knowing the original image. The secret message that has to be hidden within the encrypted image is encrypted using data hiding key K_w so that the secret message cannot be detected when the message is embedded within the image that has been encrypted. As the next step, every available pixel of this encrypted image is scanned from left to right and also from top to bottom. This step is done so as to embed the hidden message into every offered pixel by substituting the MSB of the pixel with a single bit of secret message.

$$p_{\text{ew}}(\text{i}, \text{j}) = b_k{*}64 + p_e(\text{i}, \text{j})\ mod\ 128 \tag{2}$$

As the value of the first pixel cannot be predicted, it is not able to mark the first pixel and thus its pixel value is not modified.

3.4 Information Extraction and Perfect Image Reconstruction

Data hiding in this technique will be always a reversible process. This means that the information or secret message that has been embedded into the image can be retrieved using the data hiding key K_w and also the original image I may be perfectly reconstructed without any loss. Here the reconstructed images I will be precisely similar to the original image I or it will be accurately near processed image I'.

In the decoding part, as discussed earlier, the secret message can be extracted by scanning the encrypted marked image from top to bottom and from left to right and from that extracting the MSB of each available pixel. Also the original image can also be perfectly reconstructed without any error. For this purpose, the encrypted marked image is decrypted to get the other LSB bits and the MSB value is predicted by the previously used method of prediction. The reconstructed image will be terribly like the initial original image. Considering this condition for perfect reconstruction, there may be three potential probabilities:

- This condition exists only when on the receiver side, the receiver has solely K_w
- Second condition arises when the receiver solely has K_e
- Third condition exists only when on the receiver side, recipient has both the keys

Consider the single condition, in which the receiver has solely K_w, the data hidden encrypted image is scanned for its pixels from top to bottom and also from left to right so as to extract the MSB of every available pixel. As a result of this process, the secret message is retrieved from these pixel values.

$$b_k = \text{p}_{\text{ew}}(\text{i}, \text{j})/64 \tag{3}$$

Here the index of the bits of the secret message is represented with $0 \le k < m \times n$ which is extracted from the encrypted marked image. The bits of hidden message thus extracted will be an encrypted form of the original data, thus using the data hiding key K_w, the message may be correctly used. Consider the second condition in which the receiver only has the encryption key K_e, then before the data extraction and decoding stage, the original image I may be perfectly reconstructed. The associated steps for the perfect reconstruction of the image are explained below:

- A pseudo-random sequence $s(i, j)$ is produced by employing the encryption key K_e which has a dimension of mxn bytes.
- In the next step, the pixel value of the original image is retrieved. For this, the encrypted image is scanned for its pixels from top to bottom and from left to right to obtain the pixel values.
- The LSB bits thus obtained are the encrypted form of the original pixel value. Thus the original LSB bits of the image I is obtained by XORing these bits with the generated pseudo-random sequence $s(i, j)$.

3.5 *Statistical Analysis Parameter*

1. Payload

   This parameter explains about the capacity of the image to contain the secret message within it. The embedding capability or payload is always represented in bpp i.e. bit per pixel. In all the prevailing strategies, the most achieved payload or embedding capacity is stated as 1 bpp. The embedding capacity should always be high i.e. more secret message bits should be contained inside the image after encryption.

$$\mathrm{EC} = N/H * W \tag{4}$$

2. Reconstructed image quality (PSNR)

   After the decryption stage, the image should be perfectly reconstructed so that the quality of the image is not degraded. The reconstructed image quality is represented in dB. The maximum achieved quality in prevailing strategy was 40 dB.

$$\mathrm{PSNR} = 10 \cdot \log_{10} \frac{255^2}{\frac{1}{m*n} \sum \sum (p(i,j) - p'(i,j))^2} \tag{5}$$

4 Experimental Results

The most efficient technique that can be introduced to reversible data hiding may be considered as the HCRDH approach (high-capacity reversible data hiding) approach. Here the main algorithm step in reversible data hiding is considered as the correction of prediction errors. For this to be accomplished, a preprocessing step for the original image is performed. This allows the perfect reconstruction of the original image after the extraction of the secret message embedded within the image I. The preprocessed image is then encrypted without any issues after this processing. In the embedding stage, as discussed in the previous sections, the MSB of each available pixel of the encrypted image is replaced with a bit of the secret message.

For data hiding in encrypted images, totally different performances have to be measured. These are amount of number of extracted bits, the embedding rate, and the quality of the final image reconstructed. The most effective trade-off between all these parameters needs to be optimized. The embedding rate is expressed in bit per pixel (bpp). The prime objective is to anticipate the increase in quantity of information that is hidden in the encrypted in the image. Also, the quality of the reconstructed image should be compared with that of the original one. For this, the peak-signal-to-noise ratio (PSNR) is employed. We first applied our approach on the same 4 original image of 512 × 512 pixels (Figs. 1, 2, 3, 4, and 5, Table 1).

Fig. 1 Original images I from the BOWS-2 database: (**a**) airplane, (**b**) barbara, (**c**) boat, (**d**) lenna

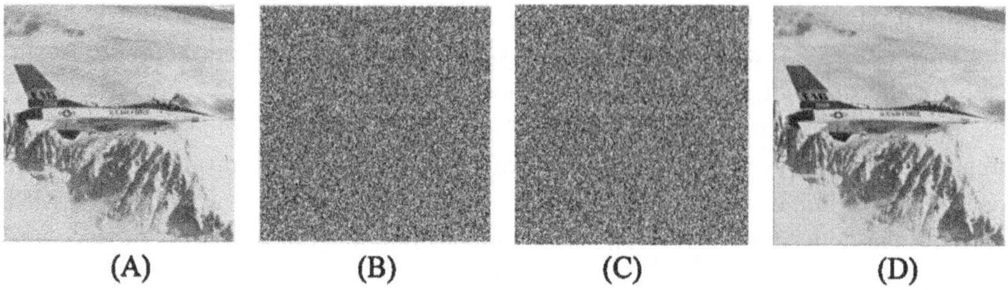

Fig. 2 Illustration of the proposed approach with the image airplane: (**a**) Original image, (**b**) Encrypted image I_e, (**c**) Marked encrypted image I_{ew}, (**d**) Reconstructed image I

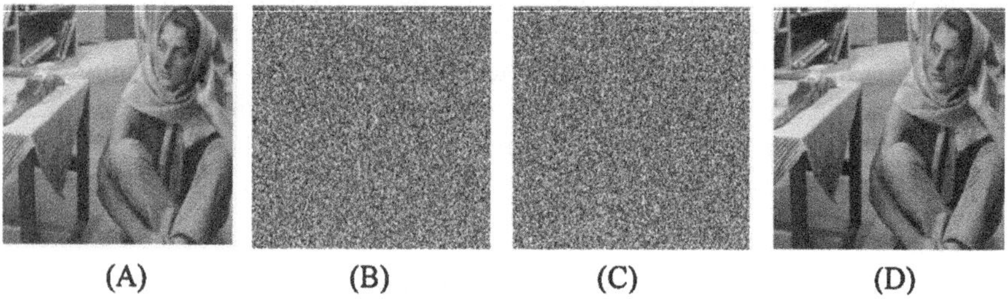

Fig. 3 Illustration of the proposed approach with the image barbara: (**a**) Original image, (**b**) Encrypted image I_e, (**c**) Marked encrypted image I_{ew}, (**d**) Reconstructed image I

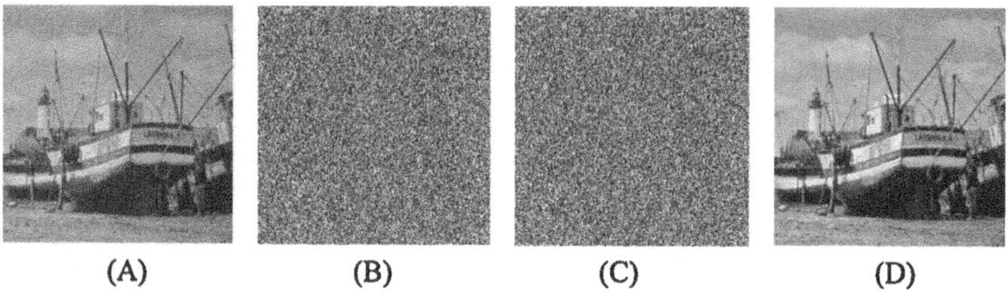

Fig. 4 Illustration of the proposed approach with the image BOAT: (**a**) Original image, (**b**) Encrypted image I_e, (**c**) Marked encrypted image I_{ew}, (**d**) Reconstructed image I

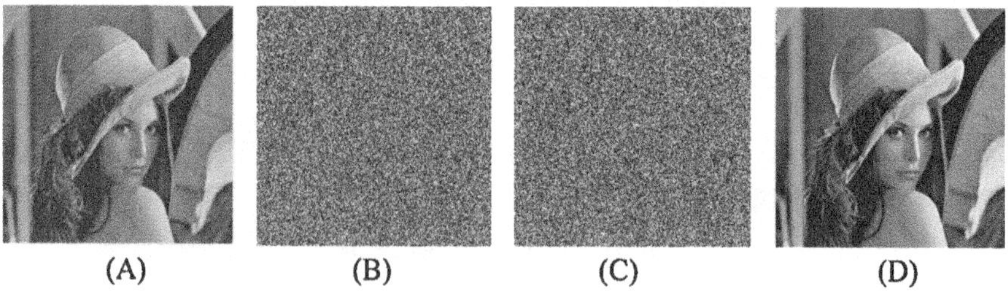

Fig. 5 Illustration of the proposed approach with the image lenna: (**a**) Original image (**b**) Encrypted image I_e, (**c**) Marked encrypted image I_{ew}, (**d**) Reconstructed image I

Table 1 Performance measurements of the proposed approach compared with [5]

		Best case	Worst case	Average
MSB prediction	Payload	1 bpp	1 bpp	1 bpp
	PSNR	$+\infty$	29.0	57.4
	SSIM	1	1	1
Second MSB prediction	Payload	1 bpp	1 bpp	1 bpp
	PSNR	$+\infty$	29.8	58.2
	SSIM	1	1	1

5 Conclusion

Here an effective technique of reversible information hiding with second MSB prediction in encrypted images with a really high payload is employed. This can be a technique that uses MSB substitution rather than using the least significant bits for a reversible data hiding technique. A high capacity or payload can be obtained due to the reason that MSB prediction may be far easier than the LSB prediction. So it can be said that MSB prediction is more efficient. Also in this case the reconstructed image quality may not be degraded. Initially the prediction errors caused due to MSB prediction are strictly analyzed and this information is stored in an error location map. Here the initial image is preprocessed i.e. some of the pixel values are modified so as to remove all the prediction errors that can be caused. By substituting the values of the MSB of each pixel within the image, it can be potential to hide a single bit per every pixel. In addition to the present value, the top value payload is obtained that leads to high quality reconstructed image. This technique provides an honest integrity and this may be used to maintain the confidentiality, integrity, and authentication of the image encrypted whereas providing the identical time legitimacy. The future work of this method can be considered as hiding over 1 bpp. It is also possible to use the next significant bits of most of the pixel of the encrypted image. By this method, the number of hidden message data can be increased, can cut back the amount of prediction errors can have enhanced reconstructed image quality during decoding with a correction to prediction errors.

References

1. Puech, W., Chaumont, M., Strauss, O.: A reversible data hiding method for encrypted images. In: Electronic Imaging 2008: International Society for Optics and Photonics, p. 68 191E (2012)
2. Qian, Z., Zhang, X.: Reversible data hiding in encrypted images with distributed source encoding. IEEE Trans. Circuits Syst. Video Technol. **26**, 636–646 (2016)
3. Puteaux, P., Puech, W.: An efficient MSB prediction-based method for high-capacity reversible data hiding in encrypted images. IEEE Trans. Image Forensics Security. **13**, 1670–1681 (2018)
4. Ni, Z., Shi, Y.Q., Ansari, N., Wei, S.: Reversible data hiding. IEEE Trans. Circuits Syst. Video Technol. **16**, 354–362 (2016)

5. Ou, B., Li, X., Zhao, Y., Ni, R., Shi, Y.Q.: Pairwise prediction-error expansion for efficient reversible data hiding. IEEE Trans. Image Process. **22**, 5010–5021 (2013)
6. Hong, W., Chen, T.S., Wu, H.Y.: An improved reversible data hiding in encrypted images using side match. IEEE Signal Process. Lett. **19**, 199–202 (2012)
7. Ma, K., Zhang, W., Zhao, X., Yu, N., Li, F.: Reversible data hiding in encrypted images by reserving room before encryption. IEEE Trans. Inf. Forensics Security. **8**, 553–562 (2013)
8. Zhang, X.: Reversible data hiding in encrypted image. IEEE Signal Process. Lett. **18**, 255–258 (2011)

Reversible Data Hiding and Coupled Chaotic Logistic Map Using Image Encryption

K. Anupama and K. Shanooja

1 Introduction

Technologists and researchers have made various efforts to solve the problems and issues generally reflected by the rapid growth of Internet, cloud computing, and the multimedia technologies in the field of data security. The issues are of mainly information security, integrality, copyrighting, hacking, etc. Information security means securing the information from unauthorized access, disclosure, modification, use, disruption, in inspection, recording, or destruction. Generally saying, the information security must secure or ensure the protection of information throughout the lifespan of the information transmission through a public domain. The data hiding can be categorized into two, Cryptography and Steganography. Cryptography is practice of converting the plain text into scribbled form called cipher text. Decryption is the opposite process of encryption, which recovers the plaintext back. The encryption and decryption are controlled by the key in each instance. In steganography, it is an art of hidden writing, conceal a secret message inside a cover medium so that it is impossible to sense the existence of the secrete message. It protects the messages and communication parties. Large varieties of steganographic techniques are there exist, in which some are more complex than others but all of them have strong and weak points. In steganography, all modern data compression, spread spectrum, information theory, any cryptography technologies are brought together for privacy on Internet. According to [1] the origin of "Steganography" derives from Greek and it means "cover writing." The commonly used method to conceal secrete message in image is steganography.

K. Anupama (✉) · K. Shanooja
MEA Engineering College, Affiliated to APJ Abdul Kalam Technological University, Malappuram, Kerala, India
e-mail: k7.anupama@gmail.com; shanooja@meaec.edu.in

M. Palesi et al. (eds.), *Second International Conference on Networks and Advances in Computational Technologies*, Transactions on Computational Science and Computational Intelligence, https://doi.org/10.1007/978-3-030-49500-8_15

Reversible data hiding (RDH) is the applications of steganography. Reversible data hiding is a process of reconstruction of the original image after the extraction of the embedded message in it without any alteration or error.

Different variety of image file formats exists in image steganography. According to [2] there are different steganographic algorithms which are used for different image file formats. Lossless and lossy compression method types are there. In both methods, for storage some space is allocated, but has different procedures for it. Lossy compression creates short files as it deletes excess image data from the original image and also the details.

Huge concentration is paid to RDH in encrypted domains, as it shows outstanding performance that the original image can be restored after the hidden data is extracted out while protecting the image content's confidentially. In this work, the cover image analysis is performed for prediction of possible errors to avoid the distortion of the image while reconstruction in decoding phase. Median edge detector (MED) is used for the prediction of problematic pixel location and formation of error location binary map is done in preprocessing step. Along with preprocessing, the image is encrypted by using the coupled logistic chaotic map and after that the problematic pixel location is inserted in the encrypted image by using MSB substitution method. The remaining available MSB of the encrypted image is scanned for image embedding. In the decoding phase, the concealed data is extracted by using the data hiding key and the cover image is reconstructed by using the encryption key without any distortion. As a result of using the coupled chaotic logistic map the visual security of encrypted image is increased. The statistical analysis is done and it is better than the previous state of the art.

2 Secure Lossless Data Hiding in Encrypted Domain

Reversible data hiding in encrypted images is one of the most leading research areas. Two phases are there; one is encoding phase and the other is decoding phase. In encoding phase, it consists of four processes mainly they are: the prediction error detection using MED, the encryption using coupled chaotic logistic map, the embedding of error location map, and message embedding using MSB substitution. Decoding phase consist of extraction of camouflaged message using corresponding key in data embedding and reconstruction of cover image without any errors using proper key. Illustration of encoding phase is given in Fig. 1.

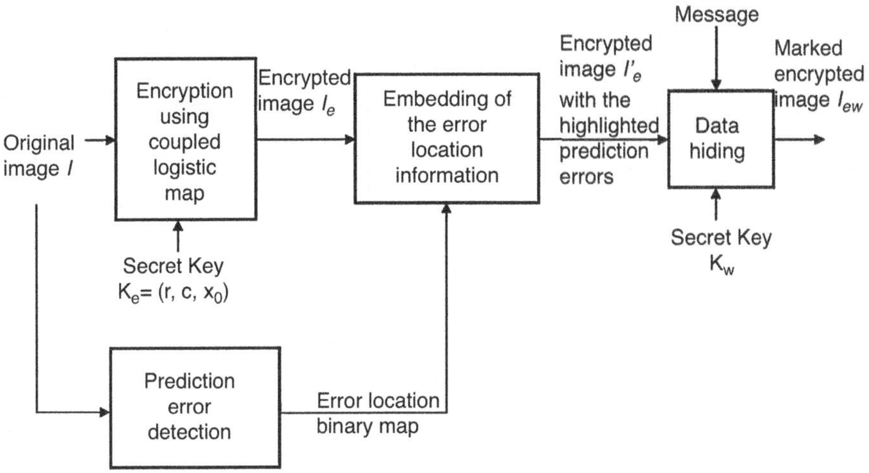

Fig. 1 Overview of the encoding phase

2.1 *Prediction Error Detection Using MED (Median Edge Detector)*

The MSB substitution method is used for data embedding, so the preprocessing of the cover image is essential for better recovery. The main requirement of reversible data hiding is minimum distortion or error free. The reversible data hiding is practice of restoring the original image back after the removal of the hidden data. Once the MSB values mislaid in the encryption process then it is difficult to get back in the decoding phase, so there need a process for recovering the lost MSB values back. Here median edge detector (MED) is used for prediction process. In MED method the prediction value of a pixel of an image is calculated using its neighbors. MED is a high performance predictor [3]. MED uses only three pixels to determine the type of pixels area which is currently predicted. Predictor decides where the pixel is in, whether it is in horizontal edge, vertical edge, or smooth area. MED predictor predicts the pixel based on local characteristics. Based on the casual area in pixel, there are three sub predictors. The causal neighbor's area is shown in Fig. 2. MED is an efficient predictor that recognizes three different types of causal areas.

- Firstly, let us consider the current pixel $p(t, l)$, with t lies between 0 and m and l lies between 0 and n. Take the inverse value of this current pixel $p(t, l)$ using the Eq. 1.

$$inv(t, l) = (p(t, l) + 128) \text{mod} 256 \quad (1)$$

- Now calculate the prediction value $pred(t, l)$ using the Eq. 2.

Fig. 2 Causal pixels for MED predictor

p(t-1, l-1)	p(t, l-1)
p(t-1, l)	p (t, l)

$$pred(t,l) = \begin{cases} \min(p(t-1,l), p(t,l-1)), & \text{if } p(t-1,l-1) \geq \max(p(t-1,l), p(t,l-1)) \\ \max(p(t-1,l), p(t,l-1)), & \text{if } p(t-1,l-1) \leq \min(p(t-1,l), p(t,l-1)) \\ p(t-1,l) + p(t,l-1) - p(t-1,l-1)), & \text{otherwise} \end{cases} \quad (2)$$

- Absolute difference between $pred(t,l)$ and $p(t,l)$ and between $pred(t,l)$ and $inv(t,l)$ is calculated and the results are recorded.
- Finally, compare the values between $(|pred(t,l) - p(t,l)|)$ and $(|pred(t,l) - inv(t,l)|)$, if $(|pred(t,l) - p(t,l)|) > (|pred(t,l) - inv(t,l)|)$, prediction error and information of position of the prediction is stored in a binary map [4].

2.2 *Image Encryption Using Coupled Chaotic Logistic Map*

Data transmission on public computer networks has created the necessity for security. Numerous encryption techniques have been emerged for data encryption. There exist symmetric, asymmetric, or hybrid encryption processes and can be applied to block or stream [5]. Image has innate uniqueness and is different from textual information content. So the traditional methods like Data Encryption Standard (DES) or Advanced Encryption Standard (AES) are not extraordinary options for image encryption [6].

Now for the image encryption chaos-based methods are used to protect the content of the image. The chaotic map encryption [7] has a complicated dynamic behavior, but it as normally simple nonlinear model. The sequence generated by using chaotic map [8] is high sensitive to the change in initial condition value and this system shows a variety of dynamics depending on the value of the bifurcation parameter p. In this work, coupled chaotic logistic map generator is used for the generation of pseudo-random binary sequence. Chaotic logistic map generator poses very good complex dynamics. By considering one more logistic map makes the generated sequence more sensitive and increases security of encrypted image. The result of the first logistic map is given to the input for the second logistic map. The key is used as the parameter for the chaotic generators. The generated pseudo-random sequence is exclusive-or (XOR) with the image pixels. The coupled chaotic logistic map using encryption step is shown in Fig. 3. The image is encrypted pixel by pixel by using the pseudo-random binary sequence $s(t,l)$ generated from the

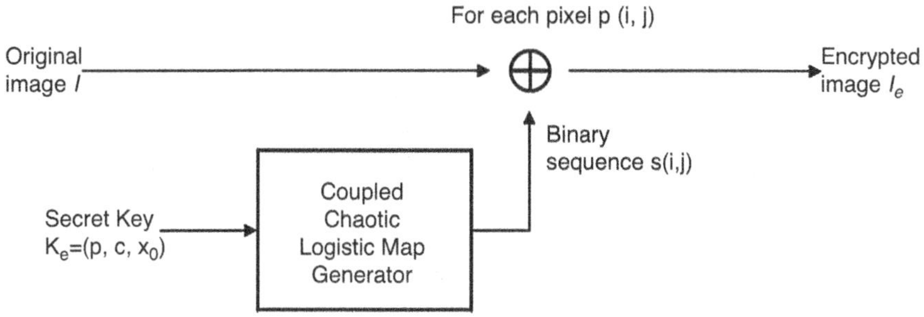

Fig. 3 Illustration of encryption step

chaotic generator. The encrypted image pixel p_e is generated by Eq. 3. Chaotic logistic map uses the pseudo-random properties of the logistic map and its equation is given below in Eq. 4.

$$p_e(t, l) = s(t, l) \oplus p(t, l). \tag{3}$$

$$X_{n+1} = pX_n(1 - X_n), \tag{4}$$

where $X_n \in [0, 4]$ and $p \in [1, 4]$. This relationship exhibits chaotic behavior for values of $p \approx 3.9$. The Eq. 4 is considered as the basic equation in the coupled chaotic logistic map and this is a discrete map function.

In this work, the encryption process is completely reversible and no overflow, hence able to recover the cover image back in decryption process.

3 Embedding of the Error Location Map and Secret Message

In this section, the predictor error location information is embedded into the encrypted image. Before the embedding process, the original image is encrypted. The encrypted image I_e is modified to keep away from prediction errors [4]. After adapting the encrypted image, the secrete message is embedded into by using the MSB substitution method. The to-be insert image is encrypted for to ensure security. The remaining available MSB value of the adapted encrypted image pixels is used for the embedding of secrete message bits b_k. The available MSB value of the adapted encrypted image pixels is replaced with the bit of to-be inserted message. A data hiding key is used for data embedding process. The embedding equation is given in Eq. 5, where p_{ew} is marked encrypted image pixel and p_e is the encrypted image pixel.

$$p_{ew}(t, l) = b_k \times 128 + (p_e(t, l) \bmod 128). \tag{5}$$

4 Cover Image Recovery and Hidden Message Extraction

The decode phase is illustrated in Fig. 4 shown below. The hidden message is decoded with the help of secret key used in the embedding time. Firstly, scan the marked encrypted image I_{ew} in line order for each pixel and the MSB value is extracted by using Eq. 6 and store it. The cover image is decrypted with the help of MSB prediction and the reconstructed image I'.

$$b_k = p_{ew}(t, l)/128, \tag{6}$$

On scanning the marked encrypted image, if the sequence of eight MSB equal to 1, then it shows the start of an error sequence. So the following pixels are not marked and scan is continued for the next sequence of eight MSB equal to 1, which shows the error sequence is over. This process of scanning for extraction is repeated until the end of image.

This approach is completely reversible, the original cover image I can perfectly reconstruct without any alteration. The Peak Signal-to-Noise Ratio (PSNR) is getting infinity and the Structural Similarity Index Map (SSIM) is getting value 1. To recover the cover image back, marked encrypted image I_{ew} must be decrypted with the secret key used in the encryption stage by using Eq. 7. $\tilde{p}$ represents the pixel of the reconstructed image I'. By using Eq. 6, it is possible to recover the correct seven LSB values only, so to predict the MSB values the MSB prediction method is utilized. The MSB values are predicted by using Eqs. 8 and 9.

$$\tilde{p}(t, l) = s(t, l) \oplus p_{ew}(t, l) \tag{7}$$

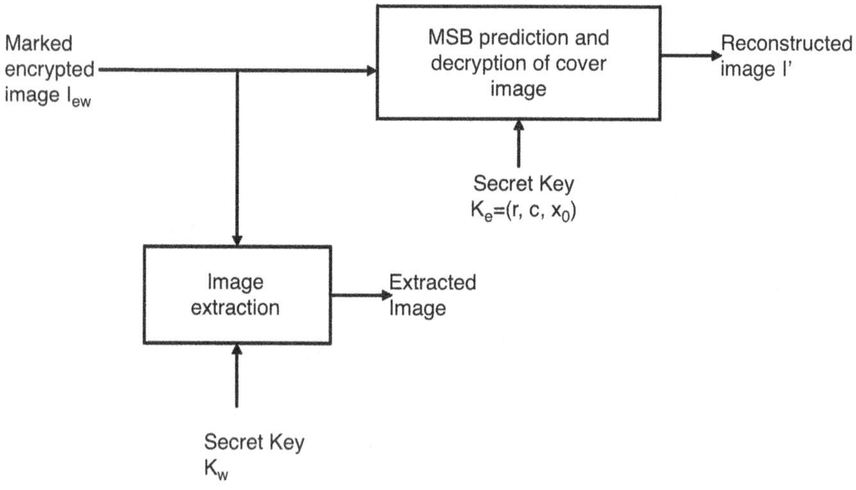

Fig. 4 Overview of decoding phase

$$\begin{cases} \Delta^0 = \left| pred(t,l) - \tilde{p}(t,l)^{\text{MSB}=0} \right|, \\ \Delta^1 = \left| pred(t,l) - \tilde{p}(t,l)^{\text{MSB}=1} \right|. \end{cases} \tag{8}$$

$$\tilde{p}(t,l) = \begin{cases} \tilde{p}(t,l)^{\text{MSB}=0}, & \text{if } \Delta^0 < \Delta^1, \\ \tilde{p}(t,l)^{\text{MSB}=1}, & \text{else.} \end{cases} \tag{9}$$

5 Experimental Results and Comparisons

For better understanding of how this work was done, the results of this work are shown below. MATLAB 2014a environment is used for the implementation of this work. Gray scale images sized 512×512 are used, standard test images and images from database [9] are taken for experiments. The key used for these experiments is $(r, p, x_0) = (3.896754, 0.123456,$ and $0.567891)$. After pre-processing the original image, there will be problematic pixel which have the most probable chance to predict wrong. So observe the amplitude of the error pixel and do necessary modification to avoid this error. So the image have to adapt before embedding the data. The problematic pixels often found in the contours. More prediction error will reduce the payload. One important process in this work is the prediction error detection. Without proper prediction error detection, it is not possible to achieve a high PSNR and SSIM value. The results of this method using the image "0.pgm" from the BOWS-2 database [9] are shown in Fig. 5. The obtained decrypted image has PSNR $+\infty$ and SSIM having 1. In some input images the PSNR and SSIM values may decrease.

The statistical analysis of the work is performed to verify the high visual security level of the encrypted or marked encrypted image. For this, diverse statistical metrics are used: horizontal correlation coefficients, vertical correlation coefficients, Entropy, Number of Changing Pixel Rate (NPCR), Unified Averaged Changed Intensity (UACI), and PSNR between original images and encrypted or marked encrypted images. The result of statistical analysis is tabulated and shown in Table 1. The entropy value is 7.981, which is very high for encrypted image and close to 8. These entropy values show that the gray-level distribution is uniform. The PSNR value of the encrypted image is ($\approx$8.21 dB), which shows that the original image and encrypted image are entirely different. The correlation of adjacent pixel of the original image is high and is shown in Table 1. The NPCR value is 100%, which is the maximal value. The UACI value is also better and close to 32.70%.

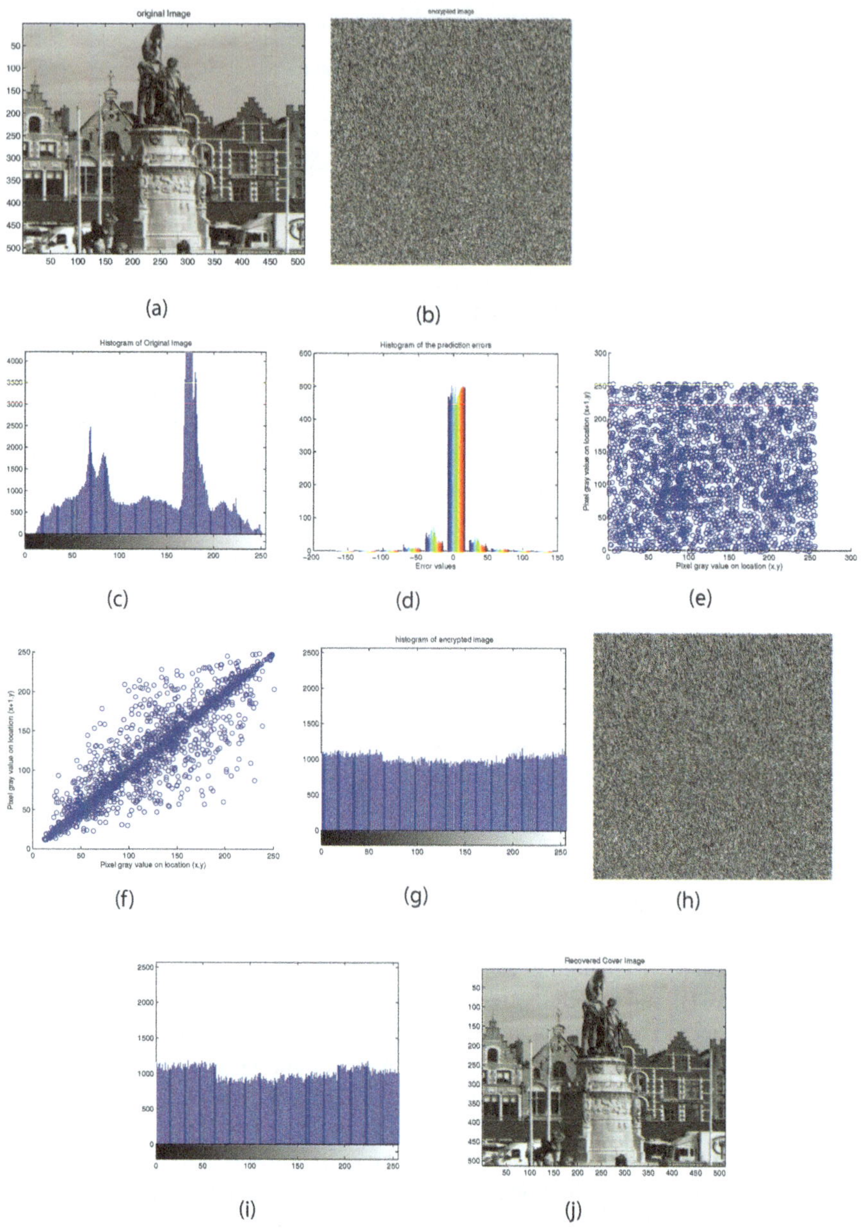

Fig. 5 Experiment result of proposed method: (**a**) cover image "0.pgm," (**b**) encrypted image, (**c**) histogram of cover image, (**d**) histogram of prediction errors, (**e**) horizontal correlation in the encrypted image, (**f**) horizontal correlation in the cover image, (**g**) histogram of encrypted image, (**h**) marked encrypted image, (**i**) histogram of marked encrypted image, (**j**) recovered cover image

Table 1 Quality evaluation of the obtained images using proposed method

Image	Horizontal correlation	Vertical correlation	Entropy	NPCR (%)	UACI (%)	PSNR (dB)
Original image Fig. 5a	0.9389	0.9437	7.3227	–	-	-
Encrypted image Fig. 5b	0.0347	−0.0431	7.9817	100	32.7022	8.276470
Marked encrypted image Fig. 5h	0.0096	−0.1431	7.9817	100	32.7022	8.276471

5.1 *Horizontal and Vertical Correlation Coefficients*

$$\mathrm{corr}_{p,p_N} = \frac{E\{|p - E(p)|\,|p_N - E(p_N)|\}}{\sqrt{V(p)V(p_N)}}, \tag{10}$$

where p_N represents the considered neighbor of pixel p, $E(x)$ is sample mean, and $V(x)$ is the sample variance.

5.2 *Shannon Entropy*

$$H(I) = -\sum_{l=0}^{255} P(\alpha_l)\log_2(P(\alpha_l)), \tag{11}$$

where I is a m $\times$ n image with 256 gray-levels $\alpha_l(0 \le l \le 256)$ and $P(\alpha_l)$ is the probability of α_l.

5.3 *Number of Changing Pixel Rate (NPCR)*

$$\mathrm{NPCR} = \frac{\sum_{i=0}^{m-1}\sum_{j=0}^{n-1} d(t,l)}{m \times n} \times 100, \tag{12}$$

where $d(t,l)$ is defined as:

$$d(t,l) = \begin{cases} 1, & \text{if} p(t,l) = p'(t,l), \\ 0, & \text{otherwise.} \end{cases} \tag{13}$$

5.4 *Unified Averaged Changed Intensity (UACI)*

$$\mathrm{UACI} = \frac{100}{m \times n} \sum_{t=0}^{m-1} \sum_{l=0}^{n-1} \frac{\left| p(t,l) - p'(t,l) \right|}{255}. \tag{14}$$

5.5 *Peak-Signal-to-Noise Ratio (PSNR)*

$$\mathrm{PSNR} = 10 \cdot \log_{10} \frac{255^2}{\frac{1}{m \times n} \sum_{t=0}^{m-1} \sum_{l=0}^{n-1} (p(t,l) - p'(t,l))^2}. \tag{15}$$

The performance of the coupled chaotic logistic map using encryption is compared with the piecewise linear map using encryption. The encrypted image is highly secure to the differential attack and is evaluated by using NPCR and UACI values. The NPCR and UACI values are compared with the existing methods stated in [4]. These values obtained by the proposed method is better than other state of the art and is tabulated in Table 2. The encryption time of proposed method is less than the time consumed by the piecewise linear map using encryption. The encryption using piecewise linear chaotic map consumed encryption time is 1.07 s and by considering proposed method, the time consumed for encryption is 0.4625 s. It means that proposed method encryption process takes only less time.

The performance of proposed approach is compared with the previous approaches using the standard test images Lena, Barbara, and Airplane which

Table 2 Comparison of NPCR and UACI value of various encrypted image

Method	Input image	Parameters of encrypted image	
Puteaux [4]	Lena	99.7902	29.368
	Barbara	99.7902	30.286
	Airplane	99.7902	31.6661
	0	99.6254	30.675
	1	99.7902	37.3576
	2	99.7902	35.4168
	3	99.7902	35.0662
	4	99.7902	30.705
Proposed method	Lena	100	29.908
	Barbara	100	33.0275
	Airplane	100	36.6963
	0	100	32.7022
	1	100	45.3956
	2	100	41.5807
	3	100	39.0431
	4	100	33.0275

Table 3 Performance analysis of NPCR and UACI values of encrypted image

Test image	NPCR (%) and UACI (%) values of encrypted image	Kyung [10]	Chin [11]	Haibin [12]	Puteaux [4]	Proposed
Lena	NPCR (%)	99.60	99.9	99.5716	99.7902	100
	UACI (%)	28.43	33.13	28.7491	29.368	29.908
Barbara	NPCR (%)	99.61	99.8	99.5182	99.7902	100
	UACI (%)	29.64	33.17	28.7107	30.286	33.0275
Airplane	NPCR (%)	99.61	-	99.61	99.7902	100
	UACI (%)	30.21	-	32.8065	31.6661	36.6963

are given in Table 3. From Table 3 it is understood that, the proposed method using embedded prediction error mechanism with coupled chaotic logistic map, has the NPCR value 100% and the UACI value is also better. The NPCR and UACI values show that the encrypted image is highly protected from statistical analysis like differential attack, etc. The SSIM value of our proposed method is 1. This means our reconstructed image is similar to the cover image. So this method found application in medical and military field.

6 Conclusion

This method allows us good embedding capacity and lossless recovery of exact cover image. Median edge detector predictor is used to improve the image recovery and to increase the payload capacity. Rather than making big modification of one value, it is better to small change some pixels to recover best image. The encrypted images are having better security to differential attacks. Coupled logistic chaotic map is used for the generation of the pseudo-random bytes for the encryption. Number of Changing Pixel Rate and Unified Averaged Changed Intensity values obtained are better than the other state-of-the-art method.

References

1. Anderson, R.J., Petitcolas, F.A.P.: On the limits of steganography. IEEE J. Sel. Areas Commun. **16**(4), 474–481 (1998)
2. Chan, C.K., Cheng, L.M.: Hiding data in images by simple LSB substitution. Pattern Recognit. **37**(3), 469–474 (2004)
3. Weinberger, M.J., Seroussi, G., Sapiro, G.: The LOCO-I lossless image compression algorithm: Principles and standardization into JPEG-LS. IEEE Trans. Image Proc. **9**(8), 1309–1324 (2000)
4. Puteaux, P., Puech, W.: An efficient MSB prediction-based method for high-capacity reversible data hiding in encrypted images. IEEE Trans. Inf. Forensics Secur. **13**(7), 1670–1681 (2018)

5. Wu, Y., Zhou, Y., Noonan, J.P., Agaian, S.: Design of image cipher using Latin squares. In: Inf. Sci. **264**, 317–339 (2014)
6. Liu, W., Sun, K., Zhu, C.: A fast image encryption algorithm based on chaotic map. Opt. Laser Eng. **84**, 26–36 (2016)
7. Ravi, R.V., Subramaniam, K.: Optimized wavelet filters and modified Huffman encoding-based compression and chaotic encryption for image data. Int. J. Appl. Eng. Res. **12**(13), 3961–77 (2017)
8. Mandal, M.K., Banik, G.D., Chattopadhyay, D., Nandi, D.: An image encryption process based on chaotic logistic map. IETE Tech. Rev. **29**(5), 395–404 (2012)
9. Bas, P., Furon, T.: Image database of BOWS-2. http://bows2.eclille.fr/
10. Seo, K.Y., Kim, D.S., Yoon, E.J., Yoo K.Y.: Improved reversible data hiding using mean values in an encrypted image. In: Proceedings of the International Conference on Security and Management. CSREA Press (2017). ISBN: 1–60132-467-7
11. Chang, C.C., Huynh, N.T., Chung, T.F.: Efficient searching strategy for secret image sharing with meaningful shadows. Int. J. Mach. Learn. Comput. **4**(5) (2014) https://doi.org/10.7763/IJMLC.2014.V4.448
12. Wu, H., Li, F., Qin, C., Wei, W.: Separable reversible data hiding in encrypted images based on scalable blocks. In: Multimedia Tools and Applications. Springer Science+Business Media, LLC, Part of Springer Nature, Berlin (2019). https://doi.org/10.1007/s11042-019-07769-w

Close Examination Camera with Automatic Quality Assessment Capability for Telemedicine

C. Sarathkumar, R. Prajith, George Varkey, and S. Aji

1 Introduction

Telemedicine is an emerging area of medicine, which uses information technology and telecommunication to empower healthcare services. Telemedicine can be defined as the usage of advanced telecommunication technologies for diagnosing, research, transferring patient data, and also for improving disease management and treatment in remote areas. Or in another way the term "telemedicine" can be used to refer to the utilization of telecommunication technology—in particular the internet—for medical diagnosis, treatment, and patient care [1, 2].

In conventional telemedicine systems, the vital signs like pulse rate (for measuring heart beat rate), SPO2 (for measuring oxygen saturation in blood), ECG, etc. are continuously transmitted to the remote specialist. Additionally video conferencing (VC) is conducted, so that the remote doctor gets a more factual picture about the patient's condition. During virtual consultation between the doctor and the patient, devices like close examination camera (CEC), electronic stethoscope, endoscope, etc. may be used. The outputs from these devices are presented to the remote specialist.

Close examination camera is a handheld device used to examine body parts like skin, eye, etc. It is usually operated by a nurse or a doctor. This is a highly expensive unit whose output is usually a video feed. But unfortunately it is seen that providing a stabilized high-quality video feed is not feasible/required in many practical cases. A few scenarios are given below:

C. Sarathkumar · R. Prajith (✉) · G. Varkey
Mobilexion Technologies Pvt. Ltd., Thiruvananthapuram, Kerala, India

S. Aji
Department of Computer Science, University of Kerala, Thiruvananthapuram, India

M. Palesi et al. (eds.), *Second International Conference on Networks and Advances in Computational Technologies*, Transactions on Computational Science and Computational Intelligence, https://doi.org/10.1007/978-3-030-49500-8_16

- The quality of CEC feed should be of higher quality than VC feed, since it is used for making diagnosis. In developing countries like India, high-speed internet is not available in most villages. Carrying a third high video channel along with video channels from patient's and specialist doctor's side may overburden the network.
- Also if less expensive cameras are used, it is seen that due to lack of good focusing system and/or image stabilization techniques, the output from these handheld devices may not always be of acceptable quality. It is left unto the nurse's skill to reduce the imperfections arising from the above problems—which many a time can lead to poor performance—particularly if the nurse is of older age.
- In many consultations, direct VC between doctor and patient is not needed. For example in diagnosing certain retinal conditions, skin lesions, etc., the CEC image is forwarded to relevant specialists, who provide appropriate diagnosis as and when they have time available.
- It is also seen that doctors and nurses particularly in government hospitals/care centers are overworking lot. Due to the high load of patients, a nurse may not be able—in most cases—to devote more than a few minutes to each patient. An automated quality assessment system is imperative in these cases so as to ensure that time taken to consult a patient does not increase significantly, if newer technologies like telemedicine are being used.

This paper presents a method to autodetect the amount of motion blur and out-of-focus blur present in an acquired image. We use a no reference quality measurement system to rank each image that is captured with an index value. At last, the top indexed values will be listed and top "*n*" number of them are stored or transmitted over the telemedicine system to the end point.

Broadly speaking, the existing image quality assessment (IQA) approaches can be classified into three classes, namely full reference (FR), reduced reference (RR), and no reference (NR) metrics. Applications like CEC where a reference image is not available to calculate the distortion belong to the NR-IQA category. Blur estimation has been studied since the past few decades, and many solutions for estimating it have been proposed in the same time. They can be classified mainly into two categories: region-based and edge-based methods. Region-based methods try to estimate the local image patches and typically produce dense blur maps directly. Edge-based methods calculate the blur at image edges and then try to propagate this information throughout the whole image.

Serra et al. proposed to estimate blur in an image using spike-and-slab prior together with an efficient variational expectation maximization (EM) inference scheme [3]. Karaali et al. proposed an edge-based method for estimating defocus blur map from a single image by calculating the reblurred gradient magnitude ratios [4]. Kerouh et al. proposed a method to estimate blur amount, by modeling the histogram of high-frequency coefficients using a suitable pdf (probability density function). This was done based on the fact that generally the blur visual impairment alters high-frequency components of an image such as edges and texture more than low-frequency components like background [5]. Sang et al. used singular value curve (SVC) to create a new no-reference blur index for still images [6].

2 Operational Overview

2.1 Platform

The operational dynamics of our system was tested as a part our stand-alone acute care system named Ubimedique Acute Care System (UMACS), which is running at several locations. This system can also be interfaced to any existing HIMS [7] to collect patient information if needed. It is a portable device that can be assembled and made fully operational in a few hours of setup time. Central to the system is a small trolley with a Wi-Fi modem and an attached tablet PC (Fig. 1).

UMACS has two software mobile applications intended to be used by the relevant stakeholders. They are the nurse's dash board (NDB) and the doctor's

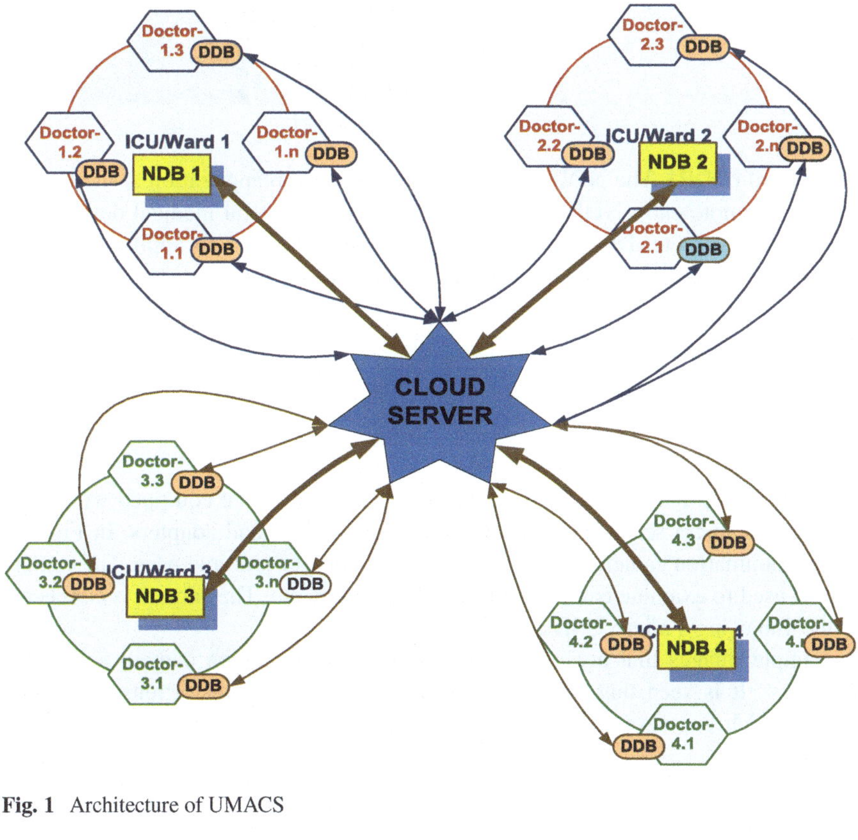

Fig. 1 Architecture of UMACS

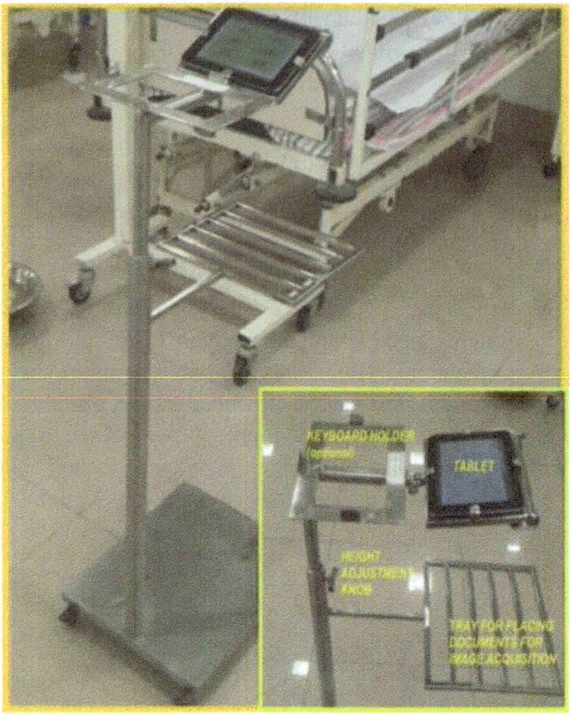

Fig. 2 UMACS trolley with tablet

dash board (DDB). The NDB represents the mobile/web application at the nurse's side (the remote end). NDB is used to control the peripheral medical devices like CEC, tele-ECG, etc. DDB similarly represents the mobile/web application at the doctor's/consultant's side. They can use the presented information to come to a diagnosis (Fig. 2).

2.2 CEC

The close examination camera is like a pen torch that can be equipped with high-resolution camera sensor along with a microscopic lens and adapters. In Fig. 3 a close examination camera is shown with its adapters for multiple uses. The camera can be used to examine ears, eye pupils, skin, mouth, etc. The camera is connected to the Tab using a micro USB cable.

Sample images that are taken from the close examination camera are shown in Fig. 4. It is seen that in practical usage, two types of distortions commonly occur, which can severely degrade the quality of acquired images. They are the following:

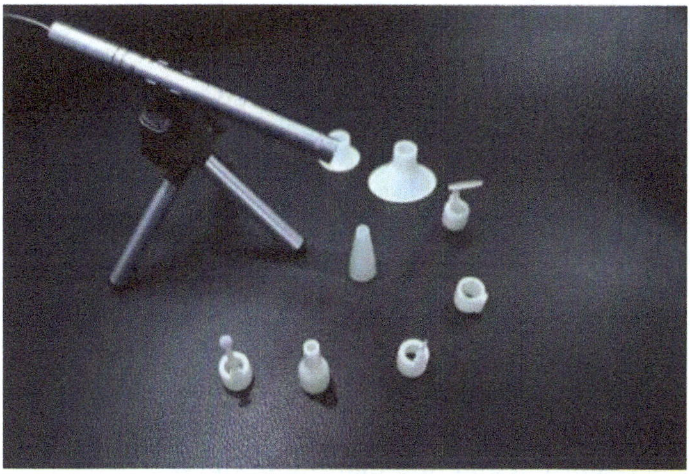

Fig. 3 CEC with adapters

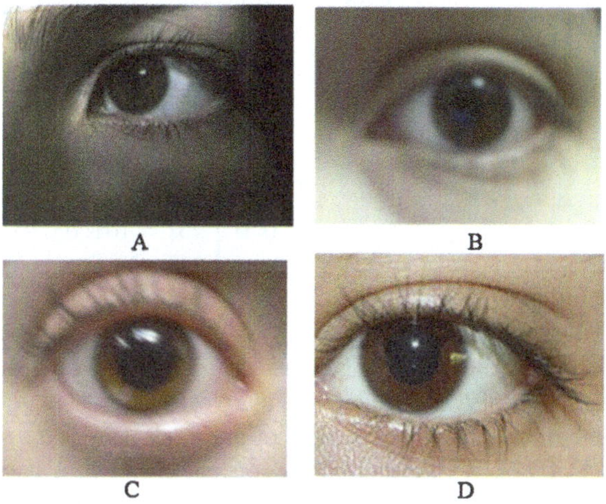

Fig. 4 Eye pupil images captured using close examination camera (**a**) with low focus image (**b**) with out-of-focus image (**c**) with motion blur (**d** quality image)

- Motion blur: This occurs while the examination camera is operated by the hands of a doctor or nurse from the patient's side. Since the camera is a microscopic camera and has 10–60× magnification, a small shake in the hand can generate a large amount of motion blur in the captured image. Also most of the low-cost cameras lack any sort of built-in image stabilization mechanisms.
- Out-of-focus blur: This usually occurs when the camera lens is manually focused. But a shake or movement towards or outwards the focusing area leads to losing of focus.

An automatic quality assessment program is then run over these images to find out if the acquired images are of acceptable quality.

3 Quality Estimation

The blur estimation technique used in our system is based on the singular value curve (SVC) method presented by Sang et al. [6]. Although SVC method was presented with a single threshold value "*c*," we have used different threshold values for different categories of images like retinal images, skin lesions, tongue images, etc. These values of "*c*" were found out using different custom datasets for each category. Each image was subjected to blur of five increasing intensities. The "*c*" value for each category of images was selected by calculating monotonicity of the IQA method using the Spearman rank-order correlation coefficient (SROCC) method.

3.1 Blur Distortion and Singular Value Curve

A real gray-scale image with dimensions $m \times n$ can be decomposed into a product of three matrices, $A = U{\cdot}S{\cdot}V^T$, where U and V are orthogonal matrices.

$$U^T U = I, U^T U = I,$$

$$s = \begin{bmatrix} S_r & 0 \\ 0 & 0 \end{bmatrix}, S_1 = (\sigma_1, \sigma_2, \sigma_3 \ldots \sigma_r) \tag{1}$$

Here "*r*" represents the rank of matrix A. The diagonal entries of matrix S are essentially the singular values of A, S_1 represents the singular value vector, the columns of U are the left singular vectors of matrix A, and the columns of matrix V are the right singular vectors of matrix A. This is the singular value decomposition (SVD) of A. To elaborate the concept of NR-IQA based on SVD, a random image and its five blurred versions from the CSIQ database [8] are shown in Fig. 5.

The amount of blur increases from Aa to Ae. These images are then subjected to SVD to obtain singular vectors S_1. A singular value curve (SVC) is now plotted with the index of the singular value vector along the x-axis and the singular value along the y-axis as shown in Fig. 6. As seen from the plotted curve, the singular values decay exponentially. It is also noticed that with larger degree of blur, the curve becomes steeper. The same law is seen if we do analysis on images from LIVE dataset [9] also.

Fig. 5 Source image A and its five increasingly blurred versions from CSIQ database; the degree of blur is sorted in ascending order from images Aa to Ae

3.2 Using Singular Value Curve to Create No-Reference Blur Index

As exemplified in Fig. 6, the shape of the SVC of natural images closely resembles an inverse power function. Let $y = S_1(i)$, $x = i$, which we can approximate roughly by the following equation:

$$y = x^{-q}, \tag{2}$$

where y is the singular value $S_1(i)$, and x is the corresponding subscript i of the singular value vector. We can use q to capture the image blur since the steepness of SVC corresponds to blur amount. Taking log

$$\ln\left(\frac{1}{y}\right) = q \ \ln x \tag{3}$$

and substituting

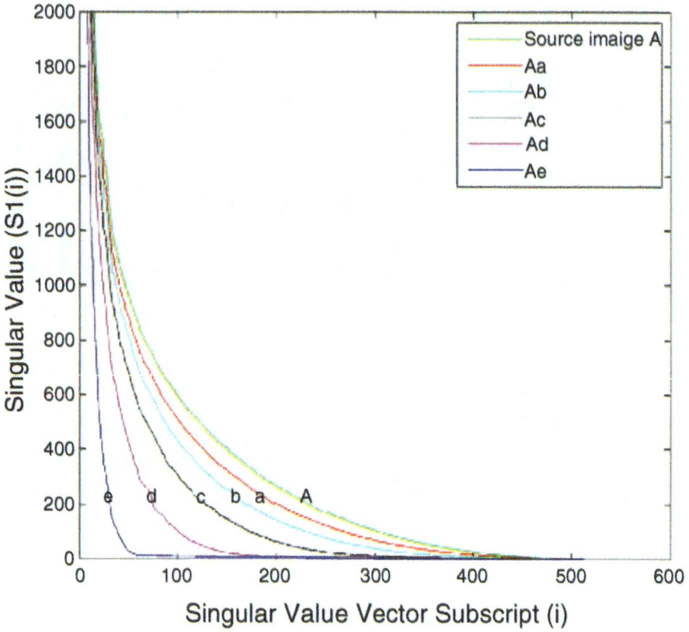

Fig. 6 Singular value curve of Fig. 5

$$M = \ln\left(\frac{1}{y}\right), N = \ln x$$

result in

$$M = q\ N. \tag{4}$$

This is a linear equation with coefficient q solvable by linear regression. We use least squares to minimize the residual sum of squares:

$$\min \sum_{i=1}^{r} e_i^2 = \sum_{i=1}^{r} (M_i - qN_i)^2. \tag{5}$$

Equating the derivative of Eq. (3) to 0, we get

$$\sum_{i=1}^{r} 2\,(M_i - qN_i)\,(-N_i) = 0. \tag{6}$$

Now q is obtained as

$$q = \frac{\sum_{i=1}^{r} N_i M_i}{\sum_{i=1}^{r} N_i^2}, \tag{7}$$

or

$$q = \frac{\sum_{i=1}^{r} \ln i \; \ln(1/S_1(i))}{\sum_{i=1}^{r} \ln i \ln i}. \tag{8}$$

When the singular values are small in value, the SVCs of blurred images are hard to discriminate. So we use only the larger singular values to estimate the blur amount. The blur quality index can be written as

$$\text{BlurPred} = \frac{\sum_{i=1}^{r} \ln i \; \ln(1/S_1(i))}{\sum_{i=1}^{r} \ln i \ln i} > c, \tag{9}$$

where S_1 is the singular value vector, i is the subscript of the singular value vector, and c is a threshold value. The "c" as earlier mentioned is calculated from reference images collected during consultation.

To create the reference dataset, we collected 200 images in each modality of examination like retinal images, skin lesions, tongue images, nail images, etc. They were subjected to five different amounts of blur—either motion or out of focus—to create a 1200 dataset for each category. The blur prediction algorithm is run over each dataset with different c values, and SROCC is calculated. The "c" value that gives the best result is then fixed for that particular category.

The c values thus obtained are stored along with their corresponding categories in a database. This will allow the system to use appropriate c value during live operation, depending on the modality (retinal images, skin lesions, tongue images, nail images, etc.) that is being examined with close examination camera at a particular time.

Sample blurred images used to create datasets are shown in Fig. 7 (eye) and Fig. 8 (tongue). The top-left image is the original image.

4 CEC in Teleconsultation

The steps used in our system to find the best images from the input feed of close examination camera from the remote patient location are described below.

4.1 Work Flow of the System

The system is composed of capture, indexing, and selection modules (Fig. 9).

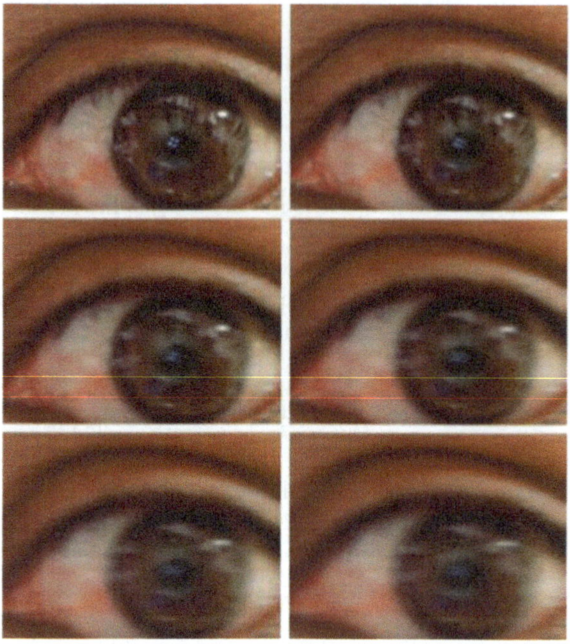

Fig. 7 Sample eye image (top-left) and its five blurred versions in the dataset

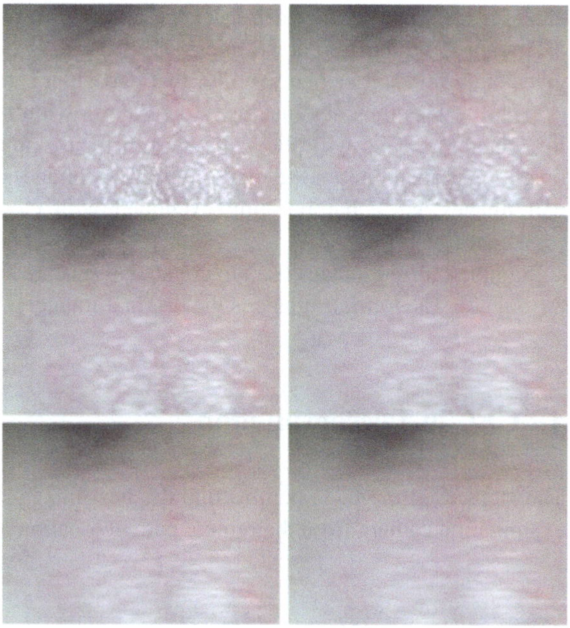

Fig. 8 Sample tongue image (top-left) and its five blurred versions in the dataset

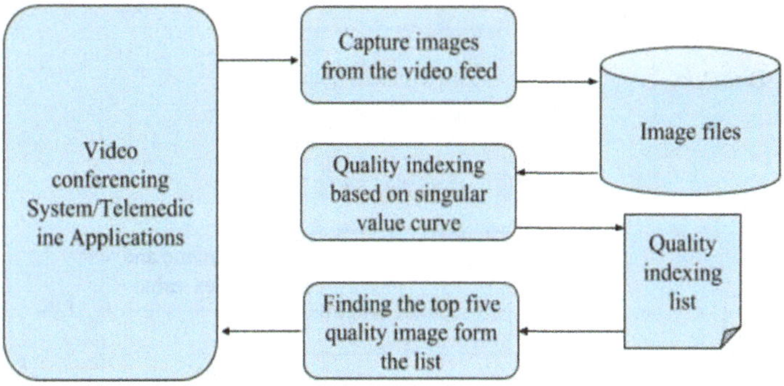

Fig. 9 Work flow of the system

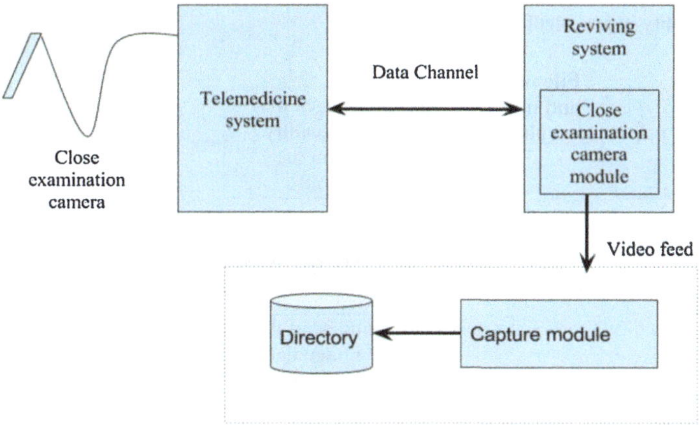

Fig. 10 Data flow to the capture module

4.2 *Capture Module*

This module captures the input images from the video feed coming from the close examination camera. This feed can be transmitted over the same channel that is used by the video conferencing system, for the reason of low bandwidth in the remote location. Using a capturing module, the input video feed is sliced frame by frame and then stored into a location that can be accessed by the indexing module (Fig. 10).

4.3 *Indexing Module*

The indexing module assesses each image that is stored in the directory and calculates an index value based on the singular value curve. Along with an

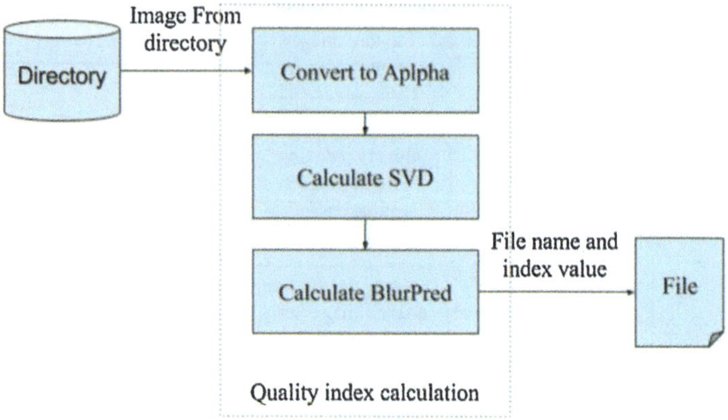

Fig. 11 Quality index calculating module

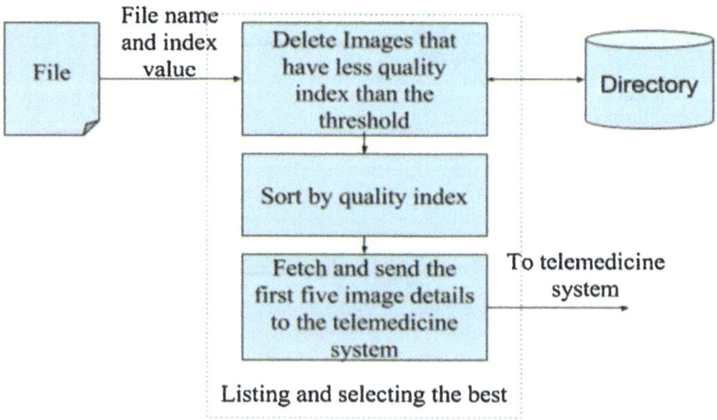

Fig. 12 Listing and selecting the best image from the list

autogenerated file name, this index value is stored in the memory for use by the selection module (Fig. 11).

4.4 Selection Module

The final stage is to eliminate the unwanted images and list the remaining images. The images that satisfy the threshold quality level fixed using the quality assessment stage will be sorted in with better images listed on top. Using empirical studies we conducted, it was seen that sending the top 5–10 images to the consultant doctor's side is sufficient for obtaining accurate diagnosis. The images coming below the threshold are removed from the database (Fig. 12).

5 Results

5.1 Determination of Threshold c

About 200 images were selected in each modality of examination like retinal images, skin lesions, tongue images, nail images, etc. These images were taken with varying light conditions and observation angles (as much as expected in a consultation room). The selected images were subjected to five different amounts of blur—either motion or out of focus—to create a dataset of 1200 images for each category. The blur prediction algorithm is run over each dataset with different c values, and SROCC is calculated. The "*c*" value that gives the best result is then fixed for that particular category. The best SROCC obtained and their corresponding c values are provided in table below (Table 1).

5.2 Test Result on Blur Datasets

Similar to creating the reference datasets as described in the previous section, a test dataset was also prepared to validate the results. 25 images in each modality of examination like retinal images, skin lesions, tongue images, nail images, etc. These images were taken with varying light conditions and observation angles (as much as expected in a consultation room). The selected images were subjected to five different amounts of blur—either motion or out of focus to create a dataset of 150 images for each category. The blur index was calculated for each dataset, and then SROCC was used to find the monotonicity of prediction. The method used provided very good results. The SROCCs obtained for the test datasets are shown in the table below (Table 2).

Table 1 Best SROCC and *c* values for reference datasets in each modality

c Value	Modality	SROCC
50	Eye	0.9163
45	Skin	0.8648
60	Tongue	0.9526
30	Nails	0.9224

Table 2 SROCC for test datasets in each modality

Modality	SROCC
Eye	0.9163
Skin	0.8648
Tongue	0.9526
Nails	0.9224

5.3 *Running Time*

Speed is an important factor for selecting an NR-IQA method in live consultations. Here it is necessary to judge the quality of an image in real time. For nonlive consultations, it is of lesser importance. The image size was kept at 512 × 512 for this computation. If the blur calculation is done on the Android system, it was seen that at an average one image took around 0.55 s to be computed. If the blur calculation is done in a nurse's laptop with i7 processor, the computation took only 0.18 s at an average. The system was developed using Android, OpenCV, and Python.

6 Conclusion

It is predicted that telemedicine-based technologies will substantially alter existing health delivery systems in the near future. But significant technical challenges are still needed to be solved for it to obtain wide acceptance among the healthcare community and the general population. The close examination camera with automatic quality assessment capability described herein is a cost-effective solution aiming to solve one of the major issues affecting telemedicine consultations. It is easy to assemble and operate, and it does not affect the workflow of the existing HIMS systems. It is modular and expandable and is minimal in its requirements of computational power. This in turn reduces the cost significantly, making it appropriate for wide variety of telemedicine applications.

Acknowledgment We are deeply indebted to the management of hospitals and institutions that helped us in this endeavor. These institutions have provided a congenial atmosphere for developing and testing out this work.

References

1. Wootton, R.: Telemedicine. BMJ. **323**(7312), 557–560 (2001)
2. Perednia, D.A., Allen, A.: Telemedicine technology and clinical applications. JAMA. **273**(6), 483–488 (1995)
3. Serra, J.G., et al.: Variational EM method for blur estimation using the spike-and-slab image prior. Digital Signal Process. **88**, 116 (2019)
4. Karaali, A., Jung, C.R.: Edge-based defocus blur estimation with adaptive scale selection. IEEE Trans. Image Process. **27**(3), 1126–1137 (2018)
5. Kerouh, F., Ziou, D., Serir, A.: Histogram modelling-based no reference blur quality measure. Signal Process. Image Commun. **60**, 22–28 (2018)
6. Sang, Q., et al.: No-reference image blur index based on singular value curve. J. Vis. Commun. Image Represent. **25**(7), 1625–1630 (2014)

7. Varkey, G., Balakrishnan, A., Prajith, R.: Presentation of patient information on the doctor's smart phone for acute care. In: Proceedings of the 10th International Conference on Theory and Practice of Electronic Governance. ACM (2017)
8. Image Coding and Analysis Laboratory, Oklahoma State University, Categorical subjective image quality. http://www.vision.com
9. Sheikh, H.R., Wang, Z., Cormack, L., Bovik, A.C.: Live image quality assessment database, release 2005. http://live.ece.utexas.edu/research/quality

Pre-Silicon Validation of 32-Bit Indigenous Processor for Space Applications

S. J. Anjana, K. Padmakumar, Joji Daniel, L. S. Syamlal, Ganta Nagendra Mourya Teja, K. Ranjani, R. Paramasivam, and M. Narayanan Namboodiripad

1 Introduction

The processor on-board the launch vehicle performs navigation, guidance, control and stage sequencing functions. Since the launch vehicle missions are designed with zero defect as the target, it is highly imperative that the processor design is fully validated in the ground before the chip is fabricated. Insufficient validation had resulted in the infamous Pentium division bug, necessitating a major recall of the Pentium computers [1]. Computer architecture refers to those attributes of a system visible to a programmer and has a direct impact on the logical execution of a program [2]. For a microprocessor, instruction set architecture (ISA) can be taken as its specification. Validation shall ensure that the correct result is always obtained for a sequence of valid instructions, and it shall also detect the potential errors and design deficiencies.

For the custom-designed 32-bit processor, hereafter referred as PROC32, the most prevalent technique in the industry, simulation-based validation is used, with additional test cases at each stage of design cycle. Another standard technique of formal verification, wherein specification can be proven to satisfy a given set of properties, was not used for this fairly simple design, where the time to delivery of product was a constraint. Elaborate test programs developed by the designers and standard validation suites for IEEE754 floating-point data were used for ISA validation. Since PROC32 design is non-pipelined and without cache, simultaneous interactions between different design elements and timing uncertainties do not affect its functioning. Thus, systematically crafted test cases can achieve complete

S. J. Anjana (✉) K. Padmakumar · J. Daniel · L. S. Syamlal · G. N. M. Teja · K. Ranjani
R. Paramasivam · M. Narayanan Namboodiripad
Vikram Sarabhai Space Centre (VSSC), Thiruvananthapuram, India
e-mail: sj_anjana@vssc.gov.in; syamlal_ls@vssc.gov.in

M. Palesi et al. (eds.), *Second International Conference on Networks and Advances in Computational Technologies*, Transactions on Computational Science and Computational Intelligence, https://doi.org/10.1007/978-3-030-49500-8_17

design validation of PROC32. The methodology presented in this paper involves testing the microprocessor design using a combination of VHDL simulator and FPGA prototype and verifying the outputs against those of instruction simulator and standard test suites. The paper discusses the test case design strategy, test platform and validations done for characterising the reliability of PROC32. The paper concludes with the summary of performance metrics.

2 Literature Survey

In a typical engineering project on system on chip (SoC), verification effort is significantly high, consuming more than 57% [3]. The key goal of system validation is to ensure that its behaviour is correct and consistent. It encompasses functional and performance verification. The designers need to start detecting the bugs early in development, continue testing till the silicon wafer is fabricated and should be able to reuse the vectors once the chip is out from the foundry. For the low-level bugs in a processor design to be brought out, the internal state of its microarchitecture is to be assessed by applying stimuli like assembly programs to cover the entire design space.

3 Processor Overview

The indigenisation elements in PROC32 include the processor architecture, hardware design, chip fabrication and system software tool suite, as depicted in Fig. 1. ISA has been defined primarily to meet the functional and safety requirements of flight software. In-house design ensures that the chips can be re-fabricated based on demand. The indigenous device manufacturing leaves no scope for bug intrusion like Trojans. The in-house system software tool suite supports customisation of user requirements.

PROC32 is designed around von Neumann architecture, ensuring deterministic performance for real-time applications. To avoid catastrophes, fraction arithmetic, which works on a closed set of values (−1, +1), is adopted and overflows are indicated through hardware arithmetic errors (AEs). PROC32 is an accumulator-based machine with microprogrammed control logic, simple addressing modes and 152 fixed-length instructions operating on fixed-point, floating-point and integer data

Instruction Set Architecture	Hardware Design
ASIC Fabrication	System Software & IDE

Fig. 1 Indigenous 32-bit processor ecosystem

types. The address space supports large code and data size for embedded real-time applications. Hardwired floating-point unit (FPU) also implements trigonometric instructions.

For faster memory access, on-chip RAM is provided for software, application data, operating system (OS) data and stack. SECDED (single bit memory error correction and double-bit error detection) protection is provided with enable/disable options. Programmable memory wait states facilitate interfacing with devices of different access times. Four programmable counters enable real-time scheduling, and a watchdog timer helps to detect and recover from any deadly situation like hanging of processor. About 256 software interrupts and 12 hardware interrupts are supported. Two on-chip indigenous MIL-STD-1553 protocol controllers serve as input–output communication interfaces.

4 Instruction Set Architecture

The programming model includes 26 registers, each 32-bit long. Two accumulator registers are used for 32-bit and 64-bit operations. Seventeen index registers enable efficient data access across multiple segments. Loop Counter facilitates iterative execution of instructions. Program counter keeps track of the sequence of instructions to be executed. Stack Pointer manages the stack. Status Register setting controls interrupts and write protection of system memory area, besides reflecting the hardware arithmetic errors. Three instruction formats permit operands in memory (Type-I), constant operands (Type-II) and operands in registers (Type-III). Two addressing modes are provided for operands with respect to program counter or index register using 20-bit signed offset field in the opcode. Little endian convention is used for data access. Generic data formats like 16-/32-bit integer (signed/unsigned), real number (IEEE754) and fraction arithmetic are supported. Two's complement arithmetic is used for fixed-point instructions. A 64-bit float has a precision of 15 decimal digits and a range of $1.79 \times 10^{+308}$ and 32-bit float has a precision of 6 decimal digits and a range of $3.40 \times 10^{+38}$.

Instruction set has been evolved to meet the real-world data processing requirements. Special instructions like multiply and accumulate (MAC) are intended for digital signal processing applications like filters. Dedicated instructions are provided for microdiagnostics (CPU/memory tests), interrupt handling and to load/read timers. Pre-scale instructions select the clock frequency of operation of the real-time counters. Instructions are defined to configure the wait states for memory-mapped peripherals and to transfer data between 16-bit peripherals and 32-bit processor. Test and set instruction enables semaphore operation by controlling access of resource by multiple processes in a multi-tasking system. Stack operations support modular software structure. Bit manipulation instructions (Set/Reset/Read) aid ON/OFF control in applications like programmable logic controllers. Special instructions process blocks of data, like array copy, initialisation and checksum computation for integrity checks.

5 Test Methodology

The processor testing is a challenging area and requires in-depth understanding of the CPU design. In microprogrammed control unit, microcode translates machine instructions into sequences of detailed circuit-level operations. Each microinstruction of PROC32 is a 55-bit long word, functionally divided into 20 fields. It generates ALU control signals and address of next microinstruction. CPU takes multiple clock cycles to execute a single machine instruction, one for each microprogram step. A dedicated FPU performs floating-point, trigonometric and data type conversion and MAC operations.

For PROC32, the software development tool suite is designed and developed in-house. Ada cross compiler translates the source programs in Ada into equivalent assembly code. Assembler converts assembly language programs to re-locatable machine code. Linker links the object code with the system libraries to generate executable code with physical addresses assigned. This is executed in the instruction simulators and target hardware, using the setup in Fig. 2. The vectors generated from the test programs can drive the RTL model in VHDL simulator running on 3.4 GHz Intel core i7 machine with 8 GB of memory. The internal state of processor, with registers, memory and system bus, is fully observable and controllable in this test scenario.

In simulation-based validation, each test is first executed in PROC32 simulator and then run on the design under test (DUT), which is the Register Transfer Logic (RTL) implementation of the microprocessor. Graphical user interface (GUI) and command-line versions of microcode and instruction simulators enable testing in PC by modelling the instruction set, registers, memory and execution flow of instructions. These generate execution trace and clock statistics. The single-stepping feature of the simulator is very useful in the initial design phase. The inputs and outputs are consolidated in a table. It is ensured that the hardware AE bits in Status Register are set in case of arithmetic overflows and software AE word bits are set in case of abnormal inputs to the software library routines. The deviations of actual outputs from references are traced using VHDL simulator, and the defects rectified after thorough analysis.

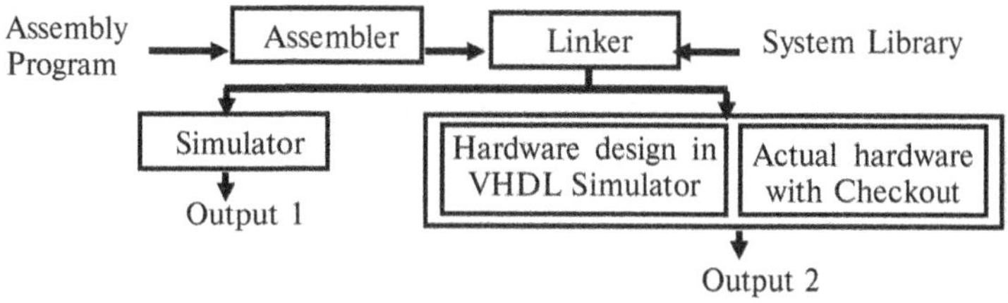

Fig. 2 ISA validation setup

6 Test Case Design Strategy

Test cases were judiciously chosen to satisfy the objectives of ensuring maximal functionality coverage and fault coverage as explained below. Statement coverage is a measure of the ability of test programs to access every executable statement in code. Code coverage of 99 % was achieved for PROC32 in a total of 3×10^8 simulation cycles. The major bugs that necessitated code changes are discussed in Sect. 7.

6.1 Functionality Coverage

Test programs are designed to exercise all the instructions for various input combinations in the different addressing modes so as to ensure the correctness of implementation of the ISA. Assembly-level processor self-test programs written for 152 instructions achieve 100% functional coverage. Corner case data values and inputs outside the functional domain are fed to the instructions, and the response is characterised. Table 1 summarises the statistics of test cases for different types of instructions.

The instructions are tested using inputs derived from equivalence partitioning, a standard technique for revealing errors with less number of test cases, reducing the test time. Boundary value testing is carried out to verify the behaviour of instructions for nominal and boundary values. Floating-point instructions are tested with additional inputs like denormal, infinity and NaN (Not a Number), and outputs are checked for overflows and underflows. It is verified that denormal inputs

Table 1 Instructions versus test cases

Instructions	Test cases
Floating-point 32-/64-bit Add/Sub/Mul/Div/Neg/Abs/Compare (12) + trigonometric functions Sin, Cos, Arctan, Sqrt, Exp and Log (12) + MAC (2) + data type conversion (10)—36 nos	6000 each—216,000 nos
Fixed-point and unsigned arithmetic and logic 16-/32-bit Add/Sub/Mul/Div/Neg/And/Or/Xor—28 nos	20 each—560 nos
32-bit integer Mul/Div/MAC—3 nos	25 each—75 nos
Logical shift 16-/32-bit (max 15/31 shifts)—4 nos	25 each—100 nos
Arithmetic shift 16-/32-bit (max 15/31 shifts)—6 nos	25 each—150 nos
Control transfer (branch/call/return/compare)—18 nos	15 each—270 nos
Load/store 16-, 32-, 64-bit (8), peripheral interfaces (2), bulk data process (3), bit operations (3), semaphore (1)—17 nos	5 each—85 nos
Register copy, initialisations, sign extension—20 nos	2 each—40 nos
Control instructions (interrupt, timer, memory protect)—13 nos	15 each—200 nos
Stack (push/pop/save/restore)—4 nos	2 each—8 nos
Microdiagnostics (CPU/memory tests)—3 nos	1 each—3 nos

Table 2 Representative test matrix

Instructions	Typical inputs
16-bit signed arithmetic	−32768, −32767, −1, 0, 1, 32766, 32767
32-bit signed arithmetic	−2147483648, −2147483647, −32768, −1, 0, 1, 32767, 2147483646, 2147483647
16-bit unsigned	0, 1, 32767, 32768, 65534, 65535
32-bit unsigned	0, 1, 32767, 65535, 4294967294, 4294967295
16-bit shift (logical and arithmetic)	0_h, 1_h, $7FFF_h$, 8000_h, $FFFF_h$, 5555_h, $AAAA_h$; No. of shifts—0, 1, 2, 3, 13, 14, 15
32-bit shift (logical and arithmetic)	0_h, 1_h, $7FFFFFFF_h$, 80000000_h, $FFFFFFFF_h$, 55555555_h, $AAAAAAAA_h$; No. of shifts—0, 1, 2, 15, 29, 30, 31
32-/64-bit floating point	+0.0, −0.0, +min, −min, +nominal, −nominal, +max, −max, +Infinity, −Infinity, +sNaN, −sNaN, +qNaN, −qNaN, denormal, −denormal
Trigonometric (32- and 64-bit floating point)	π/2, π/3,π/4, π/6, −π/2, −π/3, −π/4, −π/6, other special cases in floating-point representation
Data type conversions	Values in the range of source and destination data types

are treated as ±0.0 and underflows result in ±0.0. The precision and rounding scheme are ensured as per design. Data type conversion and MAC instructions are tested with out-of-range values, besides others. Floating-point MAC instructions are verified through matrix multiplication outputs too. Standard FPU verification test suite [4] was downloaded from the internet, and 117,470 test cases were executed. Programs are designed to test different sequences of valid instructions too. The register and memory contents are verified before and after execution of each instruction, to ensure the behaviour as per specifications. The functional outputs from VHDL simulator and Status Register contents, indicating arithmetic errors if any, are verified against the reference outputs, along with the execution clock cycles. Table 2 features the representative test inputs.

The library routines, which invoke appropriate processor instructions after input pre-processing, are also tested in the target. Trigonometric functions are implemented using CORDIC (coordinate rotation digital computer) algorithms in the processor design. For verifying these, test programs are written in Ada and Assembly and the linker-generated binaries are executed in the VHDL simulator. Reference outputs are generated using C++ programs that invoke the corresponding standard mathematical libraries available in PC compiler and also log the floating-point status flags. The CORDIC functions implemented in microcode simulator serve as another reference.

6.2 Fault Coverage

Test vectors are generated for the fault coverage of memory. Programs are written to access all memory locations (load/store) with patterns intended to detect 'stuck-at' faults, facilitating toggle coverage (the percentage of bits that toggle at least once from 0 to 1 and at least once from 1 to 0 during the simulation). The exercise was repeated to ensure coverage of all registers too. These vectors were also delivered to the foundry where the processor chip is fabricated, for clearing functionality. Tests using Automatic Test Pattern Generation (ATPG) are used subsequently by the chip manufacturer to expose any electrical and manufacturing defects within the chip.

7 Validations Carried Out

7.1 CPU Testing

CPU test instruction checks the ALU microcode operations (3 Arithmetic and 5 Logical operations), flags (Overflow, Carry, Zero and Negative) and external registers (Status Register, Memory Address Register, Memory Data Register and Micro-Index Register). Self-testing of 152 instructions is carried out as detailed in the previous sections.

7.2 Error-Handling Capabilities

Arithmetic overflows, floating-point operations with NaN and divide by 0 conditions result in hardware AE bit in the Status Register, leading to the consolidated exception bit (EX), triggering EX interrupt. OS can then take appropriate error-handling actions in the Interrupt Service Routine (ISR). Status Register also indicates underflow, inexact representation and denormal inputs for floating-point data, as shown in Fig. 3.

7.3 Interrupts and Exceptions

PROC32 supports 12 hardware interrupts, of which only 9 are connected. These include illegal opcode interrupt, SECDED-related double-bit, single bit and check bit memory interrupts, two timer interrupts and two 1553B protocol controller errors. The interrupted program context is saved in system memory area by a

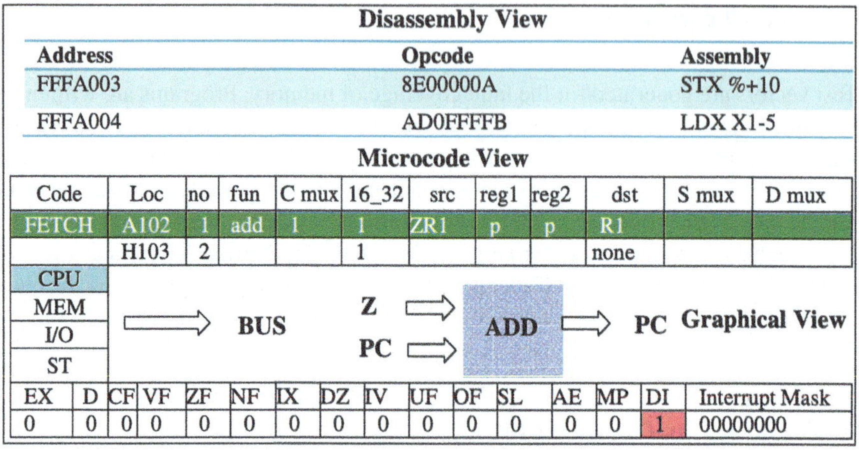

Fig. 3 Microcode simulator snapshot

hardware process transparent to the application task executing at that instant. Test programs are written to inject errors to trigger the hardware interrupts. About 256 software interrupts are tested by invoking instructions. Exception conditions are generated by selecting inputs that would lead to arithmetic saturation. It is verified that PROC32 handles the abnormal situations gracefully, giving control to the software to initiate recovery actions.

7.4 Memory and Peripherals

Assembly instruction is available to check the integrity of boot code in PROM. CRC (cyclic redundancy check) checksum and XOR checksum checks are implemented for PROM and tested. RAM health is tested by means of March pattern and Walk pattern tests. SECDED error detection/correction unit based on modified Hamming code for internal RAM is tested by injecting error after disabling SECDED and verifying after enabling it. Memory wait states are configured for different blocks through instructions and memory access times measured. For testing input/output interfaces, all patterns ranging from 0000_h to $FFFF_h$ are sent from checkout system to PROC32, received back through 1Mbits/sec 1553B link and compared. Real-time counters are verified by timer test programs executing for a fixed duration, the counter frequency being derived from processor operating clock frequency using pre-scale options.

7.5 Flight Software Evaluations

The confidence on the processor increases when it is proven in the actual domain for which it is primarily designed. Thus, launch vehicle flight software, with a representative mix of fixed and floating-point instructions, is thus a natural candidate for assessing the performance of the processor. Flight software is developed in Ada Language for Flight Software Application (ALFA) and tested using PC-based Ada compiler. Real-Time Executive (REX) software, comprising of boot code, static task scheduler and application interfaces, is developed in Assembly language and ALFA. Ada modules are cross-compiled to generate assembly code files. System Integrator utility integrates these modules with the REX software to generate a single assembly language file. The assembled flight code is linked to generate the binaries and downloaded to the memory of a Xilinx FPGA prototyping board, which is programmed with the VHDL design of PROC32. Figure 4 represents the proto-board for tests.

Simulated input profiles (SIPs) were fed through a checkout system, and the flight software execution outputs were compared with the reference outputs of a trajectory simulation program running in PC. The software for different launch vehicles like PSLV, GSLV and GSLV MkIII were tested, and the outputs for the nominal, +3σ and -3σ cases were matching with the reference outputs of trajectory simulation software.

Fig. 4 FPGA prototyping board for tests

7.6 Processor Timing Performance

The timing performance has improved by around 75%, compared with the 16-bit on-board computer. This can be attributed to increase in word length of the processor, hardware FPU, more index registers, powerful instructions and higher clock speed of PROC32. The number of assembly instructions per Ada statement has also reduced considerably, due to the improvements in the architecture and associated compiler optimisations, resulting in a reduction in flight software code size by around 40%.

8 Observations and Corrections

Some typical observations during validation and corrections implemented are listed. The results indicate that the test suite can detect design errors and data-specific bugs.

8.1 Floating-Point Instructions

Observation Related to Accuracy for Trigonometric Instructions

The accuracy requirements for sine, cosine, arctangent and logarithmic functions are 6 and 15 decimal digits for single and double-precision floating-point numbers, respectively. But, the mantissa of the output was not exactly matching with the expected values for small input angles. [(Actual − Expected)/Expected] exceeded 10^{-6} and 10^{-15} for small values of 32- and 64-bit floating-point numbers, respectively.

Analysis and Correction

The accuracy loss is due to inappropriate setting of the number of CORDIC iterations in the algorithms. Also, the values of the constants used in the algorithms were not large enough. The constants were appropriately modified for accuracy.

The constants used to represent iterations in the CORDIC algorithms were appropriately modified for the single and double-precision floating-point functions of sine, cosine, arctangent and logarithm to meet the accuracy requirements.

Observation in Arctangent Instructions

During corner case tests for 32-bit and 64-bit arctangent instructions accepting two arguments with an operational domain of [0..1], errors in Table 3 were observed:

Table 3 Observation in arctangent instructions

Instruction	Inputs	Reference outputs	Hardware output
32-bit	00800001_h ($1.17e^{-038}$), 1.0	00800001_h	40000000_h (2.0); No error bit set
64-bit	8010000000000001_h ($-2.22e^{-308}$), 1.0	$80100000\ 00000001_h$	$C000000000000000_h$ (−2.0); No error bit set

Table 4 Observation in 64-bit floating-point negate/absolute instructions

Inputs	Reference outputs	Hardware output
8000000000000001_h ($-4.94e^{-324}$)	8000000000000001_h; Denormal bit set	0000000000000001_h (denormal); No error bit set
8040000000000001_h ($-1.78e^{-307}$)	0040000000000001_h	8040000000000001_h; Denormal bit and EX bit set

Analysis and Correction

Arctangent function with two arguments involves division operation (y/x). Root cause for the observation is absence of small angle approximation. The instruction is modified to accept a single argument, with small angle approximations. Division of inputs is performed by software before invoking arctangent function.

Observation in 64-Bit Negation/Absolute Instructions

While testing instructions on target, denormal bit was not set in the Status Register for some denormal inputs. The bit was wrongly set and instruction not executed for certain normal numbers, with exception bit set, as shown in Table 4.

Analysis and Correction

For double-precision numbers, single precision negate/absolute operation implementation was reused. Denormal check was implemented improperly, without considering the differences in representation of exponent between single precision (8-bit exponent) and double-precision (11-bit exponent) numbers. Negation and absolute operations are implemented by manipulating only the sign bit, and no other operation, like check for denormal numbers, is performed.

Observation Related to Denormal Numbers

The FPU design conforms to FTZ (flush to zero), which flushes denormal results to zero when the application is in the gradual underflow mode. Instructions involving denormal numbers as inputs are not processed by the FPU. Output of an operation

Table 5 Observation in denormal floating-point numbers

Operation and inputs	Reference outputs	Hardware output
Denormal + normal	Sum	Denormal; ‘D’ and ‘EX’ bits set
Normal + denormal	Sum	Normal; ‘D’ and ‘EX’ bits set
Atan2 (1.0, 00000001_h)	Π/2	Π/4; ‘D’ and ‘EX’ bits set

with denormal number input is same as the first operand. ‘D’ bit was set leading to ‘EX’ bit setting in Status Register, which is further used by software. Also, correct control flow was not ensured after comparison operation involving at least one denormal value. Typical example is given in Table 5.

Analysis and Correction

Denormal inputs are not processed by the instructions. The first input is returned as the result, with the exception also raised. The control flow takes the path corresponding to ‘Less’ in a floating-point comparison. DAZ (denormal numbers-are-zero) concept is introduced, which treats denormal inputs to floating-point instructions as zero. Also, ‘D’ bit set in such cases is prevented from setting ‘EX’ bit.

Observation Related to NaN

FPU returns qNaN as outputs for instructions involving qNaNs and sNaNs. Invalid (‘IV’) bit is set leading to ‘EX’ bit setting in Status Register, which is used by software for further error handling. Also, correct control flow is not ensured for comparison operations with NaN.

Analysis and Correction

As per IEEE754 standard, comparison between NaN and any floating-point value x yields a result NaN $\neq$ x. The FPU causes the control flow to take the path corresponding to ‘Less’ in a floating-point comparison. The design is modified to set Control Invalid bit (‘CIV’) bit in Status Register, in addition to ‘IV’ and ‘EX’ bits, so that software can take appropriate recovery action.

Observation for Truncation Instruction

FPU was returning infinity as output for truncation instruction, used to convert a double-precision floating point to single precision, for certain inputs. For example, for $3.40e^{+038}$ ($47EFFDFFFDFFFFFF_h$), hardware returned infinity ($7F800000_h$) with 'OF' and 'EX' bits set, instead of the expected value of $7F7FF000_h$.

Analysis and Correction

The result is wrong when unbiased exponent of result is 127, and mantissa needs to be rounded. The output saturation happens irrespective of the mantissa bits. Exponent of result is incremented only when mantissa overflow happens.

8.2 Non-Floating-Point Instructions

During signed and unsigned instruction tests on target, certain deviations were observed. After corrections, instructions were re-tested and outputs were as expected.

Observation 1 and Correction

In Shift Left Arithmetic 16-bit instruction, MS 16 bits were not made 0 when number of shifts is 0. MS 16 bits of the result are forced to 0 in case of no shifts.

Observation 2 and Correction

In 32-bit unsigned Multiply and Divide instructions, Negative bit in Status Register was set when MS bit of 32-bit result is 1. Unsigned Multiply and Divide instructions are corrected so as not to affect the Negative (N) flag in Status Register.

9 Conclusion

The single-core 32-bit processor design follows simple fetch-and-execute von Neumann architecture, which restricts simultaneous access of an instruction and data. The single instruction stream and single data stream processing in PROC32 facilitates exhaustive evaluations of the design with a reasonable effort. This is unlike other state-of-the-art multi-core pipelined processors, with multiple instruction streams and multiple data streams, where the verification requires focus on the instruction-level parallelism. Systematic verification and validation activities were carried out for demonstrating the adequacy and suitability of PROC32 for meeting the workloads of flight software computations. The excellent performance in the benchmarking flight applications establishes the correctness of the design decisions made. The enhanced throughput and reduction in code footprint have resulted from the architectural improvements, coupled with compiler optimisations, in PROC32, over its predecessor.

The systematic testing undergone on the hardware description language (HDL) design enables post-silicon validations to be conducted in a reduced timeframe. The test vectors already developed can be reused for assessing the functionality of the chip from the foundry. The computer system realised using the processor core fabricated as Multi-Chip Module (MCM)/Application Specific Integrated Circuit (ASIC) will undergo extensive validations in various test beds and also several qualification tests at different environmental levels. Once inducted in various launch vehicle missions of ISRO, this processor will find a place in similar safety-critical applications of other R&D organisations.

References

1. Halfhill, T.R.: The truth behind the Pentium bug. Byte, New York (1995)
2. Stallings, W.: Computer organization and architecture: designing for performance. Pearson, New York (2000)
3. Hrishikesh, M.S., Rajagopalan, M., Sriram, S., Mantri, R.: System validation at ARM enabling our partners to build better systems (2016)
4. Berkeley TestFloat. http://www.jhauser.us/arithmetic/TestFloat.html. Accessed 20 June 2019

Towards Stock Recommendation and Portfolio Management Systems Using Network Analysis

Susan George, Hiran H. Lathabai, Thara Prabhakaran, and Manoj Changat

1 Introduction

Stock market development has a discernible reputation of being instrumental in the growth of a nation and as an indicator of economic activity. Stock market can be viewed as a system that comprises fundamental units called stocks/shares and their economic transactions between various traders. The resulting systemic behaviour cannot be simply explained using some major players in the stock market. In order to understand the static and dynamic behaviour of complex system, we need a fundamental insight into the interaction of its functional units. Unfortunately the structural complexity of stock market and its dynamics makes it incomprehensible and inexplicable through paradigms lesser than network paradigm. Network analysis can be used to tackle the complexity of stock market analysis and to some extent, its properties can be used to make predictions about market dynamics. Network approach to stock market analysis commenced with the work of Mantegna [1], in which network was created using stocks as nodes and correlation between stocks as edges. Onnela et al. studied New York Stock Exchange and constructed asset graphs and asset trees based on the split-adjusted daily closure price correlations and discussed their properties and differences [2]. Vizgunov et al. [3] studied Russian stock market and Boginski et al. [4] investigated US stock market using cliques [5]. Clustering of companies within a specific stock market index, like the Dow Jones (DJ) or S&P 500, by using the Potts super paramagnetic method can be found in [6]. Study on topological stability of the China stock market correlation network was carried out by Huang et al. [7].

S. George · H. H. Lathabai (✉) · T. Prabhakaran · M. Changat
Department of Futures Studies, University of Kerala, Thiruvananthapuram, Kerala, India

M. Palesi et al. (eds.), *Second International Conference on Networks and Advances in Computational Technologies*, Transactions on Computational Science and Computational Intelligence, https://doi.org/10.1007/978-3-030-49500-8_18

Degree centrality (number of links) reflects only the direct interaction of a stock in the market. Though abundance of direct interactions and strength of direct interactions are good as indicators of influence, moderate or less abundant interactions and strengths do not imply that such stocks have low influence. In other words, metrics that could reflect direct as well indirect influence have to be used for identifying influential stocks. Motivated by this, in [8], stock market network was explored using one such metric—*Lobby index* or l index. Inspired from h-index [9] for measuring scientific research output, Korn introduced Lobby index as a centrality measure. Unlike degree, l index of a node depends on the connectedness of its neighbours too. Thus, a node with high lobby index will be having many highly connected neighbours. Therefore, lobby index of a node that reflects lobbying power of a stock can be a better influence score. As we discussed earlier, strength of interactions has also to be considered before making decisions about the influence of stocks in the market. Therefore, in this work, we investigate the influence of stocks in such a way that a node can be said to be of high influence if its weighted lobbying power is high. In other words, if it have many highly connected *strongly interacting* neighbours. For identifying the weighted lobbying power, we use weighted lobby index wl. Thus, in this work we intend to explore certain structural properties of market by modelling it as market networks and to identify the highly influential stocks in terms of weighted lobby index too. Portfolio analysis of the top stock based on l index conducted in [8] and is compared with the one done in this work using wl index, in order to identify the insights for portfolio management and to identify the suitable metric for portfolio analysis.

2 Stock Market Networks

Every year a huge volume of data is generated from stock market and it can be used to construct a network which reflects the market behaviour. For retrieval of information from stock market through network analysis we have to model the stock market as market networks with nodes that represent the stock name of each company and edges as relationships/interactions between stocks. Edges are created using the correlation values of stock price returns over a selected time frame. Mantegna [1] is credited with the first attempt of stock network analysis based on stock price correlations. Several attempts in the literature to study stock markets using network approach are briefed in Sect. 1. This work examines the daily stock prices in the US Stock Market for the year 2016 as a time series and establishes pair wise connections between stocks. For extracting the topological information underlying the stock market network, the similarities between 3781 stocks are compared for 1 year. For that, it is necessary to retain the most relevant links of the network and filter out the less relevant links. For each pair of stocks (nodes), we evaluated the cross-correlation of the time series of their daily stock price returns [4, 10]. Let $p_i(t)$ be the closing price of stock i on day t, then the log-return of price of stock i on day t, denoted by $r_i(t)$, is defined as $r_i(t) = \ln[\frac{p_i(t)}{p_i(t-1)}]$. Suppose r_i and r_j are the daily prices or price returns or trading volumes of stock i and stock j, respectively, over the period $t = 1$ to N. The similarity between two elements (without any time delay) is quantified by linear correlation, ρ_{ij} as given in [11] is $\rho_{ij} = \frac{|r_i r_j| - |r_i||r_j|}{(\sqrt{|r_i^2| - |r_i|^2}\sqrt{|r_j^2| - |r_j|^2}}$ and r_i is computed as $r_i = \frac{\sum_{i=1}^{N} r_i(t)}{N}$, where r_i is the average return of the stock i over N days,

By computing correlation between every stock with every other stock, we could obtain a matrix known as the correlation matrix. This matrix plays an important role in modern finance as it is highly sought in risk analysis and portfolio management. If the cross-correlation of the time series of the daily stock price returns for a pair of stocks is greater than a threshold (θ), we may establish a connection between the pair. In this treatment, 'links' are binary (existing or not existing). Along with the stock market networks created in this way, we also create weighted networks using the stocks and links whose weights that exceeds threshold value. Our work aims to study some unexplored properties of stock network and to determine the most influential stocks and industrial sections in the market. One of the many advantages of network analysis is that it offers ample tools for analysis and visualization, which in turn might help to mine the stock market for retrieving insightful implications. One of such tools is centrality analysis that helps to provide unique attributes that may indicate importance of each node like degree centrality, betweenness centrality [12, 13], etc. As discussed in Sect. 1, most of the works considered degree centrality as a local measure of influence. The potential of a relatively recent metric, namely l index or lobby index is explored in this work for the purpose of portfolio management. Also as discussed earlier, depending on the significance of overall effect of strength of interaction, weighted lobbying power can be used for portfolio management. A check criteria with decision indices for determining the significance of the effect of weights is also developed and is a part of the methodology.

3 Methodology

The procedure followed in this work consists of the following steps—(i) Data collection and pre-processing, correlation matrix, and network creation, (ii) structural/topological property analysis of the market network, (iii) Computation of degree and also weighted degree of each stock item, (iv) Identification of significance of the effect of weights, and (v) Computation of l index alone or l and wl indices together based on the result of step iv, i.e., if weights are of decisive importance or not. In this work, we analyse the relationship among various stock items in the US stock market from NASDAQ and NYSE. We opt to take the stock details for the year 2016 in order to analyse the latest topological scenario. Stocks which were active for the entire 1 year are considered. We use Standard Industry Classification (SIC) for industrial sections of market in which stocks belong (neither multiple sectional attribution of stocks nor change in sectional attribution of stocks is there). Data pre-processing, correlation matrix, and network creation, etc., are done with the help of Python and Matlab scripts. The structure and properties of stock networks that are constructed using cross-correlation of stock prices are dependent upon the choice of the threshold, θ. The total number of connections increases with lower values of threshold θ and as θ approaches zero, the network becomes entirely connected so that its underlying structure takes the form of a complete graph. Link between nodes i and j is created if the correlation coefficient of the corresponding stocks ρ_{ij} is greater than θ. For weighted network creation, correlation value is given as the weight of each edge. On stock market network (unweighted), structural properties of the market is retrieved through graph mining using software package Gephi [14]. Influential stock items based on direct influence in the market is identified through degree centrality and influential stock items with consideration of strength of direct influence is found out through weighted degree. Both these concepts are briefly explained as follows: Degree centrality and lobby centrality are taken separately for unweighted and weighted graphs.

3.1 Degree and Weighted Degree

Network approach presumes that the power of an individual node is dependent on their relationship or connectivity with other nodes. Different measures had been proposed in the literature for the nodes and edges in the network as well as for the network as a whole. These mostly represent the measure of potential importance or influence in the network. Degree centrality suggested by Freeman measures a node's central position according to the number of connections to the other nodes It can be computed as the marginals of the adjacency matrix. Degree is a basic indicator and often used as a first step when studying networks [15, 16]. Using the adjacency matrix, the degree of node i, represented as k_i is computed as $\deg_i = \sum_{j=1}^{N} A_{ij}$, where i is the node examined, j represents all other nodes, N is the total number of nodes, and A is the adjacency matrix, in which element A_{ij} takes value as 1 if node i is connected to node j and A_{ij}=0 otherwise. For analysing weighted networks, degree can be generally been extended as *weighted degree*, which is the sum of weights of edges associated with a node and is regarded as node strength. This is computed as $\text{wt.deg}_i = \sum_{i=1}^{N} W_{ij}$, where W is the weighted adjacency matrix, in which w_{ij} is greater than 0 if the node i is connected to node j, and the value represents the weight/strength of the edge and 0 otherwise. Since node strength takes into consideration the weights of edge, this has been the preferred measure for analysing weighted networks in works like [17] and [18]. As degree and strength can both be the indicators of the level of involvement of a node in its immediate neighbourhood, it is sensible to incorporate both these measures when studying the centrality of a node. In our case, as link weights are the correlation weights in [θ,1], weighted degree of the node will be $\leq$ degree. If weighted degree is close to degree for most of the edges, then there is a chance for weighted degree to provide same insights about the influence of stocks as that of the degree. Therefore, before proceeding to the identification of lobbying power and weighted lobbying power, the significance of overall effect of weights in the network has to be determined. This can be done in the following way.

3.2 Criteria for Identification of the Significance of Effect of Weights

We define an effect index φ_i for each stock i such that

$$\varphi_i = \frac{\text{deg}_i - \text{wt.deg}_i}{\text{deg}_i} \tag{1}$$

$\varphi_i \simeq 0$, if $\text{wt.deg}_i \simeq \text{deg}_i$. For a stock i's wt.deg_i to be exactly equal to deg_i, then all links associated with i has to be of weight = 1 and in such cases $\varphi_i = 0$.

Suppose all the links associated with i be of weight = θ, then,

$$\varphi_i = \frac{\text{deg}_i - \text{wt.deg}_i}{\text{deg}_i} = \frac{\text{deg}_i - (\theta \times \text{deg}_i)}{\text{deg}_i} = 1 - \theta$$

.

Thus, the maximum value of φ_i, $\varphi_i(\max) = 1 - \theta$. For a network with K stocks, the overall effect index $\varphi_{avg.}$ is computed as the average of all the φ_i values of K stocks as given in Eq. 2.

$$\varphi_{avg.} = \frac{\sum_{i=1}^{K} \varphi_i}{K} \tag{2}$$

If $\varphi_{avg.} \longrightarrow 0$, then weights do have a considerable effect and if $\varphi_{avg.} \longrightarrow \varphi_i(\max) = 1 - \theta$, it is not the case. For determining the significance of effect of interactions, a significance score $\mathcal{S}$ is introduced as:

$$\mathcal{S} = \frac{\varphi_{avg.}}{\varphi_i(\max)} = \frac{\varphi_{avg.}}{1 - \theta} \tag{3}$$

Now $\mathcal{S}$ takes a value in [0,1]. For decision purposes, especially in automated systems, a tolerance value, τ can be assigned to check if the effect of weights is significant or not. We recommend the value of τ as 0.05. Stringent analysts who may need more precise insights about influence through strength of interactions of stocks may set τ to lesser values like 0.01. Thus, if $\mathcal{S} \geq \tau$, overall effect of strengths is significant and we have to proceed for the computation of lobby index (l) and weighted lobby index (wl), otherwise computation of l will suffice.

3.3 Lobby Index and Weighted Lobby Index

Korn [19] proposed a general index to node centrality in 2008 which is named as lobby index. He argued that lobby index contains a mix of properties of other commonly used centrality measures. The l index or lobby index of a node x is the largest integer k such that x has at least k neighbours with a degree of at least k. A node has high lobby index if it has large number of neighbours and these neighbours again have to be of large number of neighbours. It is a better local measure than degree centrality because it represents not only the influence of that node, but the influence of its neighbours too. It requires much less time compared with global centrality measures and carries more local information. In this work we compare l-index in weighted and non-weighted stock market network. The degree is denoted by $\deg(x)$ for nodes and the l index is defined as follows. Let us consider all y_i neighbours of x so that $\deg(y_1) > \deg(y_2)\ldots$; then,

$$l(x) = \max\{k : \deg(y_k) \geq k\} \tag{4}$$

The lobby index can be adapted to a weighted network [20] as weighted network lobby index of node x, denoted as $wl(x)$ is defined as the largest value q such that node x has at least q neighbours with node strength at least q.

$$wl(x) = \max\left\{q : \text{ wt.deg}(y_q) \geq q\right\} \tag{5}$$

4 Results, Analysis, and Discussion

We aggregated daily time series data of 3781 stocks along with necessary information about every company in 2016 and a network was constructed by computing the cross-correlations among their daily returns time series. Previous attempts to correlation analysis in literature had showed how external forces across the market influence correlation of stocks [1, 6]. Individual stocks are taken as nodes and a threshold θ is set to create links. For links with $\rho_{ij} > 0$, based on the threshold, we added a link between i and j. As structural properties of market depends on cross-correlation of stock prices, the choice of the threshold θ determines the size and density of the resulting network. When the threshold θ is high, the number of edges in the network gets decreased which in turn results in low average degree, short diameter, and low average path length. We can see that the total number of connections increases with decreasing θ and as θ approaches to 0 the network's underlying structure gets closer to that of a complete graph. We selected the case where θ is high (greater than 0.7) for analysis of structural properties in order to consider only sufficiently strong interactions. Chi et al. [10] argued that properties such as scalefreeness can be found only if large value of θ is chosen. General network measures of NASDAQ stock market during 2016 for various threshold values computed were computed in [8] and influence analysis using degree and lobby index was also conducted. Portfolio analysis of stocks with highest lobby index revealed the industrial sections that are crucial drivers of the stock market dynamics. As done in [8], we have selected $\theta = 0.7$ and created stock market network and weighted stock market network in order to ensure the presence of stocks with sufficiently large interaction strengths.

As stated earlier, degree analysis and lobby analysis of the stock market were conducted in [8]. In this work, as an extension of the same, we conduct weighted degree and weighted lobby index in the weighted network. Firstly, comparison of degree and weighted degree analysis is done.

4.1 Significance of Effect of Weights

For checking whether weights (strength of interactions) have significant effect using the check criterion mentioned in Sect. 3.2, we have to calculate the effect index φ_i for all the stocks. After that, average effect index $\varphi_{avg.}$ is calculated using Eq. 2 as shown below.

$$\varphi_{avg.} = \frac{\sum_{i=1}^{K} \varphi_i}{K} = 0.20477$$

As $\theta = 0.7$, $\varphi_i(\max) = 1 - 0.7 = 0.3$. Now, the significance of effect of weights can be identified using Eq. 3 as given below.

$$\mathscr{S} = \frac{\varphi_{avg.}}{\varphi_i(\max)} = \frac{\varphi_{avg.}}{1-\theta} = \frac{0.20477}{0.3} = 0.68256$$

Now, the check criteria is $\mathscr{S} > \tau$. Even with $\tau = 0.05$, the criterion is satisfied and therefore weights (strength of interactions) have a significant effect. Therefore, identification of lobbying power with the consideration of strength of interactions (weighted lobbying power) is highly necessary and can be of decisive use in portfolio management and stock recommendation applications.

4.2 Lobby and Weighted Lobby Analysis

In [8], portfolio analysis of stocks with highest lobbying power is done with the help of l index or lobby index. Highest value was found to be 341 and there were 115 such stocks. While analysing the stocks with the top lobby value (341), it is found that there are 84 stocks from Major Banks, 8 from savings institutions, 3 each from Property-Casualty, transportation services, investment bankers/brokers/services, and 2 each from Engineering & construction, EDP services section. Industrial machinery/Components, Industrial specialties, Construction/Ag. Equipment/Trucks, Telecommunication equipment, Investment managers, Professional services, Major chemicals, and Life insurance have one stock each. Now, the second highest lobby value is 340 and there are 52 stocks in it. Though influence of high lobbying stocks in the market is undisputed, stock recommendation and/or portfolio management systems require a diligent mechanism of imposing a cutoff/filter to select most suitable ones. Apart from the fact that lobby value does not reflect the strength of interactions with stocks in its range of influence, the choice of cutoff value that determines the inclusion of other investment options (that offers more or less the same potential benefits) is also difficult with lobby index. While doing this it is also to be noted that, suggestion of too many options is not advisable. In this case, portfolio obtained with top lobby value might create an impression that apart from Banking, finance, and insurance, other sections are of much less importance with less stock items to be considered for investment. On a structural point of view, as k-core [21] gives the induced subgraph of the network such that all the stocks have at least k connections, with increase of k, analyst can focus on more denser part of the network. With highest value of k, one could find the densest part of the network. Degree of Occupancy (% of representation) of highest lobbying stocks ($l = 341$) in most densest part reflects the ability of highest lobby index to provide neither too low nor too much investment options, i.e., as a metric for portfolio management and stock recommendation. For our stock market network, maximal core is obtained at $k = 286$, consists of 324 stocks and 51,255 edges. Important part of the distribution list of various top lobbying stocks in the maximal core is given in Table 1 and maximal core showing top lobbying stocks ($l = 341$) is (shown in Fig. 1a).

From Table 1, we can find that only 35.49% of the maximal core is occupied by the highest lobbying stocks. This indicates that, despite being an essential indicator for portfolio management and stock recommendation, for providing better solutions to analysts or users, need for improvement to lobby index based approach is felt. As the strength of interactions is found to have a significant effect on the market, portfolio analysis using weighted lobby index (wl index) might retrieve suggestive investment options in industrial sections that occupy the first few positions in the portfolio chart. Now, we found out weighted lobby values of all the 1709 stock items in the market using weighted network. Top weighted lobby value is found to be **295**. About 198 stocks are of having $wl = 295$. Portfolio chart representing the section wise distribution of stocks with $wl = 295$ is shown in Fig. 2. Might of industrial sections like Major banks, savings institutions, and investment bankers/brokers/service in driving the market is more apparent from Fig. 2. It can also be seen that, sections like life insurance, industrial machinery/components, etc., do not seem too much unimpressive anymore as at least 4 stock recommendations in each of these sections can be made. Now the presence of stocks with various wl indices in maximal core (shown in Fig. 1b) is studied and Table 1 is created and analysed.

From Table 1, it can be seen that stocks with $wl = 295$ form the 61.11% (198 out of 324) of the most densest part of the network. This is pretty good of a representation (also apparent from visualization of the maximal core in Fig. 1b),

Table 1 Distribution of stocks according to lobby values and weighted lobby values in maximal core ($k = 286$) according to decreasing order of their representation

	Lobby values in max.core		Weighted lobby values in max.core	
Sl. No.	l-index	% of representation	Wl-index	% of representation
1	341	35.49	295	61.11
2	340	16.05	294	8.33
3	338	5.25	293	4.32
4	335	4.63	284	3.7
5	339	4.32	287	2.78
6	337	2.78	288	2.78
7	329	2.47	292	2.47
8	336	1.85	286	2.16
9	325	1.85	289	1.85
10	317	1.85	290	1.85
11	319	1.54	285	1.54
12	330	1.23	291	1.54
13	326	1.23	282	1.23
14	333	1.23		
15	327	1.23		
16	334	1.23		

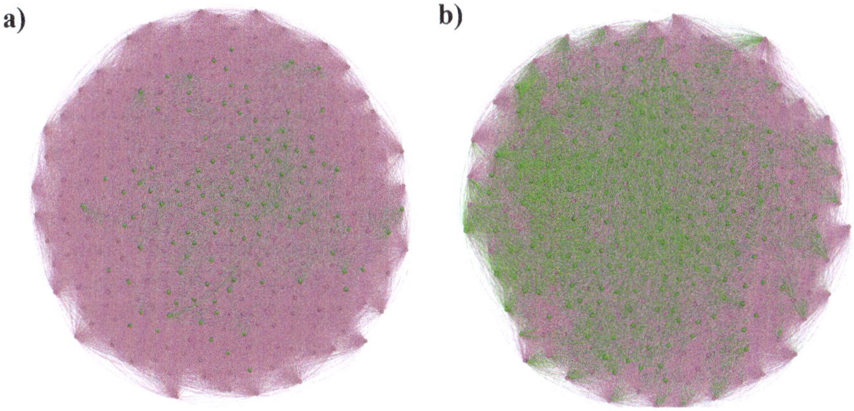

Fig. 1 (**a**) Top lobby stocks ($l = 341$) (green vertices) in maximal core (**b**) Top weighted lobby stocks ($wl = 295$) (green vertices) in maximal core

and is sufficient for a prompt recommendation as preferred by most of the users and analysts/consultants. Now analysts who use top wl for portfolio analysis can turn their attention to stock items for stock recommendation. Stocks with $wl = 295$ and their respective industrial sections (sections consisting of at least 2 stocks) are displayed in Table 2. All these results point towards the immense potential of weighted lobbying power and lobbying power of stocks for applications such as portfolio management and stock recommendation.

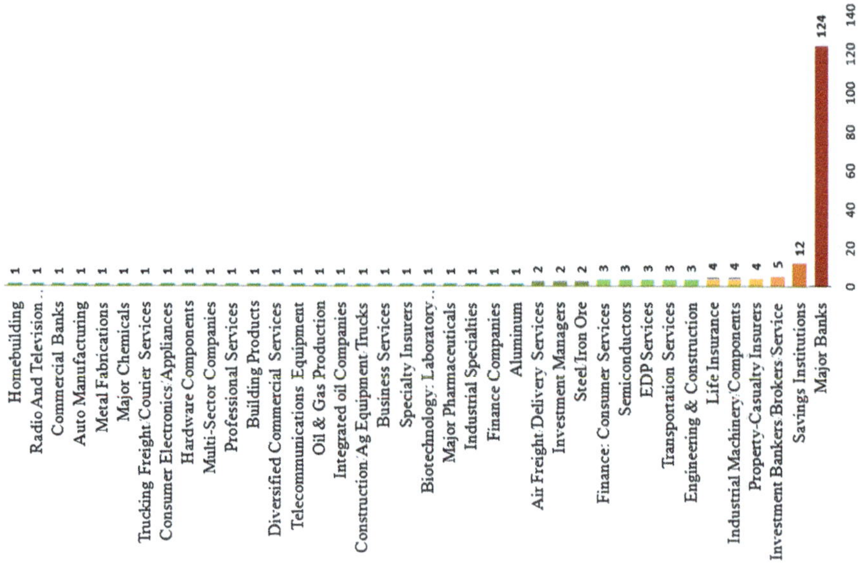

Fig. 2 Section wise distribution of top weighted lobby ($wl = 295$) stocks (sections as per SIC classification)

5 Conclusion and Future Work

Stock market is regarded as a complex system due to the its inherent complex dynamics. A prominent reason for this complexity is the inter-dependency of various stocks on others. Due to this complexity, various decisions such as investment and trade are risk prone unless done diligently. Thus, need for data mining to unveil important insights about the market dynamics for various applications including stock recommendation and portfolio management has increased manifold. As networks are found to exhibit the capability to model complex systems and inherent relationships of its various components, they can be effectively used for stock market mining too. Previous studies have suggested that density of the network itself hints the possibility of heavy dependency of stocks on other stocks. Most of the studies related to stock market based on network approach were oriented towards identification of various key players in the market based on their dependency to more number of stocks. A few of them advocated that strength of relationships/dependencies has to be used for better understanding. To our knowledge, none of the previous attempts in literature has stipulated any criteria or given any thumb rules for the benefit of analysts to assess how much significant a role the effect of weights, i.e., effect of strength dependency plays on the market. This gap is attempted to be filled in this study with the proposition of a significance score $\mathscr{P}$ and a check criteria. Once the analyst finds that strength of relationships has a significant effect, weighted network analysis can be chosen to extract more

Table 2 Stocks with highest weighted lobby value ($wl = 295$) and their respective industrial sections

Industry	Ticker symbol
Major banks	AMNB, AROW, ASB, BAC, ANF, BANR, BBT, BDGE, BHB, BHBK, BHLB, BMTC, BNCN, BOH, BPFH, BSRR, BUSE, CACB, CBF, CBU, CCBG, CCNE, CFG, CHCO, CHFC, CMA, CNBKA, CNOB, COBZ, COLB, CSFL, CTBI,CUBI, CUNB, CVBF, CZNC, EBSB, EFSC, EWBC, FBNC, FBNK, FCBC,FCF, FCNCA, FFBC, FFKT, FHN, FISI, FITB, FMBI, FNB, FNLC, FRME, FULT, GABC, GBNK, GNBC, GSBC, GWB, HAFC, HBAN, HBCP, HBHC, HEOP, HFWA, HTH, HTLF, IBCP, IBOC, INDB, JPM, KEY, LBAI, MBVT, MBWM, MOFG, MSFG, MTB, NBHC, NBTB, NTRS, OKSB, ONB, OSBC, PCBK, PEBO, PFBC, PNC, PPBI, PSTB, QCRH, RBCAA, RF, SASR, SBCF, SGBK, SIVB, SRCE, STBA, STL, TBK, TCB, TCBI, TCBK, THFF, TMP, TOWN, TRMK, TRST, TSC, UBNK, UBSH, UCBI, UCFC, UMBF, UVSP, WABC, WASH, WBS, WSBC, WTBA, WTFC, YDKN, ZION
Savings institutions	AAF, AMTD, BNCL, BRKL, CFFN, ETFC, HTBI, LTXB, NFBK, OCFC, PFS, WAFD
Investment bankers/Brokers/Service	COWN, ENH, GS, MS, PJC
Industrial machinery/Components	AIMC, BRKS, MSM, VECO
Life insurance	LNC, NWLI, PRU, VOYA
Property-casualty insurers	AFG, AGO, ESGR, GBLI
EDP services	MENT,PDFS,PEGA
Engineering & Construction	AGX, EME, PWR
Finance: consumer services	NNI, SLM, SYF
Semiconductors	MSCC, STM, TSRA
Transportation services	CLC, CSX, ERA
Air freight/Delivery services	LUV, UAL
Investment managers	EVR, MC
Steel/Iron ore	CMC, STLD

insights from the network. Also relationship with more number of stocks need not always be taken as a sole indicator of influence. Some of the recent works suggest that relationship to stocks which have high relationship with others can be regarded as a better way to express local dominance. The principle of lobby index is based on this and hence we used lobby index for identification of key players based on their relations with other key players. Players in industrial sections such as major banks, savings institutions, investment bankers/brokers/service, etc., are found to have high lobbying power and thereby likely to exercise relatively high control over the market. Taking into account the effect of strength of the relationships along with the lobbying power, weighted lobbying power can be identified and give much clearer picture about the potential of some sections that might have been underestimated if lobbying power is considered as a sole indicator of influence. Thus, in cases where strength of interactions plays a significant role, weighted lobbying power could make better resolution of stock influence or industrial section influence in the market. A comparison of portfolio analysis based on lobbying power and weighted lobbying power of stocks indicated that if the analysts are looking for a metric with ability to reveal the industrial sectional portfolio of stocks that well represent the core of the market with minimum analytic effort, weighted lobbying power is the best option. This contribution further enriched the literature of complex networks and intelligent systems. Network based data mining approach in this work is found to reveal more important inputs for decision-making than its predecessor and hence could ensure an empowered recommendation and portfolio management system.

Community detection methods and their usage for portfolio analysis is intended to be pursued as an extension of this topic. Dynamic study of the structural properties of the market using network approach is investigation worthy. Other complex network parameters and their implications on stock market can also be further investigated. Statistical mechanics of stock market networks with special emphasis on properties of lobby and weighted lobby index and their empirical validation on different stock markets is also considered as a future endeavour.

References

1. Mantegna, R.N.: Hierarchical structure in financial markets. Eur. Phys. J. B-Condensed Matt. Compl. Syst. **11**(1), 193–197 (1999)
2. Onnela, J.P., Chakraborti, A., Kaski, K., Kertesz, J., Kanto, A.: Dynamics of market correlations: Taxonomy and portfolio analysis. Phys. Rev. E **68**(5), 056110 (2003)
3. Vizgunov, A., Goldengorin, B., Kalyagin, V., Koldanov, A., Koldanov, P., Pardalos, P.M.: Network approach for the Russian stock market. Comput. Manag. Sci. **11**(1–2), 45–55 (2014)
4. Boginski, V., Butenko, S., Pardalos, P.M.: Statistical analysis of financial networks. Comput. Stat. Data Analy. **48**(2), 431–443 (2005)
5. Harary, F.: Graph Theory. Addison-Wesley, Boston (1969)
6. Kullmann, L., Kertész, J., Mantegna, R.N.: Identification of clusters of companies in stock indices via Potts super-paramagnetic transitions. Phys. A Stat. Mech. Appl. **287**(3–4), 412–419 (2000)

7. Huang, W.Q., Zhuang, X.T., Yao, S.: A network analysis of the Chinese stock market. Phys. A Stat. Mech. Appl. **388**(14), 2956–2964 (2009)
8. George, S., Changat, M.: Network approach for stock market data mining and portfolio analysis. In: 2017 International Conference on Networks & Advances in Computational Technologies (NetACT) (pp. 251–256). IEEE, Piscataway (2017)
9. Hirsch, J.E.: An index to quantify an individual's scientific research output. Proc. Nat. Acad. Sci. **102**(46), 16569–16572 (2005)
10. Chi, K.T., Liu, J., Lau, F.C.: A network perspective of the stock market. J. Empir. Finance **17**(4), 659–667 (2010)
11. Cohen, J., Cohen, P., West, S.G., Aiken, L.S.: Applied multiple correlation/regression analysis for the behavioral sciences (3. press). Taylor & Francis, Milton Park (2003)
12. Freeman, L.C.: Centrality in social networks conceptual clarification. Soc. Netw. **1**(3), 215–239 (1978)
13. Freeman, L.C.: A set of measures of centrality based on betweenness. Sociometry **40**, 35–41 (1977)
14. Bastian, M., Heymann, S., Jacomy, M.: Gephi: an open source software for exploring and manipulating networks. In: Third International AAAI Conference on Weblogs and Social Media (2009)
15. Freeman, L.: The Development of Social Network Analysis. A Study in the Sociology of Science, vol. 1. Empirical Press, North Charleston (2004)
16. McPherson, M., Smith-Lovin, L., Cook, J.M.: Birds of a feather: homophily in social networks. Ann. Rev. Sociol. **27**(1), 415–444 (2001)
17. Barrat, A., Barthelemy, M., Pastor-Satorras, R., Vespignani, A.: The architecture of complex weighted networks. Proc. Nat. Acad. Sci. **101**(11), 3747–3752 (2004)
18. Opsahl, T., Colizza, V., Panzarasa, P., Ramasco, J.J.: Prominence and control: the weighted rich-club effect. Phys. Rev. Lett. **101**(16), 168702 (2008)
19. Korn, A., Schubert, A., Telcs, A.: Lobby index in networks. Phys. A Stati. Mech Appl. **388**(11), 2221–2226 (2009)
20. Zhao, S.X., Rousseau, R., Fred, Y.Y.: h-Degree as a basic measure in weighted networks. J. Informetr. **5**(4), 668-677 (2011)
21. Seidman, S.B.: Network structure and minimum degree. Soc. Netw. **5**(3), 269–287 (1983)

A Survey on VANETs Routing Protocols in Urban Scenarios

Anishka Abraham and Rani Koshy

1 Introduction

VANET [1] stands for vehicular ad-hoc network. It falls under the category of mobile ad-hoc network (MANET). Ad-hoc network is an infrastructure less network, therefore nodes connect to the network randomly. Moving vehicles create a network by becoming network nodes or routers. There are also fixed nodes in the network called as Road Side Unit (RSU) which can be street lights or traffic signal. Road safety and better comfort services are the reasons for the evolution of VANETS.

As shown in Fig. 1 [2], there are three types of communication in VANETs. They are:

1. Inter-vehicle communication or vehicle to vehicle communication also called as V2V communication. It aims at giving road safety.
2. Vehicle-to-infrastructure communication also called as V2I communication. It occurs between a vehicle and RSU. It focuses more on providing better traffic control.
3. Inter-road side communication. It takes place between RSUs or with RSU and base station. It aims for infotainment (information and entertainment) applications by accessing the Interne.

A. Abraham (✉) and R. Koshy
College of Engineering Trivandrum, Thiruvananthapuram, India

M. Palesi et al. (eds.), *Second International Conference on Networks and Advances in Computational Technologies*, Transactions on Computational Science and Computational Intelligence, https://doi.org/10.1007/978-3-030-49500-8_19

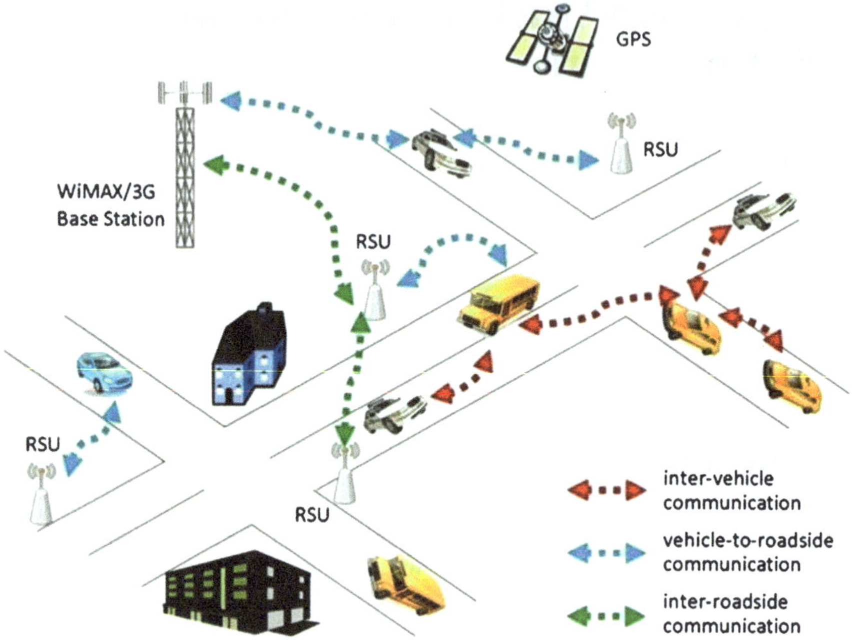

Fig. 1 VANET explanation

1.1 VANETs Working

VANET communications use a 5.9 GHz Dedicated Short Range Communications (DSRC) [1] band. DSRC produce a very low cost communication. It is similar to WiFi and works along with GPS. The safety control messages are broadcasted periodically in every 100–300 ms and cover 1000 m range of vehicles. Wireless Access in Vehicular Environments (WAVE) protocol, an IEEE 802.11p standard is used by DSRC. There are three components for each node—Event Data Recorder (EDR), On Board Unit (OBU), and Tamper Proof Device (TPD). A wireless transmitter and receiver unit is attached to each OBU which enables the vehicular communication in the network. The secret information about vehicle is stored in EDR such as its position, speed, transmitted and received messages, trip details, etc. Secure data such as driver identity and cryptographic material is stored in TPD. The cryptographic operations are done by the cryptographic material by signing, processing, and verifying the exchanged messages.

Sensors are also attached in VANET which aims in gathering data for processing it or sharing it. The information in a beacon message includes vehicle's details such as its current position, speed, its acceleration, transmission state, steering wheel angle, vehicle control information (brake status, path history), etc.

1.2 Characteristics of VANETs

Characteristics [1] based on network topology and communication mode are:

1. High mobility of nodes: Unlike any other network, VANET uses moving cars as nodes which makes high mobility of nodes in network.
2. Rapidly changing network topology: In VANET, vehicles travel continuously and that too at very high speed. Thus the vehicle's position changes repeatedly leading to frequent topology changes.
3. Short time of connection: The connection time is very short because of the vehicles high speed.
4. Wireless communication: In VANET, the communication is of wireless communication since the nodes are moving vehicles.
5. Frequent exchange of information: Because of the ad-hoc nature of VANET, frequent information exchange occurs between vehicles and also with RSUs.

1.3 Security Requirements of VANET

Secure communication of vehicular nodes is achieved by privacy and security [3]. They are:

1. Source authentication: It prevents outsider's attacks by checking legitimacy of each node.
2. Message integrity: This verifies that the messages are received without any alteration at receiver side.
3. Identity privacy preservation: Pseudo identities, which are generated from real identities must be used instead of real identities so that it is kept secretly.
4. Traceability: The real identity of the attacker vehicle should be traced by trusted authority.

1.4 Applications of VANETs

There are safety applications and non-safety applications. Reduction of road accidents and improvement of road safety are achieved by safety applications [1]. Non-safety applications aim for infotainment (information and entertainment) applications.

Safety applications include traffic management, collision avoidance, post-accident warning, curve speed warning, traffic signal violation warning, work zone warning, emergency vehicles warning, and road condition warning applications.

Non-safety applications include payment services, music and video sharing, nearest restaurants, weather, petrol stations, hotels, internet services, games, parking applications, emails.

2 Routing Protocol Classification

Many routing protocols have been evolved for communication in VANETS. Routing is an arduous task because of the frequent change in network topology that occurs because of the high mobility of vehicles. VANET routing protocols are divided into position based, topology based, cluster based, geocast and broadcast based protocols [4]. Classification of routing protocols is shown in Fig. 2.

2.1 *Topology Based Routing*

Routing table stores source to destination data. It is classified into proactive and reactive routing protocols [5].

Proactive Routing Protocol All the routing information is stored in the routing table. Next-hop node information is contained in each entry of the routing table. E.g. Destination Sequence Distance Vector (DSDV).

Reactive Routing Protocol It is also called as on demand routing protocol. Understanding of the complete network topology is not required. Source sends a request message when it needs to connect with the destination. Any intermediate

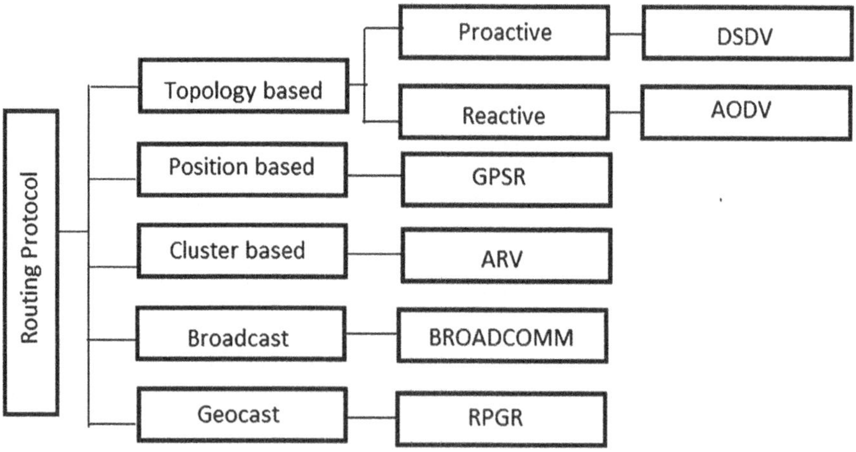

Fig. 2 Routing protocols classification

node that is present on this route receives this request message and sends a route acknowledgment message using unicast communication back to the source node. These are appropriate for large sized ad-hoc networks. E.g. Ad-hoc On demand Distance Vector (AODV).

2.2 Position Based Routing Protocol

Geographic location information of nodes is used [6] for the routing process. The source node sends information based on the geographic location of the nodes. Here, every vehicle has Global Positioning System (GPS) which helps to identify the position of other vehicle. This routing scheme uses GPS, street maps, and OBU for data transmission. E.g. Greedy Perimeter Stateless Routing (GPSR).

2.3 Cluster Based Routing Protocol

Here the total geographic area is divided into different clusters. Groups are formed by vehicles that are closer. Every cluster has a cluster head and is responsible for intra-cluster and inter-cluster communication. E.g. Aggregate Relative Velocity (ARV).

2.4 Broadcast Routing Protocol

Here the messages are broadcasts among vehicles. It uses the technique called flooding. Flooding guarantees that messages will be received by every nodes in the network. It is easy to implement and works comparatively well for a small number of nodes. It focuses on sharing safety information such as traffic, weather, road conditions, and emergency information.

2.5 Geocast Routing Protocol

Geocast routing protocol is a multicast protocol based on location. The messages are delivered in a specific geographic region from source to destination. Selective flooding technique is used here. E.g. Road Perception Based Geographical Routing Protocol (RPGR). Comparison of the routing protocols is listed in Table 1.

There are various topological and geographical routing protocols implemented in VANETs urban scenarios [6, 7]. Compared to topological based routing, geographical based routing is most suitable for VANETs because of the dynamic network topology mainly in urban scenarios [8–10].

Table 1 Routing protocols classification

Classification of routing protocol	Merits	Demerits
Proactive routing protocol	The routes are predefined Route discovery does not take place The delay is avoided	Routing table update happens frequently due to dynamic topology Unused nodes maintenance leads to high network load. Bandwidth utilization is high
Reactive routing protocol	Decreased network traffic and bandwidth is saved Routing table helps to maintain the route from sender to receiver	Route searching delay is high Network flooding causes suspension of nodes
Position based routing protocol	No maintenance of routes and routing tables Avoidance of route discovery delay	Need GPS, location and position finding service Suitable for urban scenarios
Cluster based routing protocol	It has high scalability Best for large networks	Cluster formation increases delay
Broadcast based routing protocol	Implementation easier Best for small network	Large network size causes high bandwidth utilization Flooding messages are received at each node at the same time creating congestion and collision
Geocast based routing protocol	Reduced network collision	Need for GPS, location and position finding services Satellite signal does not reach the tunnel

3 Types of Geographic Based Protocol

The first protocol came is Greedy Perimeter Stateless Routing (GPSR). Routing reliability is affected in GPSR [11] because of Greedy Forwarding (GF) which is its main forwarding strategy. In this protocol, the next-hop selection is done by choosing the neighbor of the sender node that has minimum distance to destination. To calculate shortest distance Euclidean distance is used. This next-hop node is normally placed at the communication range border because of which the packet loss ratio increases. Furthermore, in the case of transmission failure which causes re-transmissions at the Medium Access Control (MAC) layer consumes large amount of the network bandwidth. In addition to it, in relatively dense networks, congestion may occur because of the selection of next-hop nodes by not considering their load of traffic and bandwidth availability. This will result in increased chance of collisions, packet losses, and packet delay.

Many improvements on GPSR came which used vehicles mobility, distance, link quality as metric parameters. Some of them are CLPWR, DGF-ETX. Multiple metrics are combined in this protocol by using the additive composite functions which may contradict each other and may not be accurate making it arduous to optimize next-hop selection.

1. CLWPR [12]—A Novel Cross-Layer Optimized Position Based Routing Protocol for VANETs proposed by K. Katsaros et al. uses the prediction of the node's navigation information and position to improve the efficiency of routing protocol. There is no route discovery, instead next hop is selected on the basis of minimal weight.

   It is a unicast, multi-hop protocol. Every node periodically broadcasts 1-hop "HELLO" messages (called beacons). The beacon messages include information of vehicle such as velocity, position, and heading. In CLWPR, the geographical distance is not calculated, instead the calculation is done as the actual distance that a vehicle requires to travel to reach destination.

   Compared with GPSR, it performs significantly better in the urban environment in terms of packet delivery ratio (PDR) and end-to-end delay (E2ED). But it does not consider link quality of the network while selecting next hop.
2. DGF-ETX—An Enhanced Directional Greedy Forwarding for VANETs Using Link Quality Estimation [13] proposed by Ohoud Alzamzami et al. uses a link quality estimation metric, Expected Transmission Count (ETX). ETX is the probable number of data transmissions, including re-transmissions, that is needed for sending a packet over that link. Forward delivery ratio (d_f) is the probability that recipient receives a data packet successfully. Reverse delivery ratio (d_r) is the probability that the acknowledgement (ACK) packet is received successfully by the sender.

$$\mathrm{ETX} = \frac{1}{d_{\mathrm{f}} \times d_{\mathrm{r}}} \tag{1}$$

   Geographic routing does not consider link quality to select the next-hop node to reach destination. This protocol added link quality also in selecting the route. It reduces the packet loss ratio while the total network throughput and Packet Delivery Ratio (PDR) are increased.

   Recently, fuzzy logic system which is an artificial intelligence method is integrated into the network and routing protocols design. This is developed because geographic protocol aims at combining multiple parameters using the additive composite functions which may contradict each other and may not be accurate, making it difficult to select best next hop. Fuzzy logic uses fuzzy rules and input membership functions to make reasonable decisions in selecting next hop that works the way the human brain functions.

   Fuzzy logic systems analyze different metrics using an intelligent mechanism. They are used in many protocols and wireless communication systems in VANETs for improving the decision-making. It is also added to geographic routing protocols in VANETs like SRR, AFMADR, FL-DGR.

3.1 Geographic Routing Protocol Based on Fuzzy Logic

1. SRR—Fuzzy logic-assisted geographical routing over vehicular ad-hoc network [14] proposed by K. Z. Ghafoor et al. It is a Stability and Reliability aware Routing (SRR). It selects from the candidate forwarders the best next hop, by exploiting fuzzy logic systems. In order to measure signal attenuation, the distance of a neighbor from the sender is used. Progress of the packet towards the destination is ensured by the relative direction. It works well in a highway scenario since the vehicle's movement is restricted by the road network and hence it is not applicable in an urban scenario.

   Routing Metrics It uses distance and relative direction as routing metrics. Tracking the distance and tracking of the relative direction is done by using the positions of the vehicles.

   Decision-Making System Design Using Fuzzy Logic It performs fuzzification of inputs and outputs. The two metric parameters are mapped to membership functions. The membership functions for distance are {Far, Intermediate, Close}. The membership functions for direction are {Less directed, Mild directed, More directed}

   Next step is done by fuzzy inference engine. It uses IF-THEN rules to find the fuzzy output. And finally it performs defuzzification to convert fuzzy output to a numerical fuzzy weight and select the next hop with minimum fuzzy weight node.

   It uses distance and direction as metric for fuzzy logic. Compared with GPSR, it gives better packet delivery ratio. But it does not consider network bandwidth and link quality.

2. AFMADR—Adaptive fuzzy multiple attribute decision routing in VANETs [15] proposed by Gen Li et al. It works based on four parameters—distance, road density, direction, and map location.

   The four key attributes have been selected to characterize candidate vehicles. The attributes are:

   - Distance: The Euclidean distance between the candidate and the final destination vehicle.
   - Direction: The direction of packet carrier and the candidate.
   - Density: Density of the vehicle on the road that the candidate runs.
   - Dead end: The localization of the candidate, running on a common road or in an intersection or dead-end road.

   A 9-point fuzzy scale is chosen in this protocol. It then calculates fuzzy performance score. The utility of the candidate vehicle is calculated. The node with highest fuzzy weight is chosen as the best next hop. It can perform well in terms of the lowest delivery delay and the highest delivery ratio. But the congestion in the network is increased. A higher congestion increases the probability of packet losses and the packet delays.

3. FL-DGR—Fuzzy Logic Based Directional Geographic Routing [16] proposed by Ohoud Alzamzami et al. is a unicast routing to select next hop. Distance, link quality, direction, and available bandwidth are used as routing parameters. For calculating link quality, the ETX metric is used. The forwarding node with more available bandwidth, highly reliable links, and getting closer to the destination with respect to direction and distance is selected. It also utilizes a carry-and-forward mechanism when there is non-uniform node distribution and network dis-connectivity.

   Neighbors Evaluation Criteria Initially a random node is taken as source and other as destination. Each node sends a HELLO packet at regular interval and updates neighbor table.

   It calculates distance to destination on the map, link quality estimation using ETX, direction in relation to destination, and available bandwidth estimation using achievable throughput metric (ATM) for each of the nodes in the neighbor table at each node.

   Next-Hop Selection Based on Fuzzy Logic Decision System First it performs fuzzification of each metric values. Distance metric is {VeryClose, Close, Intermediate, Far, VeryFar}, ETX metric is {Good, Average, Bad}, direction metric is {MovingAway, GettingCloser}, ATM metric is {Low, Medium, High}. Next step is rule-base and inference mechanism. Fuzzy output represents the weight of each neighbor and IF-THEN rules is used to map the fuzzy values into it. Weight metric is defined as {Perfect, Acceptable, Unpreferable, VeryGood, Good, VeryBad, Bad, Worst}. The final step is defuzzification. A crisp numerical value is produced after defuzzification with respect to output membership functions. The node having the least fuzzy weight is taken as the next hop.

   It performs well in terms of end-to-end delay, packet delivery ratio, and total network throughput. The comparison of each protocol is mentioned in Table 2.

4 Performance Comparison and Discussions

The geographic based routing protocol and fuzzy logic based routing protocol performances are compared using the quality of service parameters such as Packet Delivery Ratio (PDR), total network throughput, and end-to-end delay. CLWPR distance only and distance+direction is used for comparison. CLWPR distance only used single metric, distance as parameter. CLWPR distance+direction uses both distance and direction as metric parameter.

Table 2 Comparison of routing protocols

Proposed protocol	Protocols compared	Author	Year	Quality of service parameters	Issues/Future
CLWPR	GPSR	Konstantinos Katsaros Rahim Tafazolli Mehrdad Dianati	2011	PDR E2ED	Does not consider link quality
DGF-ETX	GF DGF	O. Alzamzami I. Mahgoub	2016	PDR Throughput Delay	Considers adaptive HELLO protocol that varies the HELLO generation rate according to vehicles' mobility and density
SRR	GPSR	K. Z. Ghafoor	2012	PDR Average packet delay Control overhead	Lesser reliability of the data packet transmission Not suitable for urban scenarios
AFMADR	SRR GPSR	Gen Li1 Maode Ma2 Yantai Shu	2015	PDR E2ED Average hops	Increase in the congestion in the network Increase in the packet delays
FL-DGR	CLWPR SRR AFMADR DGF-ETX	Ohoud Alzamzami Imad Mahgoub	2019	PDR E2ED Throughput	Technique to reduce network congestion by considering number of nodes in the network

4.1 Simulation Setup

The protocols are simulated in NS-3.25 network simulator [16]. A Manhattan grid is used for simulating urban network. To generate random movement of vehicles, Simulation of Urban Mobility (SUMO) is used. The vehicle speed ranges between 20 and 40 mph. The wireless technology used is IEEE802.11p and the HELLO packet interval is set to 1 s.

4.2 Discussion

The network density effect on PDR, delay, and throughput is given in Figs. 3, 4, and 5, respectively [16]. From Fig. 3, PDR gets increased when the network density increases for FL-DGR and its PDR is high compared to other protocols. The least PDR is for SRR. For AFMADR only the PDR gets decreased when density increases. From Fig. 4, end-to-end delay is lowest for FL-DGR when network

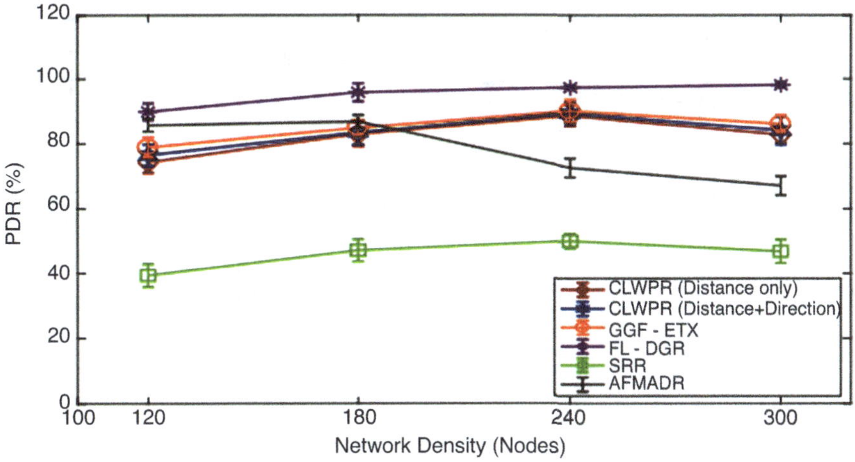

Fig. 3 Network density effect on PDR

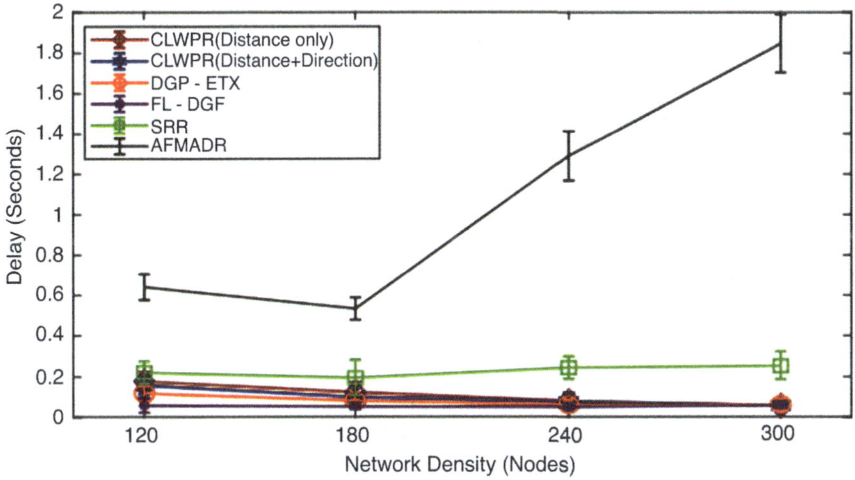

Fig. 4 Network density effect on average end-to-end delay

density is lesser and the delay becomes same for CLWPR, DGF-ETX, and FL-DGR when network density is higher. The delay gets increased and is highest for AFMADR. From Fig. 5, total throughput is highest for FL-DGR and gets increased as the network density increases. Except AFMADR, all the protocols throughput gets increased when density increases. The least throughput is for SRR.

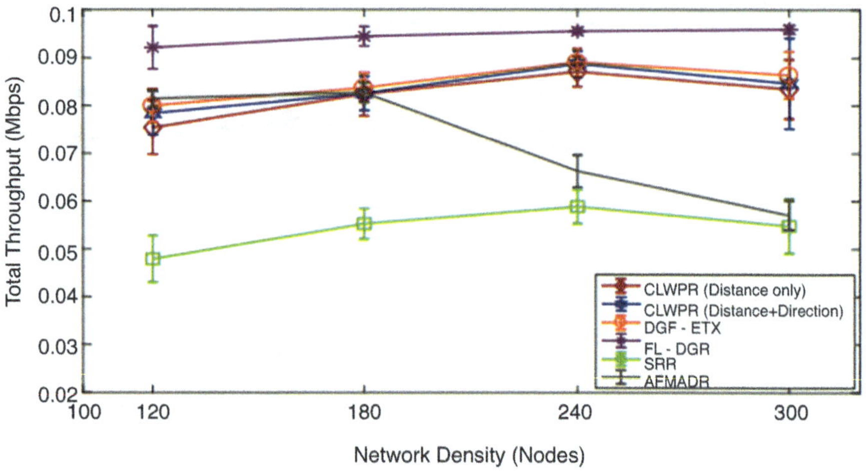

Fig. 5 Network density effect on total throughput

5 Conclusion and Future Work

Here, several geographic based routing protocols that are suitable for VANET environment in urban scenarios are covered. For urban scenarios, fuzzy logic based geographical routing is more appropriate considering end-to-end delay, latency, and packet delivery ratio since it uses an artificial intelligence in selecting best next hop. Compared to the protocols—CLWPR, SRR, DGF-ETX, AFMADR, and FL-DGR, FL-DGR is better according to PDR, end-to-end delay, and total throughput.

Every vehicle broadcasts safety messages at regular interval of time. If the network density increases, it can increase the load in the network and hence leads to network congestion. Apart from that, if the message broadcasting time interval is lesser, then there is larger network load and higher network congestion. Thus, it can affect the total throughput of the network. Hence as a future work, the network congestion can be reduced by reducing the number of packets transmitted by increasing the message packet interval when network density increases. Also, more metric parameters such as vehicle speed can be added as routing metric parameter.

References

1. Ghori, M.R., Zamli, K.Z., Quosthoni, N., Hisyam, M., Montaser, M.: Vehicular Ad-hoc network (VANET)-review. In: 2018 IEEE International Conference on Innovative Research and Development (ICIRD) (2018). https://doi.org/10.1109/TITS.2018.2867177
2. Explanation of VANET. https://www.researchgate.net/figure/A-typical-urban-VANET-scenario_fig2_252028708

3. Mishra, R., Singh, A., Kumar, R.: VANET security-issues, challenges and solutions. In: 2016 IEEE International Conference on Electrical, Electronics, and Optimization Techniques (ICEEOT) (2016). https://doi.org/10.1109/ICEEOT.2016.7754846
4. Bengag, A., Boukhari, M.E.: Classification and comparison of routing protocols in VANETs. In: 2018 International Conference on Intelligent Systems and Computer Vision (ISCV) (2018). https://doi.org/10.1109/ISACV.2018.8354025
5. Devangavi, A.D., Gupta, R.: Routing protocols in VANET–A survey. In: 2017 International Conference On Smart Technologies For Smart Nation (SmartTechCon) (2017). https://doi.org/10.1109/SmartTechCon.2017.8358362
6. Lochert, C., Hartenstein, H., Tian, J., Fussler, H., Hermann, D., Mauve, M.: A routing strategy for vehicular ad hoc networks in city environments. In: IEEE IV2003 Intelligent Vehicles Symposium. Proceedings (Cat. No.03TH8683) (2003). https://doi.org/10.1109/IVS.2003.1212901
7. Sutariya, D., Pradhan, Dr.S.: Evaluation of routing protocols for VANETs in city Scenarios. In: 2011 International Conference on Emerging Trends in Networks and Computer Communications (ETNCC) (2012). https://doi.org/10.1109/ETNCC.2011.6255911
8. Lochert, C., Mauve, M., Fussler, H., Hartenstein, H.: Geographic routing in city scenarios. ACM SIGMOBILE Mobile Comput. Commun. Rev. **9**(1), 69–72 (2005). https://doi.org/10.1145/1055959.1055970
9. Jerbi, M., Senouci, S.M., Rasheed, T., Doudane, Y.G.: Towards efficient geographic routing in urban vehicular networks. IEEE Trans. Vehicular Technol. **58**(9), 5048–5059 (2009). https://doi.org/10.1109/TVT.2009.2024341
10. Fonseca, A., Vazao, T.: Applicability of position-based routing for VANET in highways and urban environment. J. Netw. Comput. Appl. **36**(3), 961–973 (2013). https://doi.org/10.1016/j.jnca.2012.03.009
11. Karp, B., Kung, H.T.: GPSR: Greedy perimeter stateless routing for wireless networks. In: Proceedings of the 6th Annual International Conference on Mobile Computing Network (MobiCom). New York (2000). https://doi.org/10.1145/345910.345953
12. Katsaros, K., Dianati, M., Tafazolli, R., Kernchen, R.: CLWPR–A novel cross-layer optimized position based routing protocol for VANETs. In: Proceedings of the IEEE Vehicular Networking Conference (VNC) (2011). https://doi.org/10.1109/VNC.2011.6117135
13. Alzamzami, O., Mahgoub, I.: An enhanced directional greedy forwarding for VANETs using link quality estimation. In: 2016 IEEE Wireless Communications and Networking Conference (2016). https://doi.org/10.1109/WCNC.2016.7564748
14. Bakar, K.A., Ghafoor, K.Z., et al.: Fuzzy logic-assisted geographical routing over vehicular ad hoc networks. Int. J. Innov. Comput. Inf. Control **8**(7), 5095–5120 (2012)
15. Li1, G., Ma2, M., Liu1, C., Shu, Y.: Adaptive fuzzy multiple attribute decision routing in VANETs. Int. J. Commun. Syst. (2015). https://doi.org/10.1002/dac.3014
16. Alzamzami, O., Mahgoub, I.: Fuzzy logic-based geographic routing for Urban vehicular networks using link quality and achievable throughput estimations. IEEE Trans. Intel. Transport. Syst. **20**(06), 2289–2300 (2019). https://doi.org/10.1109/TITS.2018.2867177

Analysis of Hybrid Data Security Algorithms for Cloud

Vikas K. Soman and V. Natarajan

1 Introduction

Cloud computing is an emerging technology having different features, service models like Infrastructure as a Service (IAAS), Platform as a Service (PAAS), and Software as a Service (SAAS), and deployment models like private, public, hybrid, and community clouds. The cloud security should cover security for service models such as Security as a Service. The security should be from physical security to infrastructure to software security at different devices and applications in software whichever necessary. Cloud data security becomes an important task in the virtualized cloud resource environment. The data in the cloud are accessed from anywhere at any time, and manipulation can be done from any terminal through cloud service provider that does not have proper security protection mechanisms. This geographically distributed data access handling and storing data in a secured manner is a challenging problem in the cloud. The user's data should be protected securely in the cloud, so that confidentiality, integrity, and availability are maintained throughout the life cycle of data. The data centers are storing the huge amount of data coming from different cloud service providers like Amazon [1], Google [2], Microsoft [3], IBM [4], and Oracle [5]. These clouds use different data security mechanisms for protecting their data at rest on data centers. The cloud computing models have three important components such as:

1. Cloud service provider (CSP): This will preserve sufficient facilities and storage space to client's data, with a huge computational power.

V. K. Soman (✉) · V. Natarajan
Department of Instrumentation, MIT Campus, Anna University, Chennai, India
e-mail: vikassoman@gmail.com; natraj@mitindia.edu

M. Palesi et al. (eds.), *Second International Conference on Networks and Advances in Computational Technologies*, Transactions on Computational Science and Computational Intelligence, https://doi.org/10.1007/978-3-030-49500-8_20

2. Client/owner: This is an individual or organization who stores large data in the cloud.
3. User: This refers to the registered user with client or owner who uses the data stored by the client or owner.

The cloud can provide data security assurance by using firewalls, intrusion detection systems, access control list, virtual private networks and other security policies with its own devices, and sometimes it requires data from another cloud owner for business purposes and these data are not only available in the cloud but also available in the third-party cloud [6]. So authorized user through proper authentication method only should access the data stored in the cloud.

The cloud data security consists of security of data at rest and at transit. The cryptographic algorithms are mainly used to protect the data at rest, and secure protocols like HTTPS, SSH, SSL, and TLS are used for protecting data in transit. The International Standards ISO/IEC 27001:2013 and ISO/IEC 27002;2013 specifies the requirements for establishing, implementing, maintaining and continually improving an information security management system within the context of the organization.

This paper proposes a cloud data security mechanism that ensures confidentiality as well as the integrity of data at rest using a combination of Secure Hash (SHA-512), Rivest–Shamir–Adleman (RSA), and Advanced Encryption Standard (AES) algorithms and then analyzes the performance with a previous research proposal [7]. The motivation of this work is to provide a good security to the data owner who has stored his or her sensitive data in the cloud.

1.1 Treat Model

An attacker in the cloud is anyone who is unauthorized to access the cloud server data. The cloud server data can be accessed by any external or internal attacker like CSP administrator, cloud administrator, or external user. One should properly handle the modification of data in the cloud server data. Hence, the issue of tampering of cloud server data should be addressed, so that integrity of data is maintained.

This paper is divided into nine sections—Sect. 1—Introduction, Sect. 2—Cloud Data Security Challenges and Solutions, Sect. 3—Related Works, Sect. 4—Problem Statement, Sect. 5—Cryptographic Process, Sect. 6—Proposed Hybrid Cryptographic Algorithm, Sect. 7—Security Analysis of the Proposed Algorithm, Sect. 8—Results and Discussion, and finally, Sect. 9—Conclusion.

2 Cloud Data Security Challenges and Solutions

When storing data securely into the cloud, the major data security challenges are identified as shown in Fig. 1 and how to handle these challenges.

One concern is that the external attacker or any unauthorized person may access the sensitive data of the cloud owner, and other is CSP himself or herself can attack inside the cloud premise. Thus, data security, the data segregation, protection, and data leak prevention are major data security challenges that should be considered for secure cloud system implementation. The location and relocation of cloud user's data as per CSP's service level agreement (SLA) is also important. If the mobility of data is very high in the cloud, the chances of the data vulnerability is also high. Hence, data security mechanisms are important to handle these situations, and the proper auditing of the data can also be done using internal audit or external audit. The reliability of the cloud data at rest is assured by using proper data security mechanisms. The NIST [8] Cybersecuirty, CSA [9] Cloud assurance, and OWSAP [10] Web-application security are different groups of people working in cloud security issues and standard formation.

The possible solutions are listed below:

Strong cryptographic encryption-like hybrid algorithms: The symmetric cryptographic algorithms key transmission is a problem. This is solved by combining with asymmetric cryptographic algorithm forms hybrid and publicly exchange the key with the data of the owner more secure. There are other algorithms like fully homomorphic algorithm [11] and attribute-based algorithms [12], which are available to make the data operations stored by the data owner secure.

Access control policies: The access control policies at the different level of users in the cloud are also an important consideration while storing data in the cloud.

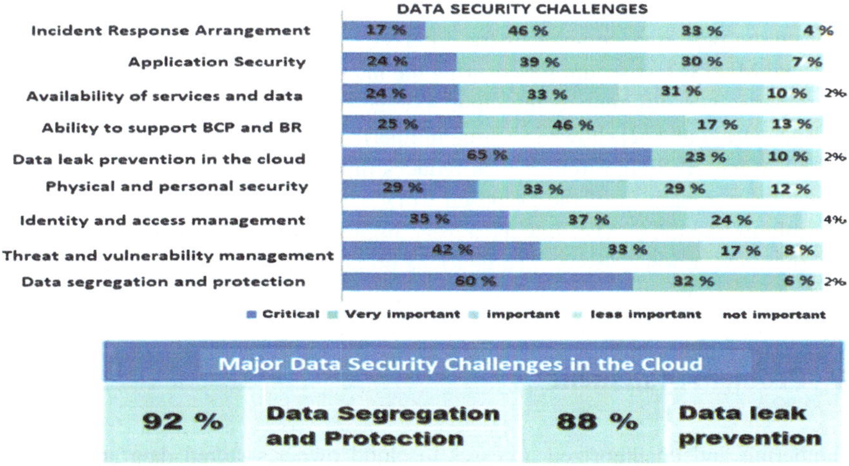

Fig. 1 The data security challenges in the cloud

Hash algorithms: These algorithms ensure the integrity of the data stored by the cloud owner. One can use SHA-512 algorithm to maintain integrity.

Key management: The proper key management is also considered for the authentication of user and data stored in the cloud. The proposed algorithm uses RSA public encryption to manage the transmission of AES private key in the cloud.

Service level agreement (SLA) [13]: This is an important agreement that provides more clarity on the data storage and other facilities provided by the CSP, data owner, and user or client in the cloud.

Good auditing mechanisms [14]: Proper auditing mechanisms also ensure the security of data. This also verifies the integrity of cloud server data both externally and internally.

Certificate-level monitoring: The certificate from trusted public key infrastructure [15] certificate authorities makes the cloud data more secure.

Trust [16]: The data owner gets trust to use his data in cloud server to store, when his data is not tampered and gets back when needs that data.

3 Related Works

Mahalle et al. [17] proposed an enhancement of cloud security using hybrid RSA and AES algorithm. The algorithm uses 128-bit AES key and 1024-bit RSA key. This proposal does not use a hash algorithm for checking the integrity of the file. As per G.P. Kanna et al. [18], the cloud outsourced data are enhanced by hybrid RSA with ECC based on identity-based encryption. This proposal does not use any hash algorithm for integrity check. Rajput et al. [19] explained how AES algorithm is used for securely storing data in the cloud. This algorithm is not suited for storing a large amount of data at the cloud, but for small data this will work very fast in the cloud; however, it has a problem in the key management. Sujithra et al. [20] showed the performance analysis for different hybrid cloud algorithms for mobile data and MD5 is less secure for ensuring the integrity of data. Yong et al. [21] gave a comparison of different secure cloud storage based on cryptographic technique. A. Bhandari et al. [22] showed secure multiparty communication in the cloud using HE-RSA and AES algorithm, which reduces time, cost and memory size. Table 1 shows the comparison of the proposed work with other works. The elliptic curve algorithms like ECC-AES having a very high-security level and high speed of algorithm execution is less compared to the proposed RSA–AES algorithm.

4 Problem Statement

Tampering and unauthorized accesses to cloud owner's stored data are possible because the security mechanisms on the stored data are not considered. Nowadays, attacks from external and internal interceptors are very high. Data tampering and

Table 1 Comparison of the proposed work with other works

Research works	Confidentiality	Integrity by using hash algorithm	Support of large amount of data	Algorithm security level	Execution speed
[17]	Yes	No	Yes	High	Very high
[18]	Yes	No	Yes	Very high	High
[19]	Yes	No	No	High	Very high
[20]	Yes	Yes (MD5)	Yes	High	High
[21]	Yes	Yes (MD5)	No	High	High
[22]	Yes	No	Yes	High	Very high
[7]	Yes	Yes (SHA-256)	Yes	Very high	High
[proposed]	Yes	Yes (SHA-512)	Yes	High	Very high

its leakage are to be prevented. The key transmission and management should be handled. The data security issues for adopting data into the cloud should be addressed, and data confidentiality, integrity, and availability should be maintained. The cost and running time of the algorithm should be reduced, so that the security of cloud data storage and the overall performance will be improved. The security of data at rest at cloud owner's servers is considered as most of the cloud owner data are at rest while adopting data into the cloud. The analysis of cryptographic algorithm is to be performed, and the security level should be compared by analyzing different issues in cloud data security.

5 Cryptographic Process

The SHA algorithm maps the large data file into fixed-size message digest [23] that is unique and is used for the integrity check of the data file in the cloud. The SHA-512 provides a 512-bit message digest of a large data file that can be stored securely in the cloud.

The RSA algorithm was publically described in 1977 [24] as follows:

Key Generation Algorithm

1. Choose randomly and secretly two large primes: p, q and compute $n = \mathrm{p} \cdot \mathrm{q}$
2. Compute $\Phi(n) = (p - 1) \cdot (q - 1)$
3. Select random integer e such that $1 < e < n$ and gcd $(e, \Phi(n)) = 1$
4. Compute d, such that $e.\ d \equiv 1 \bmod \Phi(n)$ and $1 < d < \Phi(n)$
5. Public key is (e, n)
6. Private key is (d, n)

Encryption Process of Key

1. Suppose Alice needs to send message M to Bob $(0 < M < n)$

2. Bob should send his public key to Alice
3. Alice will encrypt M as $c = M^e \bmod n$ and will send c to Bob

Decryption Process of Key

1. Bob will decrypt the received message as $M = c^d \bmod n$

The AES algorithm uses 128 or 192 or 256 bits based on 10 or 12 or 14 rounds. AES uses different rounds in which each round having different stages, which provides security through transformation, substitution, permutation, mixing, and key adding each round except that uses four transformations.

The secure data storage in cloud includes four main processes, namely key generation algorithm, hash code generation, AES encryption, and AES decryption. The 128-bit secret key is generated securely using random key generation algorithm of AES. This key is encrypted using RSA public key algorithm. The hash of the file is generated using the SHA-512 algorithm. The AES encryption is done with the private key generated using RSA key pair generator and the data are stored in the cloud server. The AES private key is encrypted and transmitted using RSA public key. This encrypted AES key and the private key of RSA are used for RSA decryption to form the decryption AES key that is used for AES decryption when the user wants to access the data (when the hash of the file matches).

6 Proposed Hybrid Data Security Algorithm

SHA-512, RSA, and AES Hybrid Data Encryption Algorithm

Input: Data message or file and key

Output: Encrypted data message or encrypted file and hash of the file and encrypted secret key

1. Cloud owner registered with their credentials, username, and password with a cloud server. The authorized owner has the access to data stored on the cloud server. The following steps are done before data stored in cloud server.
2. The 128-bit secret AES secret key, K, is generated randomly, and RSA key pairs are generated.
3. The message M and its hash code H (M) are generated using SHA-512 algorithm.

The message M is then encrypted using AES encryption with 128-bit secret key generated randomly and the encrypted message and hash are stored on the cloud server. The AES secret key is then encrypted using RSA public key encryption and transmitted. Ciphertext C (M) = E_{AES} (M, K), where K is secret key that is again encrypted E_s = E_{RSA} (*K, public key*) for secure transmission of secret key. The communication process is shown in Fig. 2.

SHA-512, RSA, and AES Hybrid Data Decryption Algorithm

Input: Encrypted data message or file, hash of the file previously stored, and encrypted secret key.

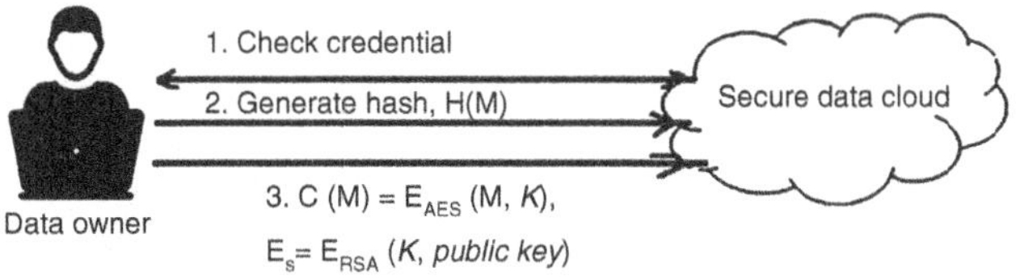

Fig. 2 Secure data file to upload into the cloud system

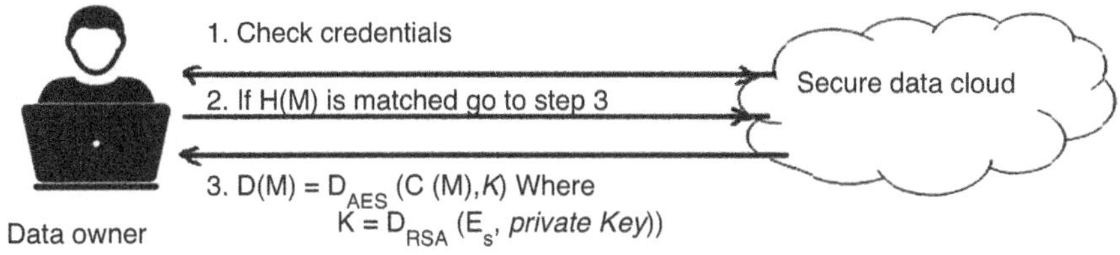

Fig. 3 Secure data file to download from the cloud system

Output: Original message

1. The cloud owner first checks whether his or her credentials are authenticated and then proceeds to step 2.
2. The server checks the hash H (M) of the previously stored message or file; if it matches, go to step 3.
3. The server does the AES decryption using RSA decrypted secret key using private key, i.e., D (M) = D_{AES} (C(M), *K*), where *K* is D_{RSA} (E_s, *private key*)) and the original data message or file is accessed securely. The communication process is shown in Fig. 3.

7 Security Analysis of the Proposed Algorithm

The security of the data at rest is more important because cloud data are at rest most of the time. So this proposed cryptographic algorithm makes the data at rest of cloud server secure. Data segregation and data leakage are prevented, and it ensures that only the authorized user can access the data at rest. The minimum key length of the AES is 128, so that if one terminal can perform 2^{56} times of key search per second, it will take at least 149 trillion years to complete AES key search and is not feasible considering the time as well as space required. AES algorithm resists against expanding algebraic expression attack because of the complexity of s-boxes expression. The RSA resists timing attack and mathematical attack. AES algorithm is strong against differential, truncated differential, linear, interpolation, and square attacks. The NIST's recommended key size comparison is shown in Table 2.

Table 2 NIST's recommended key size comparison

Security (bits)	Symmetric key algorithms	Hash functions	Public key size (bits)	
			RSA	ECC
80	–	SHA-1	1024	160
112	3DES	–	2048	224
128	AES128	SHA-256	3072	256
192	AES192	SHA-384	7680	384
256	AES256	SHA-512	15,360	512

The proposed algorithm resists from brute-force attack and birthday attack, and the complexity of security is higher than single algorithm. It also resists unauthorized access of cloud server data at rest, tampering of data stored at rest, the internal attack by CSP, etc.

1. *Brute force attack.*

   The data while adopting into the cloud may be intercepted by unauthorized methods by trying all key combinations, but the proposed method is dual encrypted and such combination finding will take several years. Such an attack is mostly useless by checking all the key combinations. The minimum AES key size is 128 bit and 256-bit key is very strong, and RSA 1024-bit key is normally used, but now 15,360 bit key is also used. Hence, a brute-force attack is much harder task to unauthorized users.
2. *Tampering of data at rest.*

   The data at rest may be accessed by unauthorized interceptor, but the proposed method uses an integrity checker like SHA-512 while uploading data into the cloud server and also used while downloading. Therefore, the integrity of the data is assured in the proposed algorithm, and tampering of data will not take place.
3. *Internal attack by CSP.*

   The data at the cloud are not under the control of data owner, and the CSP may leak the sensitive data for helping other parties. So one cannot trust the CSP blindly. The internal attack by the CSP is also resisted because the proposed method uses very strong encryption mechanism that resists decryption by brute-force attack, and also the integrity is assured by SHA-512. Hence, this attack is also resisted.
4. *Unauthorized access and Unauthorized cloud server*

   The unauthorized access to the data at rest and any interceptor who uses an unauthorized cloud server to access the owner data are prevented using the proposed algorithm. So the adopted data to the cloud are very well securely preserved by this algorithm from an interceptor.

8 Results and Discussion

The study of the cryptographic applications of data encryption and decryption algorithm in cloud computing security is analyzed, challenges are identified in the cloud data security, and the proposed algorithm enhanced the security of cloud data at rest. The RSA algorithm alone cannot encrypt large data file, but combining with other like proposed hybrid algorithm can encrypt large data files. The different file size data at a real time is analyzed using this algorithm. This may vary depending on the cloud or local resource used at the time of execution. The execution speed may vary depending on the number of processors or virtual machines. Large files such as .pdf, .doc, .ppt, .xls, .csv, and image files can be securely stored on cloud. The algorithms are executed in the OpenNebula Cloud [25] using java9 [26] and executed in a local Ubuntu machine and also data at rest are stored in the cloud server in OpenNebula and local Ubuntu machine. The comparisons of the encryption and decryption time analysis are shown in Figs. 4 and 5. The SHA-512, RSA, and AES hybrid is performing faster encryption and decryption in the cloud as well as in the local terminal while comparing with SHA256, ECDSA, and AES hybrid algorithm. But the performance efficiency is high in ECDSA, SHA256, and AES

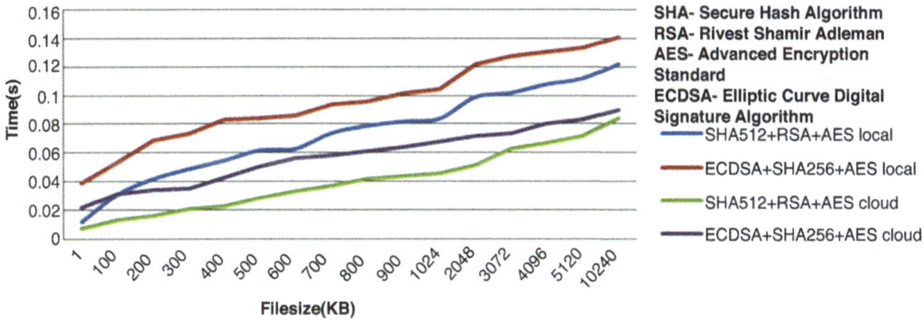

Fig. 4 Hybrid encryption time analysis local and cloud

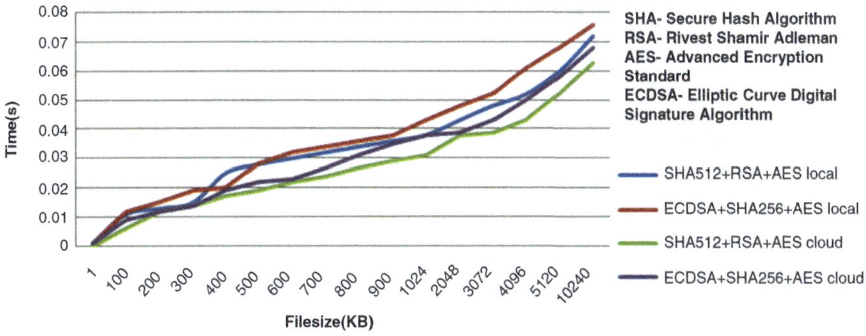

Fig. 5 Hybrid decryption time analysis local and cloud

hybrid encryption and decryption in local as well as in the cloud terminal. The algorithms' performance analyses on large data files are shown in Figs. 6 and 7.

9 Conclusion

In this paper, the data security issues faced by cloud owner when adopting sensitive data in the cloud system are analyzed. Tampering and unauthorized access of cloud server data are prevented using the proposed algorithm. Data security in the cloud data at rest can be analyzed using the proposed hybrid data security cryptographic algorithm. Performance comparison is also done using different large file sizes. This proposed algorithm performs faster than other algorithms. The security level of the

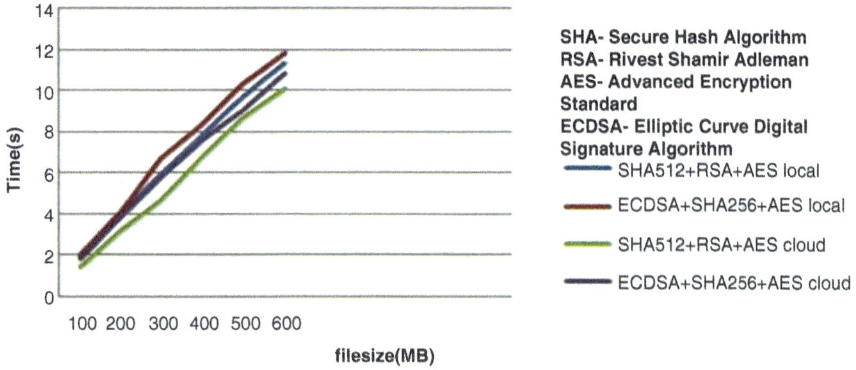

Fig. 6 Hybrid encryption execution time analysis on different large files in the cloud and the local machine

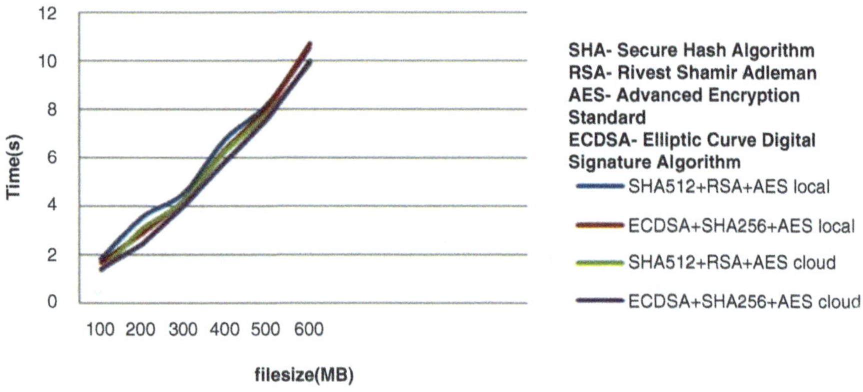

Fig. 7 Hybrid decryption execution time analysis of different large files in the cloud and the local machine

algorithm is also high so as to ensure a good secured cloud server data. Using this algorithm, a new protocol for data security can be developed in the cloud.

References

1. Amazon EC2: https://aws.amazon.com/ec2/. Accessed 18 Nov 2018
2. GoogleApp Engine: https://cloud.google.com/appengine. Accessed 20 Nov 2017
3. Microsoft Azure: https://azure.microsoft.com. Accessed 15 Nov 2018
4. IBM Cloud: https://www.ibm.com/cloud/. Accessed 25 Nov 2018
5. Oracle Cloud: https://cloud.oracle.com/home. Accessed 28 Nov 2018
6. Julisch, K., Hall, M.: Security and control in the cloud. Inf. Secur. J. Glob. Perspect. **19**(6), 299–309 (2010)
7. Soman, V.K., Natarajan, V.: An enhanced hybrid data security algorithm for cloud. In: 2017 International Conference on Networks & Advances in Computational Technologies (NetACT), pp. 416–419. IEEE (2017). https://doi.org/10.1109/NETACT.2017.8076807
8. Mell, P., Grance, T.: The NIST Definition of Cloud Computing. NIST, Special Publication, 800–145. NIST, Gaithersburg, MD (2011)
9. Cloud security alliance, security guidelines for critical areas of focus in cloud computing v4.0. https://cloudsecurityalliance.org/download/securityguidance-v4. Accessed 11 Apr 2017
10. Open Web Application Security Project (OWASP): Home page. https://www.owasp.org/index.php/Main_Page. Accessed 15 Nov 2017
11. Gentry, C., Halevi, S.: Implementing Gentry's fully-Homomorphic encryption scheme. In: Paterson, K.G. (ed.) Advances in Cryptology – EUROCRYPT 2011, Lecture Notes in Computer Science, vol. 6632. Springer, Berlin, Heidelberg (2011)
12. Gorbunov, S., Vaikuntanathan, V., Wee, H.: Attribute-based encryption for circuits. J. ACM. **62**(6), Article 45 (2015). https://doi.org/10.1145/2824233
13. Mirobi, G.J., Arockiam, L.: Service level agreement in cloud computing: an overview. In: 2015 International Conference on Control, Instrumentation, Communication and Computational Technologies (ICCICCT), Kumaracoil, pp. 753–758 (2015). https://doi.org/10.1109/ICCICCT.2015.7475380
14. Wang, C., Chow, S.S.M., Wang, Q., Ren, K., Lou, W.: Privacy-preserving public auditing for secure cloud storage. IEEE Trans. Comput. **62**(2), 362–375 (2013). https://doi.org/10.1109/TC.2011.245
15. Cloud Certificate Authority: Public key infrastructure. http://cloudpatterns.org/mechanisms/certificate_authority. Accessed 18 Jan 2018
16. Huang, J., Nicol, D.M.: Trust mechanisms for cloud computing. J. Cloud Comput. Adv. Syst. Appl. **2**, 2–9 (2013)
17. Mahalle, V.S., Aniket, K.S.: Enhancing the data security in cloud by implementing hybrid (RSA&AES) encryption algorithm. In: Power, Automation, and Communication (INPAC), 2014 International Conference on. IEEE, Piscataway, NJ (2014)
18. Kanna, G.P., Vasudevan, V.: Enhancing the security of user data using the keyword encryption and hybrid cryptographic algorithm in cloud. In: 2016 International Conference on Electrical, Electronics, and Optimization Techniques (ICEEOT), pp. 3688–3693. IEEE (2016). https://doi.org/10.1109/ICEEOT.2016.7755398
19. Rajput, S., Dhobi, J.S., Gadhavi, L.J.: Enhancing data security using AES encryption algorithm in cloud computing. In: Satapathy, S., Das, S. (eds.) Proceedings of First International Conference on Information and Communication Technology for Intelligent Systems: Volume 2. Smart Innovation, Systems and Technologies, vol. 51. Springer, Cham (2016). https://doi.org/10.1007/978-3-319-30927-9_14

20. Sujithra, M., Padmavathi, G., Narayanan, S.: Mobile device data security: a cryptographic approach by outsourcing mobile data to cloud. Procedia Comput. Sci. **47**, 480–485 (2015). https://doi.org/10.1016/j.procs.2015.03.232
21. Yong, P., Wei, Z., Feng, X., Zhong-hua, D., Yang, G., Dong-qing, C.: Secure cloud storage based on cryptographic techniques. J. China Univ. Posts Telecommun. **19**(2), 182–189 (2012). https://doi.org/10.1016/S1005-885(11)60424-X
22. Bhandari, A., Gupta, A., Das, D.: Secure algorithm for cloud computing and its applications. In: 2016 6th International Conference - Cloud System and Big Data Engineering (Confluence), Noida, pp. 188–192. IEEE (2016). https://doi.org/10.1109/CONFLUENCE.2016.7508111
23. Stallings, W.: Cryptography and Network Security: Principles & Practices, 5th edn. Prentice Hall, Upper Saddle River, NJ (2010)
24. Rivest, R., Shamir, A., Adleman, L.: A method for obtaining digital signatures and public-key cryptosystems. ACM Trans. Commun. **21**, 120–126 (1978)
25. OpenNebula: https://opennebula.org/. Accessed 15 Mar 2019
26. Oracle java9: https://oracle.com. Accessed 18 Nov 2018

A Deep Learning Approach to Malayalam Parts of Speech Tagging

M. K. Junaida and Anto P. Babu

1 Introduction

Parts of Speech Tagging is one of the elementary and consequential tasks in any Natural Language Processing (NLP) pipeline; it consists of allocating unique grammatical categories (Parts of Speech Tags) to words in a sentence. It is a well explored problem in English and other South Asian languages, but in Indian languages especially south Indian language like Malayalam, still having room for exploration because it is a challenging task for languages with rich system of morphology, inflections, and free word order. In addition, POS tagging of Malayalam is more complex due to the characteristics such as sandhi, lack of capitalization, and gender information, etc.

Sequence labelling tasks such as POS, NER, Chunking are some of the initial tasks carried on natural language processing using deep learning approach and it has been well examined and attained state-of-the-art results in last decade [1]. The recent approaches based on end-to-end Deep Neural Networks (DNN) models uses both word and character-level informations together, by using RNN, CNN and CRF, performs better results on many sequence labelling tasks. Now, the focus has shifted to designing and implementing effective neural network architectures than feature engineering [2–4].

Before neural sequence labelling models, traditional machine learning models are used to solve sequence labelling problems, e.g.: HMM, CRF, Structural SVM, etc. [5] and these are based on statistical modelling of feature values. The problem with these models are highly dependent on hand crafted features and time consum-

M. K. Junaida (✉) · A. P. Babu
Department of Information Technology, Kannur University, Kerala, India

M. Palesi et al. (eds.), *Second International Conference on Networks and Advances in Computational Technologies*, Transactions on Computational Science and Computational Intelligence, https://doi.org/10.1007/978-3-030-49500-8_21

ing. It is difficult to represent high dimensional features in real world applications because it is not practically possible by a human expertise always.

To overcome these draw backs neural network based models have been proposed with the recent popularity and success of word embedding (low dimensional features) and variants of neural network models have achieved superior results on sequential labelling task as well as most of the NLP related task as compared to traditional machine learning models. Also deep learning based models have shown promising results for sequence tagging in many languages such as English, much less work has been done on neural models for POS tagging of Indian languages.

In this paper, we apply a neural sequence labelling system for POS tagging of Malayalam language using neural sequence labelling framework NCRF++ [6] and demonstrate the effectiveness of leveraging character level knowledge for language representations. This paper is organized as the following—Sect. 2 explains the previous work and Sect. 3 discusses details about the system architecture and different methodologies employed. The corpus details and experimental setups are explained in Sect. 4, and Sect. 5 presents results and analysis of our experiment. We conclude our paper with conclusion and future work in Sect. 6.

2 Related Work

It is found in literature several works have been done towards the development of POS tagger using different approaches, which have performed in satisfactory way and allowing that good results for Indian languages. However, Generation of efficient POS tagger with large training corpus for different languages is most challenging task in this research field. It is quite a difficult task for Indian languages, especially on Dravidian languages. Very little work has been done on Malayalam because of scarcity and quality of annotated data.

Manju [7] proposed stochastic HMM-based model for Malayalam POS tagging and claimed that it was able to assign tags to almost all the words in the test case but only 1400 words are used for training which was main drawback of their work. A SVM (Support Vector Machine) based Malayalam POS tagger [8] used tagged corpus of more than one lakh words and trained with tagset of 29 tags. One more work was reported using SVM with IIIT-Hyderabad tagset and compared the performance of SVM with TNT for Malayalam POS tagging [9]. A Memory-Based Language Processing (MBLP) approach [10] and a hybrid method with rules and bigrams for Malayalam POS tagging [11] were also reported. Most of the works cited here reported above 90% accuracy, but it uses less amount of annotated corpus with small tagsets and hand crafted features.

Recently, Kumar [12] proposed a deep learning based POS tagger for Malayalam tweets. They consisted of 9915 Malayalam tweets with 17 coarse tags and achieves 90% accuracy for GRU(Gated Recurrent Unit) based deep learning sequential model.

3 System Architecture

We use a similar architecture to that used in sequence labelling models for fundamental NLP tasks such as POS tagging, chunking, and named entity recognition (NER) [13]. Which comprises three layers: a character sequence representation layer, a word sequence representation layer, and an inference layer.

3.1 Character Sequence Representation Layer

For tasks like POS tagging and NER, character embeddings has significance to look at within word morphological and shape information such as characters and it helpful to produce better results on morphologically rich languages [14]. We have used LSTM and CNN to extract character sequence information. Character CNN using one layer CNN structure with max-pooling to capture character level representations [13] are used in our experiments. Character LSTM utilizes a bidirectional LSTM for each word character sequence and concatenate the Forward LSTM (left-to-right) and the Backward LSTM (right-to-left) as character sequence representation. Figure 1 below depicts the neural character sequence representations of both CNN and LSTM.

3.2 Word Sequence Representation Layer

Unlike character embedding layer word sequence layer encompasses only the Recurrent Neural Networks (RNN). RNN is commonly used for word level classification. Young [2] proposed a bidirectional LSTM, which is a one of the variants of RNN and offers state-of-the-art result for POS tagging and many of their applications. Here, we also used bidirectional LSTM, which captured arbitrarily

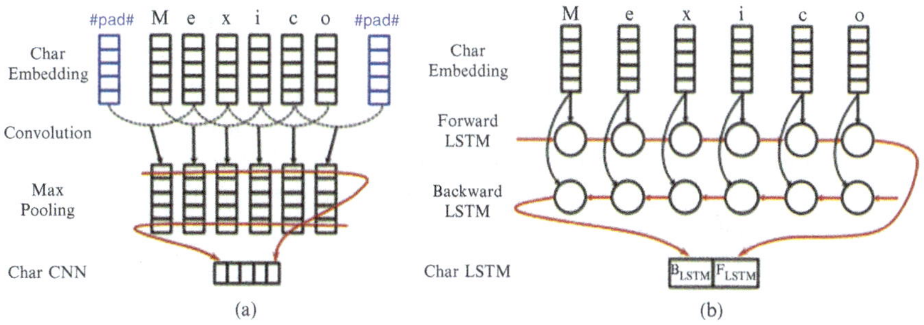

Fig. 1 Neural character sequence representations [13]. (**a**) Char CNN (**b**) Char LSTM

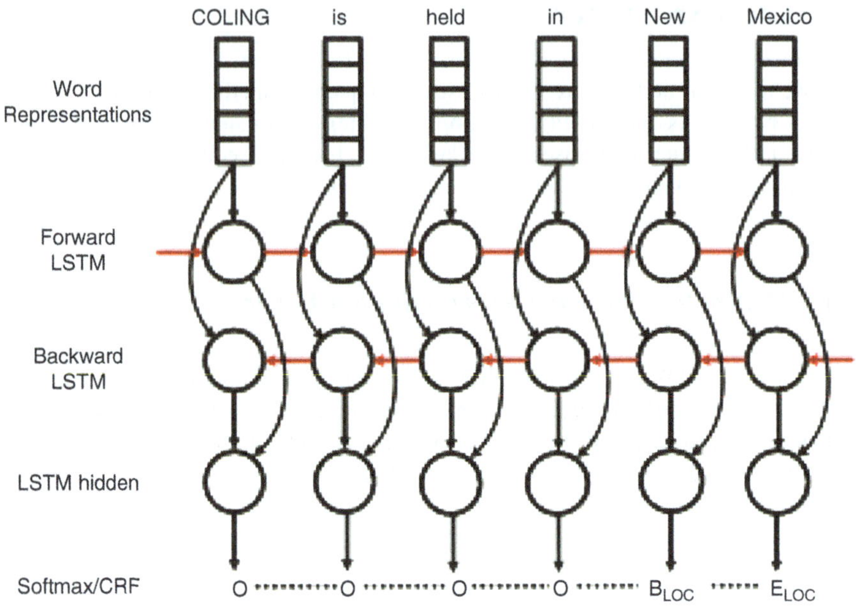

Fig. 2 Neural word LSTM sequence representations [13]

long context information around the target word to overcome the limitation of a fixed window size. As shown in Fig. 2, the word sequence information are captured by forward LSTM and backward LSTM from left to right and right to left direction, respectively. To give the global information of the whole sequence the hidden states of the forward and backward LSTMs are concatenated at each words in a sequence.

3.3 Inference Layer

The inference layer allocate labels to each words by extracting word sequence representations as features. Recently, Young [14] reported that by using CRF in inference layer boosted the performance of sequence labelling tasks. In our experiments we employ both Softmax and CRF. Softmax is a full distribution over the potential label of what the actual input could be, through the network of pre-activation and by using the soft-max non-linearity. CRF captures dependencies among labels by adding transition scores between neighboring labels. Due to the support of parallel decoding, Softmax is much more efficient than CRF for the task Parts of Speech (POS) Tagging [15].

4 Experimental Setup

4.1 Corpus Details

We used publicly available corpus through TDIL website, under the project Indian Languages Corpora Initiative phase II (ILCI Phase II), initiated by the MeitY Govt. of India, Jawaharlal Nehru University, New Delhi. The corpus contains a Monolingual POS tagged sentences from different domains, approximate 31,000 sentences.

We had done corpus cleaning by removing untagged sentences and shuffled the sentences from all domains and converted to neural sequence labelling tool data format. From the available corpus, we had split Training, Development, and Testing corpus according to Table 1.

Most of the cited work in previous section used a smaller tagset with limited corpus size. The tagset used for this task, BIS[1] (Bureau of Indian Standards) is a hierarchical tagset uses grammatical categories and their sub categories along with other morpho-syntactic attributes. This tagset is mostly used and recommended for the Indian language Parts of Speech Tagging as common tagset. Our dataset uses 35 different tags that have been detailed in BIS tagset.

4.2 Tool Used

The implementation of neural network model for Malayalam POS tagging was done by using publicly available Neural Sequence Labelling tool NCRF++.[2] It is a PyTorch based Open-source Neural Sequence Labelling Toolkit with flexible choices of input features and output structures.

4.3 Hyper-Parameters

In order to obtain better performance of neural networks, numerous dimensions need to be made regarding the hyper-parameters used. In our experiment, to fix the

Table 1 Train Dev Test split

Type	Sentences	Tokens
Train	24,000	544K
Dev	3495	78K
Test	3342	76K

[1]http://tdil-dc.in/tdildcMain/articles/134692Draft%20POS%20Tag%20standard.pdf.

[2]https://github.com/jiesutd/NCRFpp.

Table 2 Model hyper-parameters

Hyper-parameter	Value
Character dimension	50
Word dimension	200
Update function	SGD
Learning rate	0.015
Learning decay rate	0.05
Dropout rate	0.8
LSTM-layer	1
Batch size	10
Iterations	15
GPU	False

optimal hyper-parameters, two trials of experiments were carried out with different learning rate, number of epochs, and batch sizes. All models are trained using early stopping, with minimum number of 10 and maximum number of 15 epochs, which is never reached. In order to avoid the overfitting, we stopped training when we obtained no improvement in three consecutive epochs. We also make use of the regularization techniques l2 regularization and dropout value of 0.5 to improves the models performance on the unseen data as well.

Table 2 summarizes the chosen hyper-parameters used in our experiments. Hyper-parameters were tuned on the development set of POS. We set the character embedding dimension at 50, the word embedding dimension at 200 and we optimize parameters using stochastic gradient descent (SGD) optimizer with mini-batch size 10. We choose an initial learning rate of $\eta_0 = 0.015$, and the learning rate is updated on each epoch of training as $\eta_t = \eta_0/(1 + \alpha t)$, with decay rate $\alpha = 0.05$ and t($t = 15$) is the number of epoch completed.

5 Results and Discussion

Token accuracy is used to evaluate the performance of POS tagger. We used the training set to learn model parameters, the development set to select optimal hyper-parameters, and the best model on the development set is used for testing. Accuracy of the selected models on the test datasets are reported.

Table 3 shows the empirical results of three CRF-based models and three Softmax based models with different character sequence and word sequence representations on ILCI Phase II corpus. In this table "NOCHAR" suggests the model without character sequence information. "CLSTM" and "CCNN" denote the models using LSTM and CNN for character sequence and "WLSTM" represents LSTM word sequence representation.

The first two rows show the performance of BiLSTM-CRF (WLSTM+CRF) model with the character feature information and next two rows show the BiLSTM-Softmax (WLSTM+SOFT) model with character feature information, and then

Table 3 POS tagging performance on ILCI Phase II Malayalam dataset

Models	Accuracy
CCNN+WLSTM+CRF	85.40
CLSTM+WLSTM+CRF	86.94
CCNN+WLSTM+SOFT	85.20
CLSTM+WLSTM+SOFT	**87.05**
NOCHAR+WLSTM+SOFT	81.32
NOCHAR+WLSTM+CRF	81.10

the two rows show the results for BiLSTM-CRF (WLSTM+CRF) and BiLSTM-Softmax (WLSTM+SOFT) models without the character information.

As the empirical results in the Table 3 show that the BiLSTM-Softmax with character information model achieved a highest accuracy of 87.05% among other proposed models. Also, the models with LSTM character level embedding (85.40%, 85.20%) regardless of the CRF and Softmax based inference layer. In particular, Malayalam POS tagging, BiLSTM-CRF-based model and BiLSTM-Soft based model without character level information obtained similar performance (81.10%, 81.32%, respectively), outperforming BiLSTM-SOFT by approximately 0.22% in absolute terms.

When incorporating character level information into two BiLSTM-CRF (CLSTM, CCNN) and two BiLSTM-SOFT (CLSTM,CCNN) based models, we found that character level information improved the performance by about 5.84%, 4.3% and 5.73%, 3.88% respectively, thus clearly indicating that character level word embedding can capture unseen word information.

Based on the above observations, The experimental results show that the models with character level representation obtained better performance than without character level representation models. This showing the advantage of word level information with character information on Malayalam POS tagging models, which suffers the lack of annotated corpus and hand crafted features.

6 Conclusion and Future Work

In this work we have applied deep learning neural network model for Malayalam POS tagging, which obtained good performance compared to state-of-the-art results in Malayalam, with less cost for building. The experiments on ILCI Phase II Malayalam POS tagged dataset acheived a highest overall accuracy score of 87.05% by the BiLSTM-Softmax model with character level feature. Empirical results show that the character level information along with word level features would benefit Malayalam language models by improving the performance. The results show the models with word and character embedding obtained, approximately 6% of improvement than the models without character embedding. Precisely, Word LSTM with character LSTM and Softmax model gives little improvement than Word LSTM with character LSTM and CRF-based models.

The accuracy of the proposed neural network models can be improved by training with pre-trained word embeddings, so as to reduce the OOV (Out of Vocabulary) words. The character and word embedding models with CRF are not specific to Malayalam language, and can be used for other languages, especially Indian languages with the issue of data scarcity and quality of annotated data. Also, it can be extended to other sequence labelling tasks such as NER, Chunking, etc.

References

1. Ahmed, M., Samee, M.R., Mercer, R.E.: Improving neural sequence labelling using additional linguistic information. In: 2018 17th IEEE International Conference on Machine Learning and Applications (ICMLA), pp. 650–657 (2018)
2. Yang, J., Liang, S., Zhang, Y.: Design challenges and misconceptions in neural sequence labeling. In: Proceedings of the 27th International Conference on Computational Linguistics, COLING (2018)
3. Ma, X., Hovy, E.H.: End-to-end sequence labeling via Bi-directional LSTM-CNNs-CRF. CoRR, abs/1603.01354 (2016)
4. Lample, G., Ballesteros, M., Subramanian, S., Kawakami, K., Dyer, C.: Neural architectures for named entity recognition. In: Proceedings of the 2016 Conference of the North American Chapter of the Association for Computational Linguistics: Human Language Technologies, HLT-NAACL (2016)
5. Nguyen, N., Guo, Y.: Comparisons of sequence labeling algorithms and extensions. In: Proceedings of the 24th International Conference on Machine Learning, ICML (2007)
6. Yang, J., Zhang, Y.: NCRF++: an open-source neural sequence labeling toolkit. In: Proceedings of ACL 2018, System Demonstrations, ACL (2018)
7. Manju, K., Soumya, S., Idicula, S.M.: Development of a POS tagger for Malayalam - an experience. In: 2009 International Conference on Advances in Recent Technologies in Communication and Computing, pp. 709–713 (2009)
8. Antony, P.J., Mohan, S.P., Soman, K.P.: SVM based part of speech tagger for Malayalam. In: International Conference on Recent Trends in Information, Telecommunication and Computing, pp. 339–341 (2010)
9. Rajeev, R.R., Jayan, J.P., Sherly, D.E.: Tagging Malayalam text with parts of speech-TnT and SVM tagger comparison. In: Proceedings of International Conference on Advances in Computer Science (2010)
10. Nisha, M., RejiRahmath, K., RekhaRajC, T., Raj, P.C.: Malayalam morphological analysis using MBLP approach. In: 2015 International Conference on Soft-Computing and Networks Security (ICSNS), pp. 1–5 (2015)
11. Anisha Aziz, T.A., Sunitha, C.: A hybrid parts of speech tagger for Malayalam language. In: International Conference on Advances in Computing, Communications and Informatics (ICACCI), pp. 1502–1507 (2015)
12. Kumar, S., Kumar, M.A., Soman, K.P.: Deep learning based part-of-speech tagging for Malayalam twitter data (special issue: deep learning techniques for natural language processing). J. Intell. Syst. (2018). https://doi.org/10.1515/jisys-2017-0520
13. Yang, J., Liang, S., Zhang, Y.: Design challenges and misconceptions in neural sequence labeling. In: Proceeding of COLING (2018)
14. Young, T., Hazarika, D., Poria, S., Cambria, E.: Recent trends in deep learning based natural language processing [Review Article]. IEEE Comput. Intell. Mag. **13**, 55–75 (2018)
15. Ling, W., Dyer, C., Black, A.W., Trancoso, I., Fermandez, R., Amir, S., Marujo, L., Luís, T.: Finding function in form: compositional character models for open vocabulary word representation. In: Proceedings of the 2015 Conference on Empirical Methods in Natural Language Processing, EMNLP (2015)

Rainbow Tables for Cryptanalysis of A5/1 Stream Cipher

Praveen Kumar Gundaram, Swamy Naidu Allu, Nagendar Yerukala, and Appala Naidu Tentu

1 Introduction

The European Telecommunications Standards Institute (ETSI) developed a standard called Global System for Mobile communication (GSM), which describes the digital cellular networks protocols used by mobile device. These protocols provide over-the-air communication privacy. GSM standard is the most widely used network system all over the world. It was the first security system containing security threats. There were no practically secure cellular systems earlier, so call theft, eavesdropping on cellular calls and phone cloning such type of criminal activities were increased drastically. Today, GSM cellular telecommunications system is the most secured system in the world. The standardized security methods are incorporated in GSM technology. GSM provides confidentiality and anonymity of the GSM subscribers. It ensures end-to-end security for the users. The aim of GSM is to protect the privacy of subscribers as well as to protect the network from unauthorized access.

P. K. Gundaram
C R Rao Advanced Institute of Mathematics, statistics and Computer Science, University of Hyderabad Campus, Gachibowli, Hyderabad, India

Acharya Nagarjuna University, Guntur, Andhra Pradesh, India

S. N. Allu · N. Yerukala (✉) · A. N. Tentu
C R Rao Advanced Institute of Mathematics, statistics and Computer Science, University of Hyderabad Campus, Gachibowli, Hyderabad, India

M. Palesi et al. (eds.), *Second International Conference on Networks and Advances in Computational Technologies*, Transactions on Computational Science and Computational Intelligence, https://doi.org/10.1007/978-3-030-49500-8_22

1.1 GSM Security Algorithms

GSM consists of three main algorithms as specified in [7]: (1) Authentication: The A3 algorithm (mobile station authentication algorithm), (2) Key Generation: The A8 algorithm (voice privacy key-generation algorithm), and (3) Encryption: The A5 algorithm (over-the-air voice privacy algorithm). All these algorithms are relatively weak and therefore have successfully been attacked in the past. In GSM [16], the design of first two algorithms (i.e., A3 and A8) is not specified, they specified only the external interface of these algorithms. The operators can independently select the exact design of the stream cipher algorithms.

The A5 algorithm is used for the encryption of data. This provides confidentiality. There are three currently used versions of A5: A5/1, A5/2, and A5/3. In 1999, both A5/1 and A5/2 algorithms were reverse engineered from and actual GSM handset by Briceno [5]. A5 algorithm takes the session key K_c and frame number F_n as inputs and produces the pseudo-random bits, called keystream bits. Finally, these keystream bits are XoRed with the message bits to get the ciphertext.

2 Related Work

The approximate architecture of A5/1 was leaked in 1994 and the exact inner workings of stream ciphers A5/1 and A5/2 were discovered by Briceno et al. [5] in 1999. Golic initially proposed two known-key stream attacks: the time-memory tradeoff (TMTO) and guess and determine attacks on the leaked design of A5/1 in 1997 [8]. In 2000 Biryukov et al. [4] significantly improved the TMTO attack proposed by Golic. This attack could retrieve the key in less than a second. This attack needed large pre-computation and four 74-gigabyte disks. Biham and Dunkelman [3] took a different approach in 2000. This attack requires known-key stream (seconds) and within 2^{40} A5/1 cycles, key was recovered. In 2001 correlation attacks were applied by Ekdahl and Johansson [6] to A5/1. Given 4 min of keystream, their attack could find the key within a minute on a PC. Correlation between the internal state of A5/1 and the output bits were discovered by Maximov et al. [12]. Using this correlation they improved the attack on A5/1, in this they used less than 5000 frames and recovered the key within 0.5–10 min on a PC. In 2008 Barkan et al. [2] proposed an ciphertext attack only that uses error correction codes used in GSM before encryption to find the encryption key of A5/2. They applied this attack to A5/1 and also presented a cipher text-only time-memory trade-off attack on A5/1 cipher. In GSM systems using A5/1 all the above attacks requires a large pre-computation steps, a high time complexity or large amount of known keystream. We compare the time, memory, and pre-computation for each and every attack on A5/1. Some attacks on A5/1 are shown in Table 1.

Table 1 Existing attacks on A5/1

Attack	Pre-computation	Time	Memory	Data	Success rate
Henricksen et al. [11]	Two unified rainbow tables	9 s	984 GB × 2	8 known-keystreames	81%
Nohl et al. [15]	40 pre-computation tables with a total about 2 TB	5 s	2 TB	2 known keystreames	90%
Ekdahl and Johansson [6]	–	5 min (on single PC)	–	140 min (ciphertext only)	76%
Ekdahl and Johansson [6]	–	4 min (on single PC)	–	99 min (ciphertext only)	33%
Biham and Dunkleman [3]	–	1.5 days (on single PC)	–	60 min (ciphertext only)	63%
Maximov [12]	–	10 min (on single PC)	–	20 min (ciphertext only)	99.99%
Biryukov [4]	2^{48} parallelizable data preparation stage	On single PC	–	2 min of the conversation	–
Barkan (improved estimators) [1]	Calculate estimators	6–10 min	–	2000 frames	91%
Ekdahl and Johansson [6]	–	< 5 min (on single PC)	–	40 first bits from about 2^{16} frames, 5 min GSM conversation	70%
Maximov et al. [12]	140 computers working together with 22 × 220 GB HDD	2^{28} by one pc	–	–	–
Krause et al. (BDD) [10]	–	Polynomial time, complexity = $n^{O(1)}2^{an}$, $(a = 0.6403)$	–	–	–
Biased birthday attack [4]	2^{48} steps	1 s	2 × 73 GB	2 min of conversation	–
Random subgraph attack [4]	2^{48} steps	1 min	4 × 73 GB	2 s of conversation	–
Barkan et al. [2]	–	A single message	200 GB Disks	3.33 min online phase of a single PC	–
SACCH [2]	–	64 s	200 GB Disks	13.33 min online phase of a single PC	–

3 A5/1 Cipher

For the protection of over-the-air transmissions, in a GSM network they are encrypted with a stream cipher. The stream cipher A5/1 [5, 14] is a mixture of three linear feedback shift registers (LFSRs) as shown in Table 2. The specifications of three shift registers are in Fig. 1. Each register is associated with one clocking bit. All three registers are clocked using a majority rule (stop and go fashion).

At each cycle, the majority bit is determined using the clocking bits of all three shift registers. If the clocking bit agrees with the majority bit, then the register is clocked. At each cycle, at least two or three registers are clocked.

3.1 *Stepwise Procedure of A5/1*

A5/1 operates on 228-bit blocks called frames. Every 4.615 ms frames are sent and received over the air. In digital audio signals, 114 bits of data are sent from MSE and remaining 114 bits of data received from Mobile Station Equipment (MSE). A5 generates the 228-bit pseudo-random number (PRAND) using the 22-bit frame counter F_n and 64-bit session key K_c. Plaintext frame of 228 bit is XOR'ed with the PRAND in order to obtain resulting 228-bits of ciphertext.

Step 1:

- Initially, all the three LFSRs are set to zeros.

Table 2 Specifications of A5/1 stream cipher

LFSR	Length in bits	Feedback polynomial	Control bit	Tap positions
1	19	$x^{19} + x^{18} + x^{17} + x^{14} + 1$	8	13, 16, 17, 18
2	22	$x^{22} + x^{21} + 1$	10	20, 21
3	23	$x^{23} + x^{22} + x^{21} + x^{8} + 1$	10	7, 20, 21, 22

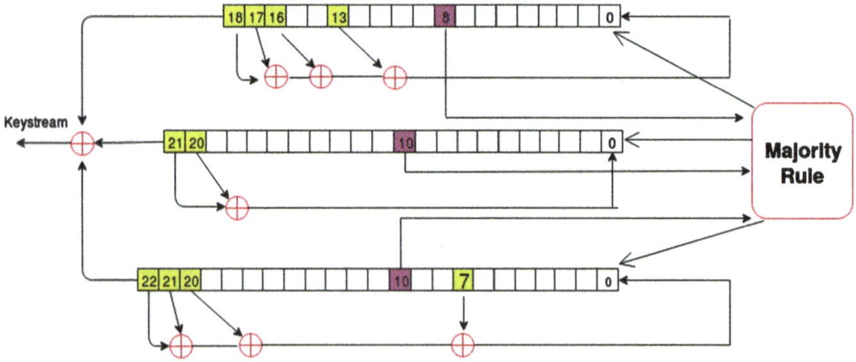

Fig. 1 A5/1 LFSRs

Step 2:

- All three LFSRs are clocked regularly for 64 times.
- Every bit in the *session key* is exclusive-ORed with the feedback bits of the LFSRs in order to obtain the new feedback bits.

Step 3:

- All registers are regularly clocked for 22 times.
- Every bit in the *frame counter* (IV) is exclusive-ORed with the feedback bits of the LFSRs in order to obtain new feedback bits.
- Each frame counter (IV) contains the number of frame that is being encrypted and frame length is 228-bits.

Step 4:

- All registers are irregularly clocked for 100 times.
- This irregular clocking follows the majority rule. Majority bits is determined based on clocking bits of the registers. If the clocking bit of the register is same as the majority bit, that register will be clocked and remaining LFSRs are not clocked. Output of the registers is discarded.

Step 5:

- After initialization of the registers is complete, the registers are clocked for 228 times with irregular clocking.
- Output of each register is XORed to produce 228 bit long keystream.

Step 6:

- Plaintext is XORed with the 228-bit keystream generated is obtained 228-bit of Ciphertext.
- To encrypt a consecutive frame, same session key will be used, but the frame counter will be incremented by 1.
- The session key changes if the mobile device has to re-authenticate itself with the network carrier.

4 Proposed Rainbow Table Methods

In any particular algorithm like A5/1, breaking depends on the calculating the inversion of one-way function. There are two methods proposed in [13]: (1) A brute force attack on an average of 2^{n-1} values can be performed until the target has been reached. (2) Pre-compute and store around 2^n input and output pairs. Sometimes only single lookup is needed for inversion of particular value. For the above two methods space and time trade-off exists. If n is large, storage and computation of 2^n entries is impractical. Requirements of storage can be reduced if we either all the encrypted texts can be decrypted or if we add more computation that is not actually stored in the table, this trade-off was first explored by Hellman. This work was later extended by Oechslin, where he proposed Rainbow Tables for TMTO attacks. TMTO attack has two phases: (1) pre-computation phase, the out of this phase is a look-up table containing key and cipher text pairs. The time required to construct the table is quite high. (2) Online phase, this phase recovers the internal state given the keystream using the look-up table built earlier.

4.1 Method-1: Rainbow Table Generation Algorithm for A5/1

The Rainbow table [9] entries are Starting (SP) and Ending Points (EP). SP is a randomly chosen 64-bit word and EP is a stored 64-bit Distinguished Point (DP), DP is a 64-bit word whose 15 LSB_s are zeros [1]. Each EP is an output generated by constructing a chain by taking its corresponding SP as an input. Generation of chain is as follows:

1. Start Point (SP) is fed into keystream generator (A5/1) as a seed value and the output is of 64 bits which becomes input to the reduction function.
2. Check whether the output of the reduction function is a distinguished point or not.
   - If the Output of reduction function is not a distinguished point, then it is fed again into keystream generator (A5/1) with the same reduction function and same process is repeated until the result is a distinguished point.
   - If the Output of reduction function is a distinguished point.
     - Output of reduction function is fed into keystream generator.
     - Select the next reduction function.
3. The chain terminates after using all reduction functions.
4. The last distinguished point in a chain is called End Point (EP).

The above procedure is repeated for all SPs.

Reduction Function In this Rainbow table construction, Reduction function is nothing but Xoring the keystream bits with one of the following numbers in the set: {1,3,...,63}, RF_{-i}: (K, 0xi)= $K \oplus 0xi$, where i: {2i+1/i∈0,1,...,31}. Thus we apply 32 Reduction functions.

Chain Generation The first column of the rainbow table contains several start points(SP's), each is of 64-bit length, which act as the seed values for the algorithm, meaning that each SP is the internal state of the A5/1, after initialization of session key(K_c) and frame number(F_n). The second column of the rainbow table contains corresponding end points (EP's), each of which is 64-bit in length. These end points are generated by the rainbow table algorithm as shown in the Fig. 2.

Attack Procedure This is a known-plaintext attack. The attacker gets the 228-bit keystream by XORing the known plaintext with the corresponding ciphertext. He

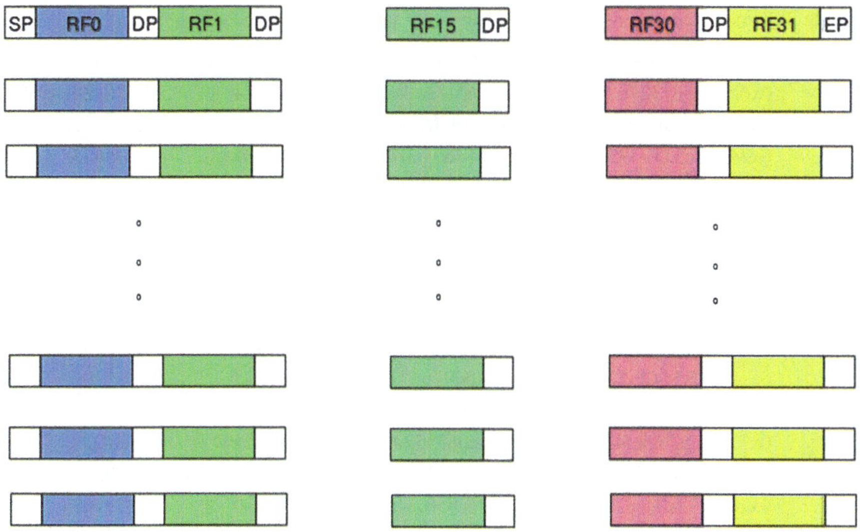

Fig. 2 Chains of rainbow table

now takes first 64-bits from a block of first 114-bits keystream of 228-bit keystream. Then he compares with EP's of rainbow table. If it matches with one of the EP's, then he takes the corresponding SP and apply the rainbow table algorithm to get the chain starting with SP and ending with EP. In this chain, the state which comes before the EP() is the internal state of the A5/1 algorithm. From this internal state, we do inverse process of initialization to get the session key (K_c). Suppose if it does not match with any of the EP's in the table, then we apply the A5/1 algorithm by seeding this 64-bit keystream and then apply 32nd reduction function and check the output with any of the EP's. If it matches, then follow the above process to get the session key. If again it does not match with any of the EP's, then again apply A5/1 followed by 31st and apply A5/1 followed by 32nd reduction function and check with EP's. Continue this process until it matches with any of the EP's in the rainbow table and follow the above process to get the session key.

Table 3 shows the rainbow table which displays (SP's, EP's) and corresponding chain length with execution time.

Method-1 Experimental Evaluation The Experimental Evaluation was done on the following systems.

Desktop Specification(DS) Ubuntu 14.04LTS (64 bit) and Processor Intel Core i7 CPU 860 @ 2.80GHz × 8.

Param Shavak(PS) System Specifications CentOS Linux 7 (Core) 64-bit, Model name: Intel(R) Xeon(R) CPU E5-2680 v4 @ 2.40GHz, Core(s) per socket:14.

Table 3 Time taken to generate rainbow table (for sample 30 pairs, SP-EP)

S.No	Start point (SP)	End point (EP)	Chain length	Execution (in s)
1	0x6b8b4567327b23c6	0xc8d205f23e8c8000	988774	7.576120
2	0x643c986966334873	0xfdf7134607518000	1224052	9.363431
3	0x74b0dc5119495cff	0x045a1d651c070000	948989	7.276483
4	0x2ae8944a625558ec	0x10753ab1374a8000	1214069	9.336693
5	0x238e1f2946e87ccd	0x0c9f4dc040b20000	1037026	7.940774
6	0x3d1b58ba507ed7ab	0x16c7f87238cc0000	900845	6.959752
7	0x2eb141f241b71efb	0x2855a0be73a68000	741602	5.675136
8	0x79e2a9e37545e146	0x45e5f7f6197f8000	988950	7.580269
9	0x515f007c5bd062c2	0xe6b39a5e4bda0000	910437	6.956055
10	0x122008544db127f8	0xa9140d94b6330000	786393	6.015109
11	0x0216231b1f16e9e8	0xde07eb27747b8000	899198	6.906403
12	0x1190cde766ef438d	0xdbbfe06755348000	960121	7.331027
13	0x140e0f763352255a	0xc6b37fa33c320000	965375	7.409501
14	0x109cf92e0ded7263	0xee82ecb502608000	981630	7.579908
15	0x7fdcc2331befd79f	0xbf0172d6f5430000	677112	5.305468
16	0x41a7c4c96b68079a	0x267bc129236c8000	982227	7.583342
17	0x4e6afb6625e45d32	0x48a910cb2dc20000	589723	4.542916
18	0x519b500d431bd7b7	0x8559f9ba39548000	1209324	9.247417
19	0x3f2dba317c83e458	0xfb0cdbacedb08000	898887	6.894113
20	0x257130a362bbd95a	0xacda565286708000	905872	7.004087
21	0x436c6125628c895d	0x9017bb1825258000	824542	6.291410
22	0x333ab105721da317	0xf5b9aad5715c8000	1611710	12.408952
23	0x2443a8582d1d5ae9	0x0db4a85a27280000	1030039	7.876851
24	0x6763845e75a2a8d4	0x8f98ece6dff88000	952045	7.249338
25	0x08edbdab79838cb2	0x9c98003f47cb8000	1240669	9.485149
26	0x4353d0cd0b03e0c6	0xf6acd64f8d0a8000	960893	7.341437
27	0x189a769b54e49eb4	0xfaac9c977c9c0000	707247	5.492332
28	0x71f324542ca88611	0x21e44f706bb78000	956025	7.415580
29	0x0836c40e02901d82	0x39f289bf81dd0000	783212	6.026209
30	0x3a95f87408138641	0xd18080609a758000	1077626	8.435817

Param Shavak is a HPC system powered with two multicore CPUs, each with 14 cores along with either one or two Xeon-phi or GPU accelerator cards and it has 5.7 Tera flops computing power with 8 TB storage.

The number of chains that were implemented in parallel are 10, 128, 181, 200, 250, 300. Table 4 depicts the speedup and compares the time (in s) for the rainbow table calculation for the existing scheme [9] and proposed scheme.

Method-1 covers 2^{15} keys in each color, 2^5 colors in each chain, and 2^{37} chains in the table shown in Fig. 2 which gives us a total of $2^{15} * 2^5 * 2^{37} = 2^{57}$ keys covered. The following are the reasons for not having all the possible 2^{64} keys:

Table 4 Execution time for different set of instances

	Time taken in seconds			
No of chains	Existing paper [9]	Our results (DS)	Our results (PS)	Memory
1	6.30	6.29	5.67282	16 Bytes
10	77.44	19.76	8.62612	160 Bytes
128	870.00	234.152	55.4599	2.0 KB
181	1217.66	335.286	77.7806	2.9 KB
200	1353.66	376.127	91.9085	3.2 KB
256	1670.04	469.186	104.675	4.1 KB
300	2036.73	499.024	126.258	4.8 KB
345	2346.70	576.154	144.922	5.5 KB

- If we cover all possible keys, then the chance of having collisions is higher, which increases calculation time.
- If we want to cover more keys, then it takes more time to generate them and also requires more space to store them.

Then the table had to be sorted by the last column value for a fast search during the attack.

4.2 *Method-2: Rainbow Table Generation Algorithm for A5/1*

Procedure is same as Method-1 which is described in Sect. 4.1. But the least significant bits of a distinguished point are zero. Generation of chain is same as described in Sect. 4.1.

Reduction Function In this Rainbow table construction, Reduction function is replacement of 17 bits of residue (r_i) in keystream generator (A5/1) output. i.e $K = (K_4||K_3||K_2||K_1)_{LSB}$ and $r_i = (-K_4 + K_3 - K_2 + K_1)\%(2^{16} + 1)$. RF_{-i}: (K, r_i)= $(K_{47-bit}||r_i)$.

Thus we apply 8 Reduction functions.

Method-2 Experimental Evaluation Table 5 shows the rainbow table which displays (SP's, EP's) and corresponding chain length with execution time. The number of chains that were implemented in parallel are 10, 128, 181, 200, 250, 300. Table 6 depicts the speedup and compares the time (in seconds) for the rainbow table calculation for the proposed scheme.

Method-2 covers 2^{17} keys in each color, 2^3 colors in each chain, and 2^{37} chains in the table shown in Fig. 2 which gives us a total of $2^{17} * 2^3 * 2^{37} = 2^{57}$ keys covered (Table 6).

Table 5 Time taken to generate rainbow table-2 (for sample 30 pairs, SP-EP)

S.No	Start point (SP)	End point (EP)	Chain length	Execution (in s)
1	0x6b8b4567327b23c6	0xe38f9412a7300000	564282	3.488360
2	0x515f007c5bd062c2	0xb0496171fe680000	532742	3.290145
3	0x579478fe749abb43	0xa43e596173de0000	536644	3.317146
4	0x7e0c57b177ae35eb	0x0ae5d6b1f5580000	740904	4.757167
5	0x57fc4fbb0cc1016f	0x5f895956a4e80000	568662	4.083596
6	0x69e7f3e52a6de806	0xd25fe82b81740000	573184	3.835382
7	0x746f2e306fde8af6	0xccdcdfa709000000	562060	3.541554
8	0x424479da1a9a9e69	0x998c68e180200000	618072	3.827082
9	0x064af49b397c46bc	0x71c60ed60a660000	527254	3.415073
10	0x4c2a71662e534a82	0x707aec6a163a0000	601787	3.839454
11	0x2db88089706b674e	0x193f489ce04c0000	526107	3.259552
12	0x66a48d1156c28e34	0xe60c0bbbf90c0000	531841	3.289621
13	0x7835626c665aca49	0xbdf1b4684a6a0000	546347	3.879913
14	0x4a9554fe392edbe4	0xf19e84524cfa0000	738682	4.592455
15	0x026baae92c02fe8c	0xf3b48d1dc39a0000	535434	3.349314
16	0x682dfed6606ed7f6	0x4ddcc86e954a0000	574886	3.576006
17	0x5992a02e29ef532d	0x699102f20fde0000	565161	3.488687
18	0x4b683d0d076e41d8	0x5cc22a529cae0000	547951	3.382093
19	0x43f8e1ac69215dfb	0xa2597faa42f60000	525893	3.287729
20	0x6163ed0d0c600e47	0x5f2dc1c292600000	778224	4.809690
21	0x102809e23b2125a3	0xac28199a3ddc0000	530291	3.292612
22	0x450b7fb6340bf64d	0xb542c9d53b240000	616428	3.820736
23	0x449f66fe557e0515	0x96f3bf5b80e20000	534434	3.634360
24	0x2c70edae44296c6d	0x0d489121b5080000	668796	4.210969
25	0x76dee9180b79d08d	0x9d65c078b2ee0000	527123	3.268758
26	0x6687f34d25415b0c	0xa73484106d9a0000	536033	3.307526
27	0x4ac9f3e20cc0a782	0xb5ac48138a640000	655710	4.102253
28	0x6d651b8d31ed2baf	0x1bf05a8bb2540000	531792	3.324015
29	0x222fc865047c27fa	0x0c7fbcfb2b900000	539336	3.348265
30	0x0acd4c11111817cf	0x35736cf6b3520000	584400	3.603085

Table 6 Execution time for different set of instances

No of chains	Time taken in seconds	Memory
1	3.452917	16 bytes
10	35.514525	160 bytes
128	465.309537	2.0 KB
181	665.71077	2.9 KB
200	733.482986	3.2 KB
256	939.309152	4.1 KB
300	1106.684391	4.8 KB
345	1274.740848	5.5 KB

5 Conclusion

Two different ways of efficient implementation of rainbow tables for the stream cipher A5/1 were presented. Time-Memory Trade-Off attack uses the rainbow table for breaking this cipher. Given ciphertext and known plain text bits, TMTO attack retrieves the internal state which is after loading K_c, furthermore deciphering of a conversation. Previous attacks need much pre-computation/memory/high time complexity. Two rainbow table constructions we presented are easy and efficient to implement in terms of time complexity. These tables can be used to attack A5/1 stream cipher by TMTO attack efficiently. Experiment results are presented for proposed constructions. Execution time is very less when compared to existing results.

References

1. Barkan, E., Biham, E.: Conditional estimators: an effective attack on A5/1. In: International Workshop on SAC 2005. LNCS, vol. 3897, pp. 1–19. Springer, Berlin (2006)
2. Barkan, E., Biham, E., Keller, N.: Instant ciphertext-only cryptanalysis of GSM encrypted communication. J. Cryptol. **21**, 392–429 (2008)
3. Biham, E., Dunkelman, O.: Cryptanalysis of the A5/1 GSM stream cipher, progress in cryptology. In: Proceedings of Indocrypt-00. Lecture Notes in Computer Science, vol. 1977, pp. 43–51. Springer, Berlin (2000)
4. Biryukov, A., Shamir, A., Wagner, D.: Real time cryptanalysis of A5/1 on a PC, advances in cryptology. In: Proceedings of Fast Software Encryption'00. Lecture Notes in Computer Science, vol. 1978, pp. 1–18. Springer, Berlin (2001)
5. Briceno, M., Goldberg, I., Wagner, D.: A pedagogical implementation of the GSM A5/1 and A5/2 "voice privacy" encryption algorithms (1999). http://cryptome.org/gsm-a512.htm (originally on www.scard.org)
6. Ekdahl, P., Johansson, T.: Another attack on A5/1. IEEE Trans. Inf. Theory **49**(1), 284–289 (2003)
7. ETSI: European Telecommunications Standards Institute- GSM Architecture. https://www.etsi.org
8. Golic, J.: Cryptanalysis of alleged A5 stream cipher. In: Proc. of Eurocrypt'97. LNCS, vol. 1233, pp. 239–255. Springer, Berlin (1997)
9. Kalenderi, M., Pnevmatikatos, D., Papaefstathiou, I., Manifavas, C.: Breaking the GSM A5/1 cryptography algorithm with rainbow tables and high-end FPGAS. In: 22nd International Conference on Field Programmable Logic and Applications (FPL), 29–31 Aug 2012
10. Krause, M.: BDD-based cryptanalysis of keystream generators. In: EUROCRYPT 2002. LNCS, vol. 2332, pp. 222–237 (2002)
11. Lu, J., Li, Z., Henricksen, M.: Time-memory trade-off attack on the GSM A5/1 stream cipher using commodity GPGPU. In: International conference on ACNS 2015. LNCS, vol. 9092, pp. 350–369. Springer International, Cham (2015)
12. Maximov, A., Johansson, T., Babbage, S.: An improved correlation attack on A5/1. In: Proceedings of SAC'04. LNCS, vol. 3357, pp. 1–18. Springer, Berlin (2005)
13. Meyer S.: Breaking GSM with rainbow tables. Preprint, arXiv:1107.1086 (2011)
14. Nagendar, Y., Prasad, V.K., Rao, A.A., Padmavathi G.: Applications of stream ciphers in wireless communications. Int. J. Comput. Sci. Eng. **6**(6), 1121–1126 (2018)
15. Nohl, K.: Attacking phone privacy. In: BlackHat 2010 Lecture Notes. Security Research Labs, Berlin (2010)
16. Stockinger, T.: GSM network and its privacy-the A5 Stream Cipher, Nov 2005

Shopping Using Augmented Reality

Bharat Suchith, Nikhitha Grace Josh, Nikitha Kurien, P. B. Yedukrishnan, and Kiran Baby

1 Introduction

Shopping is an action wherein a client peruses the accessible merchandise or services displayed by at least one retailer with the potential plan to buy a reasonable choice of them. In the late nineteenth century, shops changed from "single-function" shops offering one kind of good, to the retail chain where enormous assortment of products where sold. A bigger business zone can be found in numerous urban areas, all the more officially called central business district, however, more generally called "downtown" in the USA, or the "high street" in Britain, and souks in Arabia. Shopping centers are accumulations of stores; that is collection of a few businesses in a small area. It comprises of a gathering of retail, amusement, and administration stores intended to serve items to the encompassing district. Typical examples include shopping malls, town squares, flea markets, and bazaars.

Today, home mail conveyance frameworks and present day innovation, (for example, TV, phones, and the Internet), in blend with electronic trade; enable customers to shop from home. There are three main types of home shopping: mail or telephone ordering from catalogs; telephone ordering in response to advertisements in print and electronic media (such as periodicals, TV, and radio); and online shopping. Online shopping is a form of electronic commerce which allows consumers to directly buy goods or services from a seller over the Internet using a web browser. Consumers find a product of interest by visiting the website of the retailer directly or by searching among alternative vendors using a shopping

B. Suchith · N. G. Josh · N. Kurien · P. B. Yedukrishnan · K. Baby (✉)
Mar Baselios College of Engineering and Technology, Trivandrum, Kerala, India
e-mail: kiran.baby@mbcet.ac.in
https://mbcet.ac.in/

M. Palesi et al. (eds.), *Second International Conference on Networks and Advances in Computational Technologies*, Transactions on Computational Science and Computational Intelligence, https://doi.org/10.1007/978-3-030-49500-8_23

search engine, which displays the same product's availability and pricing at different e-retailers.

Internet shopping has turned into a noteworthy game changer in the retail business. Buyers would now be able to scan for item data and put in item requests crosswise over various areas while online retailers convey their items straightforwardly to the buyers' home, workplaces or any place they need. The B2C (business to consumer) process has made it simple for customers to choose any item online from a retailer's site and to have it delivered generally rapidly. Utilizing web based shopping strategies, purchasers do not have to expend vitality by physically visiting physical stores, yet spares time and travel expense. In only 10 years, the shopping pattern has moved from a direct, retail-engaged model, to the present iterative, advanced driven model of client conduct. Retailers, today, are attempting to give clients an omnichannel involvement. Clients experience consistent shopping, regardless of whether they are shopping on the web from a work area or cell phone, by phone, or in a physical store.

1.1 Problem Definition

Retailers should address the issues of their purchasers by meeting both the online and offline experiences—a methodology that is backed by information revealing that the expense of attention was the most dramatic increase in operational expense in the previous 25 years. Customers value a more extensive choice and plenitude of data, and keeping in mind that they accumulate data online about the items they want, customers relish the in-store involvement of touching, seeing, and trying on a product. In spite of the fact that the objective is to offer clients a consistent shopping background over various channels, the key is to locate the most ideal approach to consolidate web based business with in-store shopping to make an interconnected, powerful retail experience. The serious issue with internet shopping is that you cannot try it before you get it.

1.2 Assumptions

- Majority customers have phones with a minimum API level 26, i.e. Android 8.0 Oreo
- AR/Vuforia is supported in most phones.
- Majority of the people love to shop online.
- People like to try out new technologies.
- Internet facility is available.
- Majority of people have good quality phones so that the app does not lag.

2 Literature Review

Researches have been conducted to study what factors drive a customer to use AR. Recent studies have shown that people, who are Gadget Lovers, or frequent online shoppers, are more likely to have tried AR. The study focused on how personality, innovation seeking behavior, tech savviness, and shopping experience seeking behavior of shoppers affect their preference to shopping using AR. The study shows that income was not a hindrance to using AR. People who tried AR were not Early Adopters or Innovators in the technology adoption life cycle [1].

Applications in e-commerce have moved from web onto mobile. Mobile applications allow touch, feel, and gestures thereby taking e-commerce to the next level. Clearly AR/VR is the next step. AR simply requires a smartphone with no additional requirements, whereas VR on the other hand requires additional equipment. Reference [2] introduces an application platform called Lify that aims to reduce development and operating costs. Several platforms like Lify have been developed to support AR thereby removing the need to know the AR encoding algorithms.

Researches show that by 2020, generation Z would constitute about 40% of consumer base. They are a digital generation that focuses on speed than accuracy. They like using AR since many of their desired values are delivered from such a shopping experience. The service based economy has now turned into an experience based economy [3].

Several companies like Lenskart have adopted augmented reality. Researchers have proposed techniques that do not require the use of markers, additional hardware or special setup in order to fit eye glasses to the face of the user [4].

Several enhancements have been suggested in AR. Researchers have suggested a product recommendation system in AR thus taking retail in AR a step higher. Customers can get personalized recommendations of product in the physical spaces where they may be used [5]. The study focuses on the impact of placing recommended content (using AI to recommend) in physical context using AR [5]. It has been found that majority customers find it better to use AR than conventional browsers so they can see it in the context where it will be used. Today, AR is solving more problems than just entertainment.

3 Existing Solutions

Today, many companies have started to realize the importance of AR in the future of retail. Companies like IKEA and Wayfair have turned to AR to solve their existing online shopping issues. IKEA, a Swedish founded multinational group that sells ready-to-assemble furniture, has implemented AR in their mobile application. IKEA aims to remove the hesitancy of customers to buy a product online by projecting true-to-scale 3D furniture. But, in order to use this feature, users have to place a copy of the IKEA catalog in a frame to help the app calculate the scale. Wayfair, an American e-commerce home goods company that sells over ten million products from over 10,000 suppliers, has made AR available to both their IOS (Apple's ARKit) and Android (Google's ARCore) users. Wayfair projects its products in 3D at full scale and anchors it to the floor. Converse Shoes is an American shoe company that has been a subsidiary of Nike since 2003. They have also released an AR application called "The Sample" which allows customers to virtually try on shoes before purchasing it. It also allows customers to share a picture with the shoes on. Dulux Visualiser is another application released by Dulux Paints, that helps the customer picture their walls before painting it. The customers, therefore, can test different colors on the walls without actually painting their walls.

4 Proposed Work

4.1 Data Collection and Sampling

This study aims to implement a system which is able to detect objects already placed in the physical environment, and place a virtual product in the present surroundings. It aims to provide customers with the option to compare products of different vendors and dynamically alter positions of a pre-existing augmented object with ease. The study was based on the survey of 49 shoppers of ages 20–40 conducted by us. Those surveyed included mainly females (63.3%). Refer Fig. 1. 5.1% of the respondents are of the age of 21, 12.2% of them are of the age of 20, and 14.3% are of the age 22. Refer Fig. 2.

Fig. 1 Distribution of gender

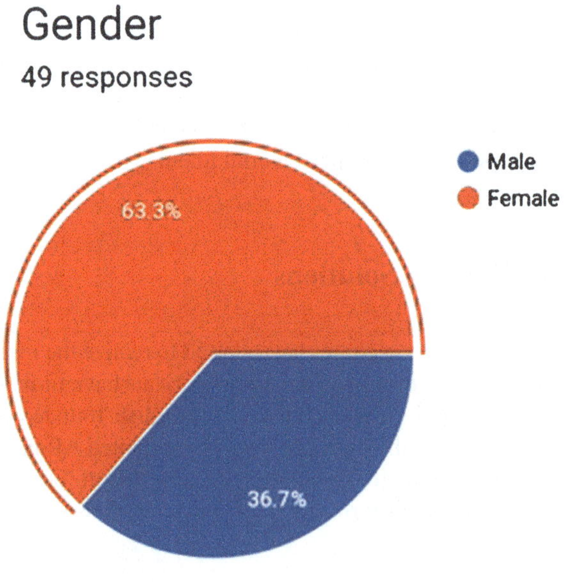

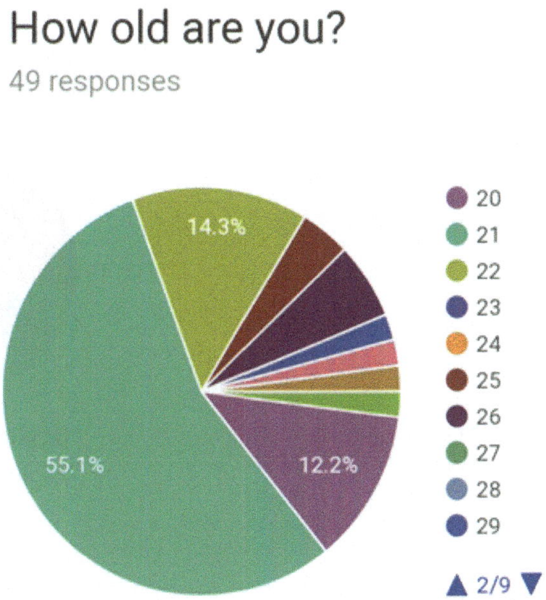

Fig. 2 Distribution of age

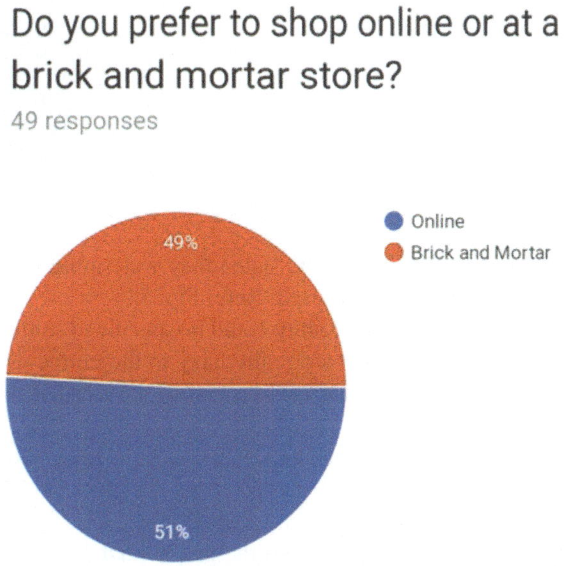

Fig. 3 Distribution of brick and mortar or online shoppers

The responses show that most people prefer online shopping (51.1%), but they are more likely to buy a product while shopping at a brick and mortar store (48.9%). Refer Figs. 3 and 4. This result clearly shows the gap that is needed to be bridged. This could possibly be overcome by the augmented reality, as it combines the effect of both the shopping online and brick and mortar store, as you could get the feel of the product and shop at ease.

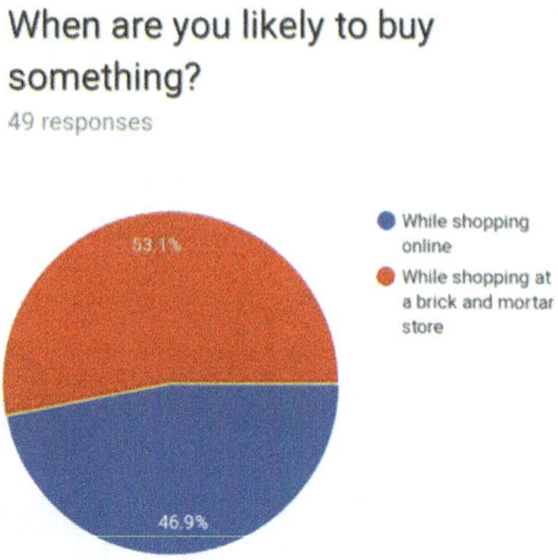

Fig. 4 Distribution of likeliness to buy online

Fig. 5 Most shoppers are likely to buy online with AR

Most (89.8%) shoppers stated they were more likely to buy a product online, if augmented reality was used. Refer Fig. 5.

Use of augmented reality could be specific to some products like shoes or clothes whose measurements are really hazy in the online world. Enhancing the ability to measure such products could increase the trust in the online store and the buyer experience.

Among the respondents, 26.5% were regular shoppers (people who are crazy about shopping), 65.3% were normal shoppers, and 6.1% did not fancy shopping. Refer Figs. 6 and 7.

It can be concluded from the graphs that people find it comfortable and prefer shopping online more than at a brick and mortar store, but at the same time, when it comes to purchasing a product, they are more likely to buy something in a brick and mortar store.

Fig. 6 Shopping interest

Fig. 7 Frequency of shopping

4.2 *Proposed Work*

An augmented reality e-commerce assistant system for furniture and other home décor commodities is designed to provide consumers with more realistic and practical product experiences and interactions. With the development of an augmented reality e-commerce portal, online consumers can bring a product into their own physical environment and even try out and visualize the product in their physical environment while shopping from their mobiles. Similar to the traditional e-commerce systems, our online shopping portal using augmented reality requires the Internet as the elementary user interaction platform. In addition to this our augmented reality online shopping portal also needs a camera to capture the consumer's physical environment and then integrate and blend it with virtual objects in real time.

Modules The modules used in this software are as follows:

- Login: This module allows customers to login to the application. The login process involves typing a valid username and password combination. All customers who have registered can login.
- Registration: This module allows the user to register to the application. It involves filling a personal profile including data such as name, address, phone number, e-mail, etc. Registered customers will be provided a valid username and password combination. Only registered customers can purchase product using the application.
- Inventory: This gives the details regarding the home decor available for sale.
- Filter: This module allows a user to filter salable products based on his/her need. For example, users may filter on price range, items, company, etc. This eases the task of choosing products that fit their need.
- Product: This module shows the details of each product like its shape, size, color, prize, warranty, etc. This module also allows an option that directs the customer to the augmented reality module.
- Augmented Reality: This module displays items in real size by projecting a visual replica though the camera on a smart phone. One can reposition the items to any angle to view the furniture in one's home space.
- Cart: This module shows the products that the user wishes to purchase. The customer also has the option to remove items in the cart later if they decide not to buy it. Items can be removed or added before payment.
- Wishlist: This module shows a list of products that the customer has liked and wishes to buy at a later time.
- Payment: This module deals with the payment options that validate the customer's purchase. The customer may proceed to payment if he/she has decided on the products in the cart. The customer may pay for the products in the cart using cash on delivery, debit card, credit card, and other payment options.

Refer Fig. 8.

4.3 Advantages over Traditional System

It allows customers to "try before they buy with a 3D product review" and enhances shopper experience as they can visit many stores at the convenience of sitting at home.

Unlike in already existing shopping applications, the users do not need any catalog in order to project products onto the real environment. Since markerless AR is used, the app itself predicts the ground plane and places the product wherever the user taps on the device.

It is a great hand of help for the differently abled customers who cannot go out for shopping. It is a combination of traditional retail experience and e-commerce. The

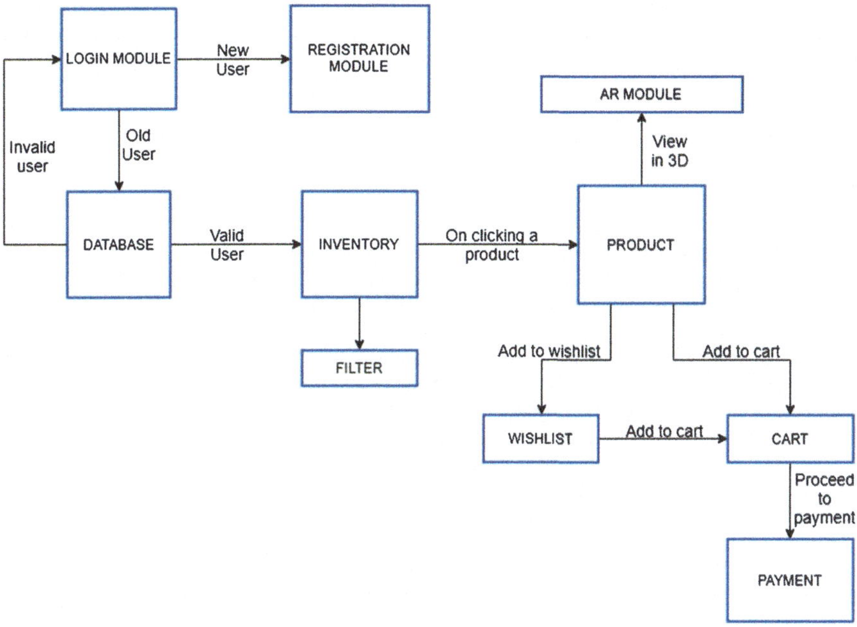

Fig. 8 Block diagram

company will have enhanced brand recognition and improved profitability as this system increases the online conversion rate and reduces returns. Shoppers will be able to purchase a genuine product by reading reviews. Thus customer satisfaction is increased.

5 Conclusion

In the current scenario, a lot of uncertainties bind consumers from purchasing their needs online. The major reasons why they go behind real and face to face shopping is the fact that a lot of confusions arise, say—cashless online transactions, doubts about the guarantee of services/products ensured online, and so on. And particularly in case of commodities like dresses, numerous complaints arouse about size fitting when the dresses are delivered. This is the time to let AR give a chance to tackle the above-mentioned uncertainties. To be precise, augmented reality supplements reality, without replacing it. As seen, AR is so far so good in setting up a real-world atmosphere virtually, thus enabling consumers to try on whatever they want to, like they go out to shops in person. Online trials of anything and everything can be easily tried based on their convenience. In addition to this, AR provides quite flexible and dynamic user interface when its application gets proposed in 3D interior designing. With this AR feature, it is unimaginably fast and comfortable to decide what furniture fits into your home ambiance, or what clothes go into your wardrobe, by just swiping through your mobile phones.

References

1. Chakraborty, S., Gupta, D.: A study of the factors impacting the adoption of augmented reality in online purchases in India. In: 2017 2nd IEEE International Conference on Recent Trends in Electronics, Information & Communication Technology (RTEICT) (2017). https://doi.org/10.1109/RTEICT.2017.8256853
2. Atalar, M., Özcan, M.: New augmented reality application in E-commerce and M-commerce. In: 2017 International Conference on Computer Science and Engineering (UBMK) (2017). https://doi.org/10.1109/UBMK.2017.8093403
3. Kapusy, K., Lógó, E.: Values derived from virtual reality shopping experience among generation Z. In: 2017 8th IEEE International Conference on Cognitive Infocommunications (CogInfoCom) (2017). https://doi.org/10.1109/CogInfoCom.2017.8268249
4. Mallik, A., Bhowmick, B.: An efficient and robust method of virtual augmentation of eye-glass for easy shopping. In: 2016 14th International Conference on Control, Automation, Robotics and Vision (ICARCV) (2016). https://doi.org/10.1109/ICARCV.2016.7838752
5. Huynh, B., Ibrahim, A., Chang, Y.S., Höllerer, T., O'Donovan, J.: A study of situated product recommendations in augmented reality. In: 2018 IEEE International Conference on Artificial Intelligence and Virtual Reality (AIVR) (2018). https://doi.org/10.1109/AIVR.2018.00013

Auto-Colorization of Images: Fuzzy c-Means and SLIC Approaches

Sanjay Kumar, Prateek Bansal, Tript Sharma, and Raveesh Garg

1 Introduction

From the black and white classic movies or historical pictures, to even image sensing and research purposes, the colorization of grayscale images has got its uses. The previous related works in this arena had drawbacks as "extra information" needed to be provided along with the original grayscale image. Like the user had to offer colorful scribbles on the target set or the user to attach a reference image to colorize the model. Our approach reduces the efforts of the users and conditionally colorizes monochrome images without any direct input.

The RGB (red–green–blue) image is given as input and pre-processed before training. The sub-squares are generated from the grayscale image, and feature extraction is done using fast Fourier transformation (FFT). The SVR (support vector regression) model is given the expected I and Q values along with features for each sub-square for each segment. MRF is applied, and we get the predicted output image.

The clustering approach of hard c-mean was extended to fuzzy c-means (unsupervised) in 1981 by Bezdek et al. [1]. FCM has broad areas of application ranging from image processing, engineering graphics to medical treatment diagnosis, geology, astronomy. Therefore, we have also introduced another approach involving the fuzzy c-means clustering algorithm.

S. Kumar (✉) · P. Bansal · R. Garg
Department of Computer Science and Engineering, Delhi Technological University, Delhi, India
e-mail: sanjay.kumar@dtu.ac.in

T. Sharma
Department of Mechanical Engineering, Delhi Technological University, Delhi, India
e-mail: tript_bt2k16@dtu.ac.in

M. Palesi et al. (eds.), *Second International Conference on Networks and Advances in Computational Technologies*, Transactions on Computational Science and Computational Intelligence, https://doi.org/10.1007/978-3-030-49500-8_24

A minor setback is that we are not able to reconstruct a full range of colors which is there in the original image like it is failing in including more shades of brown and yellow color. The reason for this could be a limited range of colors present in the dataset as it mainly consists of shades of green and blue. Increasing training dataset leads to better results. Also, such models could be highly exploited for specific domains of application.

As now the model would learn to output images comparable to real-world ones. However, if the system is redesigned considering the adversarial network or applying CNN (convolutional neural networks), accuracy, as well as results, would improve.

This paper is organized as follows: An overview of related work is presented in Sect. 2. In Sect. 3, the technical details of our approaches have been described. Results are demonstrated in Sect. 4. Finally, the conclusion and future works are outlined in Sect. 5.

2 Related Work

Welsh et al. [2] proposed the method which transferred colors based on the value and variance of the luminance of pixel. This method was improved upon by Levin et al. [3] who proposed the idea of scribble-based colorization, in which the colored scribbles are made available along with the target set. They developed the approach in which the pixels are colored based on the neighboring pixels with similar intensities. Least-square optimization is applied to propagate the information of the colored scribbles to the target image. Huang et al. [4] further contributed by proposing an algorithm based on adaptive edge detection for reducing the bleeding color artifact around the borders of the region.

The second idea involves the user to attach a reference image for the colorization of the grayscale image. The "color information" provided by the reference image is then transferred onto the grayscale (input) image. The image regions and pixels are matched by luminance parameters. A patch-oriented colorization algorithm was put forward by Bugeau et al. [5] that takes square patches into account (around each pixel).

3 Proposed Method

Our method here is mainly influenced by the work of F. Liu et al. [6]. He predicted a depth channel of the given monocular images using deep-convolutional neural networks. Their work deals with an MRF over the given image's superpixels. The estimation of the field potentials is carried out using a deep CNN. In Sect. 3.1, we are applying a marginally different approach where SVRs are exploited to provide local estimates, and an MRF is modeled over the superpixels in the image provided

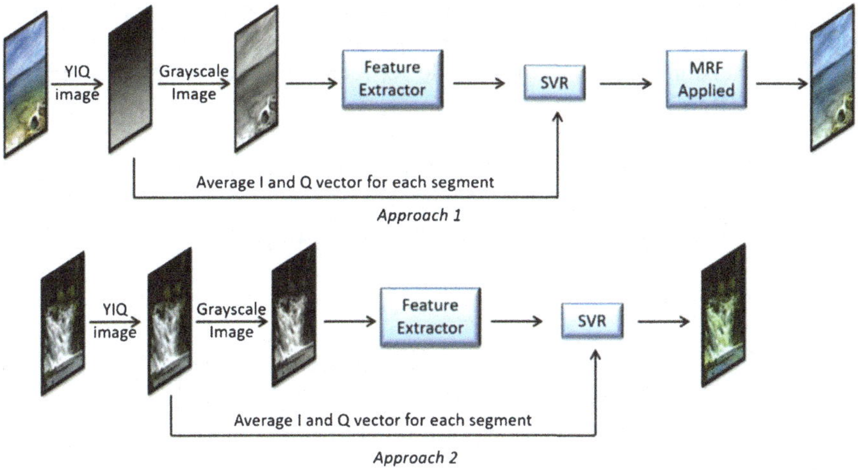

Fig. 1 Proposed approach

as input [7]. A similar approach is followed in Sect. 3.2 with a more sophisticated clustering algorithm.

Figure 1 (Approach 1) shows the process of predicting I and Q chrominance values of each pixel. The input RGB value is converted to the YIQ image, and superpixels are generated using the SLIC algorithm. The average values of I and Q channels along with the centroids of each superpixel are calculated. The centroids are used to create sub-squares from each superpixel in the "Y" channel of the corresponding image. The sub-squares are used to identify the features which are fed to the SVR along with the average I and Q vector. Finally, the predicted color is improved using MRFs. The second approach is slightly different. It involves a much more sophisticated clustering algorithm that results in the centroids of each cluster. Each pixel is mapped to a membership value corresponding to each centroid. These clusters are used for feature extraction and ultimately, identification of the colors using SVRs. The approaches have been elaborated in the following subsections.

Dataset

Yellowstone National Park images were used for training and testing the classifier. The photos were downloaded from Flickr. Pictures with "yellowstone" and "landscape" tags were selected. Images of animals or people were pruned out. The dataset images were then scaled to a uniform width of 200 pixels. Finally, the dataset was split into training and testing sets with a ratio of 80:20.

Image Representation

Our algorithm runs on several images to train our model. In the YIQ color space, the Y component represents the luminance information with I (orange-blue) and Q (purple-green) representing the chrominance information. Our algorithm takes luminance information (Y channel) of an image as input, whereas the chrominance channels, which are I and Q channels, are used in predicting output image.

The conversion formulas used to toggle between the two-color spaces are:

From RGB to YIQ

$$\begin{bmatrix} Y \\ I \\ Q \end{bmatrix} = \begin{bmatrix} 0.2989 & 0.5870 & 0.1140 \\ 0.5959 & -0.2744 & -0.3216 \\ 0.2115 & -0.5229 & 0.3114 \end{bmatrix} * \begin{bmatrix} R \\ G \\ B \end{bmatrix}$$

From YIQ to RGB

$$\begin{bmatrix} R \\ G \\ B \end{bmatrix} = \begin{bmatrix} 1.0002 & 0.9560 & 0.6211 \\ 1.0001 & -0.2720 & -0.6470 \\ 1.0000 & -1.1060 & 1.7030 \end{bmatrix} * \begin{bmatrix} Y \\ I \\ Q \end{bmatrix}$$

The working model is trained not over a single image but rather on a corpus of images. The main objective is to train the model independent of the extra input or the supporting feature information.

3.1 *Approach 1*

Image Segmentation

SLIC superpixel algorithm [8] is used to segment the images in different sections. The SLIC algorithm is used because of its effectiveness in forming compact and uniform segments, and it was implemented by us in scikit-image [9]. Figure 2 illustrates this effect of SLIC clustering and its results on a test image. Fig. 2b shows how pixels with a similar value in a given localized section form a bigger pixel.

Traditional models predicted color pixel-by-pixel. The model is formulated by estimating two real values (I and Q channels) of chrominance for each image segment. Moreover, the assumption that the two channels (i.e., I and Q) are independent has been made. Given a vector $y \in R^m$ and training vectors $\varphi(x^{(i)}) \in R^s$ (where $s = p^2$ and $1 \leq i \leq m$), the following minimization problem is specified by a support vector regression, where C and ϵ are chosen constants:

$$\min_{w,b,\varepsilon,\varepsilon^*} \frac{1}{2} w^T w + C \sum_{i=1}^{n} \left(\varepsilon_i + \varepsilon_i^*\right) \tag{1}$$

$$y_i - w^T \phi\left(x_i\right) - b \leq \epsilon + \varepsilon_i \tag{2}$$

Fig. 2 (**a**) Grayscale image and (**b**) superpixels formed using SLIC training

$$w^T \phi(x_i) + b - y_i \leq \epsilon + \varepsilon_i \text{ Subjected to } \varepsilon_i,\ \ \varepsilon_i^* \geq 0, i = 1, \ldots, n \tag{3}$$

This optimization problem can be accomplished productively by making use of the SVR function of scikit-learn, which carries out Epsilon-support vector regression optimization. In this method, training of two different SVRs is done for each channel of output. We take a square of 10x10 pixels to take out feature vectors in every segment of the image. The square is made around each centroid. After this, a 2D fast Fourier transform (FFT) is carried out on the squares taken before, and this gives us our feature vectors $\varphi(x^{(i)})$ [10]. These feature vectors are used in both regressions as an input, and the expected values of U and V of $x^{(i)}$ segment are used as outputs. The default Gaussian kernel is used to train SVRs.

Image Testing and ICM Smoothing
The feature extraction process and segmentation operation have been performed for predicting the chrominance parameters on the image test set. The two support vector regressors are executed over the fragments, providing the initial color prediction. An algorithm based on MRFs (Markov random fields) has been implemented to regularize the color shades of similar superpixels. One field per estimated channel is used in the model. The local potentials of a superpixels hidden chrominance value are represented as a Gaussian $\sim N(\mu_i, \sigma)$, where μ_i is the predicted value of the SVR on the corresponding superpixel. Moreover, using a distance-function, the potentials between hidden chrominance values of neighboring superpixels are also represented. The total energy of a superpixel $x^{(i)}$ having hidden chrominance c_i can be represented using the following equation:

$$E_i = \frac{||c_i - \mu_i||_2^2}{2\sigma^2} + \gamma \sum_{x^{(j)} \in N(i)} ||c_i - \mu_i||_2^2 \tag{4}$$

where γ weighs the relative importance of neighboring pixels compared to single pixels, and $N(i)$ denotes the set of adjacent superpixels to $x^{(i)}$. If the two pairs of neighboring superpixels $x^{(i)}$ and $x^{(j)}$ are considered, then there is inclusion of $x^{(j)}$ in set $N(i)$ provided $||\varphi(x^{(i)}) - \varphi(x^{(j)})||^2$ is less than the threshold value. In order to lower the total energy of MRF and minimize it, iterated conditional modes (ICM) are performed until convergence.

Now since we have our actual Y channel and also the estimated channels (U and V) for any target image, these channels are converted to RGB space, which gives us our final colorization estimate.

3.2 Approach 2

Fuzzy c-means (or FCM) is a clustering method that enables a data piece to fit more than one cluster. The expected distance between cluster centers (for each cluster) and the data points determines the clusters formed. The membership degree (U_{pq}) becomes the deciding factor and decides that the data item belongs to which cluster. The following coefficients have been specified that tell about the membership degree (U_{pq}) to be in the rth cluster [11].

where d_{pq}—distance of pth item from qth cluster; d_{pr}—distance of pth item from rth cluster; m—value of fuzzification factor.

The fuzzification factor is defined between 0 and 1 (both inclusive) by the user. This coefficient dictates the degree (or level) of fuzziness. When the value of m tends to 1, the method functions as a fresh partitioning approach, and for higher m values, it has been noted that the clusters overlap more. The primary goal is to segregate or partition the data into groups in such a manner that the likeliness of features within each cluster is optimum, whereas the likeliness of data items for two or more different clusters is reduced. Moreover, it quantifies the measures of partitioning (dividing a dataset in C—clusters).

The following objective function is minimized in this method:

$$J_m = \sum\nolimits_{p=0}^{N} \sum\nolimits_{p=1}^{C} u_{pq}^m \left\| x_p - c_q \right\|^2 \tag{5}$$

where $m \in \mathbb{R}$, $m \geq 1$, U_{pq} is the membership degree of x_p in the cluster q; x_p is the pth of d-dimensional measured data; c_q is the d-dimension center of the cluster; $||*||$ is L-2 norm expressing the similarity between any measured data and the center.

The objective function shown above is optimized through an iterative approach (fuzzy partitioning) where the membership U_{pq} and the cluster centers c_q are updated by

$$U_{pq} = \frac{1}{\sum_{r=1}^{c} \left(\frac{||x_p - c_q||}{||x_p - c_r||} \right)^{2/m}} \tag{6}$$

$$C_q = \frac{\sum_{p=1}^{N} u_{pq}^{m} . x_p}{\sum_{p=1}^{N} u_{pq}^{m}} \tag{7}$$

The iteration stops when

$$\max_{pq} \left\{ \left| u_{pq}^{(r+1)} - u_{pq}^{(r)} \right| \right\} < \epsilon' \tag{8}$$

where ϵ' is the termination condition ($0 < \epsilon' < 1$), whereas r signifies the number of steps of iteration. This method converges to a local minimum. It is also known as a saddle point of J_m.

4 Results

It is seen that the models perform well on the test set as can be seen from the results in Fig. 3. The first column represents the original images, while the second and third columns are the predicted images of SLIC and fuzzy approaches, respectively. Although, not all colors match with our predicted results correctly, in general, give reasonable coloring outputs.

4.1 Approach 1

The model successfully colors the features of the environment differently. This can be seen in the third example, where our model distinctively predicts colors for trees and grass. It is found that running the ICM smoothing reduced the error by 74.6%, which was earlier 68.1% when only SVR was implemented. Thus, it states that anomalies in the predicted image are decreased.

The major problem seen in this model is that shades of one section of the image mostly bled into the neighboring sections. This can be seen in the fifth example where the patch of sky in the left side is yellow colored as the green color of the tree bled into its neighboring pixels giving the respective segment a yellow color. This means that for small superpixels, the 10×10 squares taken around the centroid of a superpixel mostly contain more pixels when compared to pixels present within the superpixel itself. Due to this, the color estimated by the model is strongly dependent on neighboring superpixels, and it can result in an error if the neighboring superpixels have sufficiently distinct shades of color. Lastly, the

Fig. 3 Original image (left), predicted image (middle), and ground truth (right)

model mostly displays images having low saturation, which can be seen in the third example. It has been assumed that a superpixel has a unique coloring, but the fact is that they can take many equally probable colorings. This assumption in the model may cause the SVRs in averaging chrominance output values, thereby reducing the saturation of the images and the color estimated. Charpiat et al. [12] have effectually incorporated a multimodal approach to enhance the saturation of the output colors.

4.2 Approach 2

In the SLIC operating method, it is observed that despite belonging to neighborhood areas, the colors predicted could be drastically different. It can be seen in examples four and five that some of the superpixels remained grayscaled, i.e., no prediction of chrominance channels is present for them. In the fourth example, the grass appears to be monochromatic. Similarly, in the fifth example, the left patch of the grass is not colored. This was observed as a problem in SLIC clustering since it assumes the image to be composed of many bigger pixels, dividing one big cluster into smaller segments called superpixels.

This problem was administered with the help of a more sophisticated clustering algorithm, fuzzy c-means clustering which significantly helped in reducing the anomalous prediction of neighboring pixels. It has overcome the shortcomings of the SLIC approach due to the creation of more coherent clusters. Furthermore, the membership function helped in accurately predicting the color at the boundary pixels of the segments.

Both the models face difficulty in reconstructing a full range of colors which is there in the original image. It also fails in including more shades of brown and yellow color. The reason for this could be a limited range of colors present in the dataset as it mainly consists of shades of green and blue. Hence if the model is trained on images having a balanced color variety, it can be improved. The trained model for the fuzzy approach received an accuracy of 75.87%. The results of the model have been presented in Fig. 3.

Hyperparameter Tuning

After evaluating the model on a different set of values of the SVR and ICM hyperparameters, selection of those values is made that resulted in the minimum possible error. The error has been defined as the expected (mean) distance between the actual RGB values and predicted RGB values. After processing, the values of parameters have been initialized as follows: $\gamma = 2$ for the ICM, $\in = 0.0625$ and $C = 0.125$ for the SVRs. Similarly, the hyperparameters for fuzzy c-means clustering were decided on two factors: clustering accuracy and computational expense. After testing, we initialized $M = 1.2$, $\in' = 0.005$ and MAX_ITERS (maximum number of iterations) = 1000.

5 Conclusion and Future Works

This paper introduces a state-of-the-art, fully automated method of colorization using simple linear iterative clustering and support vector regression to reduce the efforts of the users an idea to conditionally colorize monochrome images without explicit user input. The output chrominance values are further refined by Markov random fields and iterative conditional modes. Moreover, the novel fuzzy c-means clustering algorithm resulted in slightly better results of the output images. We seek the development of a system that is capable of colorizing grayscale photos by a simple process of stating the relevant tags for a target image.

The efficiency of the model can be improved using several measures. The model can be designed further to accept numerous images of the training set by combining the different colors and also the feature spaces of the various pictures. Application of state-of-the-art techniques, like CNN's, could further increase the accuracy and improve our results.

The need is to estimate the distribution of the values for each and every color here and not just a single value. Currently, support vector regression is used, which is inadequate for this work.

Post-processing methods, like total variance minimization, can be used to achieve more effective results. Moreover, currently, the focus is mainly on step-downing the loss of cross-entropy on per pixel scale. If the system is redesigned considering the adversarial network, then results could improve, as now the model would learn to output images comparable to real-world ones.

References

1. Bezdek, J.C., Ehrlich, R., Full, W.: FCM: the fuzzy c-means clustering algorithm. Comput. Geosci. **10**(2–3), 191–203 (1984)
2. Welsh, T., Ashikhmin, M., Mueller, K.: Transferring color to greyscale images. ACM Trans. Graph. **21**(3), 277–280 (2002)
3. Levin, A., Lischinski, D., Weiss, Y.: Colorization using optimization. ACM Trans. Graph. **23**(3), 689–694 (2004)
4. Luo, W., Huang, F., Huang, J.: Edge adaptive image steganography based on LSB matching revisited. IEEE Trans. Inf. Forensics Secur. **5**(2), 201–214 (2010)
5. Bugeau, A., Ta, V.T.: Patch-based image colorization. In: Proceedings of the 21st International Conference on Pattern Recognition (ICPR2012), pp. 3058–3061. IEEE (2012)
6. Liu, F., Shen, C., Lin, G.: Deep convolutional neural fields for depth estimation from a single image. In: Proceedings of the IEEE Conference on Computer Vision and Pattern Recognition, pp. 5162–5170 (2015)
7. Greiffenhagen, M., Ramesh, V., Comaniciu, D., Niemann, H.: Statistical modeling and performance characterization of a real-time dual camera surveillance system. In: Proceedings IEEE Conference on Computer Vision and Pattern Recognition. CVPR 2000 (Cat. No. PR00662), vol. 2, pp. 335–342. IEEE (2000)
8. Achanta, R., Shaji, A., Smith, K., Lucchi, A., Fua, P., Süsstrunk, S.: SLIC superpixels compared to state-of-the-art superpixel methods. IEEE Trans. Pattern Anal. Mach. Intell. **34**(11), 2274–2282 (2012)

9. Van der Walt, S., Schönberger, J.L., Nunez-Iglesias, J., Boulogne, F., Warner, J.D., Yager, N., Yu, T., The Scikit-Image Contributors: Scikit-image: image processing in Python. PeerJ. **2**, e453 (2014)
10. Itti, L., Koch, C., Niebur, E.: A model of saliency-based visual attention for rapid scene analysis. IEEE Trans. Pattern Anal. Mach. Intell. **11**, 1254–1259 (1998)
11. Singh, T., Mahajan, M.M.: Performance comparison of fuzzy C means with respect to other clustering algorithm. Int. J. Adv. Res. Comput. Sci. Softw. Eng. **4**(5), 89–93 (2014)
12. Charpiat, G., Hofmann, M., Schölkopf, B.: Automatic image colorization via multimodal predictions. In: European Conference on Computer Vision, pp. 126–139. Springer, Berlin, Heidelberg (2008)

A Survey on Time-Series Data Prediction Models Using Recurrent Neural Networks

Jeril Lalu and Binu Jose A.

1 Introduction

Time-series data is one-dimensional and has a temporal property, i.e. time is integral to the data sequences. Such data are obtained from sources that usually measure a quantity or perform operations that output data in a sequence.

Data can span large time intervals, and the granularity of data can be as fine as a minute or as broad as a day. The granularity partly determines the accuracy of predictions as hidden data patterns can be detected by recurrent neural networks.

A recurrent neural network (RNN) is a deep learning neural network that uses the past and present information to predict the future.

However regular RNNs suffer from a vanishing gradient descent, i.e. if the weights being updated in each time instance are very small, the network will take a very long time to converge at the optimal weights. Also, the previous output information can be lost if the time sequence is very large.

Long short-term memory (LSTM) is a modification of RNN that includes memory storage in the form of three gates: the input gate, the forget gate, and the output gate. Data flow is akin to a conveyor belt. LSTMs solve the vanishing gradient descent problem to a great extent. However, LSTMs take a long time to be trained, and this can be a disadvantage when extremely large datasets are used as inputs.

Gated recurrent units (GRUs) are a modification of LSTMs such that they reduce the number of gates from three to two, i.e. update gate and reset gate. GRUs are designed to train faster. Usage depends on problem situations.

J. Lalu (✉) · Binu Jose A.
Department Of Computer Science and Engineering, Mar Baselios College of Engineering and Technology, Thiruvananthapuram, India
e-mail: binu.jose@mbcet.ac.in

M. Palesi et al. (eds.), *Second International Conference on Networks and Advances in Computational Technologies*, Transactions on Computational Science and Computational Intelligence, https://doi.org/10.1007/978-3-030-49500-8_25

2 Regular LSTM Models

2.1 Hotel Reservation Forecasting

Jian et al. [1] proposed a system to forecast hotel reservations. Two LSTM models were used. One was to predict the number of days a room would be booked, starting from an arbitrary day, and the second was to predict average room rates. Around 5 years of historical data was available for the cities Atlanta, Chicago, and Jacksonville, whereas Boston only had around 3 years of data available.

For the sake of comparison with other methodologies, the averages of the estimation columns (considering Boston separately due to difference in dataset duration) were taken as final average performance values. MSE and RMSE per city and LSTM model are shown in Table 1.

The time-based LSTM obtained an average estimation MSE and RMSE of 0.0004 and 0.0196, respectively. The time-rate-based LSTM obtained an average estimation MSE and RMSE of 0.0007 and 0.0252, respectively.

2.2 Power Load Forecasting

Weicong et al. [2] proposed a short-term load forecasting model to adjust power production levels throughout the span of a day. Daily half-hour readings were obtained from 69 customers via smart meters, chosen based on ownership of water heaters. Ninety-two days of time-series data were used, and this data was fed into an LSTM model that predicted short-term load readings. Table 2 lists the resultant MAPE values obtained by each LSTM configuration.

Table 1 Time-based and time-rate-based LSTM results per city [1]

	Time-based		Time-rate-based	
City	Estimation MSE	Estimation RMSE	Estimation MSE	Estimation RMSE
Atlanta	0.0002	0.0145	0.0003	0.0180
Boston	0.0004	0.0199	0.0006	0.0244
Chicago	0.0008	0.0282	0.0016	0.0400
Jacksonville	0.0003	0.0162	0.0003	0.0177

Table 2 Average MAPE values per time step [2]

LSTM and time step taken	Average MAPE of individual forecasts (%)	Average MAPE of aggregate forecasts (%)	Average MAPE forecasting the aggregate (%)
LSTM 2 steps	44.39	8.18	9.14
LSTM 6 steps	44.31	8.39	8.95
LSTM 12 steps	44.06	8.64	8.58

Heng et al. [3] proposed a deep recurrent neural network powered by LSTM cells for household power consumption. Their methodology consisted of two stages, i.e. load profiles pooling where household load profiles were pooled together to avoid overfitting and household short-term load forecasting where the PDRNN was trained and tested by loading profile batches randomly.

The dataset used was electricity consumption (kWh) sampled half-hourly spanning a time period from July 1, 2009 to December 31, 2010. From this, 920 customers were randomly selected and 92 groups each containing ten customers were formed. The PDRNN obtained a value of 0.4505 in RMSE and an MAE of 0.2510.

2.3 Financial Market Prediction

Thomas et al. [4] proposed LSTMs for financial market predictions, and a multi-stage approach was used to address the problem, which involved trading and training dataset generation, feature and target extraction, LSTM network generation, benchmarking, forecasting, ranking, and trading.

The data input was the constituent stocks of S&P 500 from 1990 to 2015. A training-trading set consisting of 750 days of training period and 250 days of testing period was defined. The LSTM model obtained an RMSE value of 0.0206.

Yujin et al. [5] proposed a new framework that consisted of an LSTM prediction module and an overfitting prevention LSTM module.

The prevention module took different input features each time. Direct information contained only the index values, while indirect information contained the stock prices. Features from both types of data were extracted.

The prediction module predicted the stock market index using the index values as the input. Sequence data with a window length of 20 was chosen. The daily closing price of the stock market index was the input, for 20 consecutive days.

The ModAugNet model used LSTMs in both modules. The resultant output was the succeeding day's closing price.

The input stock market indices dataset was from S&P 500 and KOSPI 200 and spanning a period from 4th January 2000 to 27th July 2017.

The model configurations tested were ModAugNet-f, a model with a prevention module and five company stocks data pre-fixed, and ModAugNet-c, same as ModAugNet-f except that the prevention module was fed with 252 combinations of five company stock data.

Using S&P 500, ModAugNet-f obtained MSE of 1665.3, MAPE of 2.0089%, and MAE of 33.849. ModAugNet-c obtained MSE of 342.48, MAPE of 1.0759%, and MAE of 12.058.

Using KOSPI 200, ModAugNet-f obtained MSE of 14.45, MAPE of 1.2397%, and MAE of 2.832. ModAugNet-c obtained MSE of 7.56, MAPE of 1.0077%, and MAE of 1.975.

2.4 *Human Activity Duration Prediction*

Kundan et al. [6] proposed a prediction system which predicted durations of human activities. Two LSTM models were used. One, a hybrid LSTM, which attempted to predict the next human activity and its duration jointly. The second, a cascaded LSTM, which only predicted the next activity.

The input dataset was obtained by a custom-made mobile application called ContextTagger, which prompted the subject to periodically record their daily activities.

Segment level activity data used the American Time Use Survey Dataset. This dataset had information on time usage of people in the USA. Data spanned from 2003 to 2015 and included activities such as paid work, religious activities, socializing, etc.

Eleven datasets were created. Seven were from subject whose individual data was collected. The remaining four datasets were chosen from four segments of population from the ATUS data, based on their occupation. Two-thousand days were randomly picked from the database, 1600 used for training and 400 used for testing. Further filtration was performed such that sleeping was the first activity of the day in each of the sets. This reduced the training set to 1467 days and test set to 363 days, i.e. 1830 days total.

For the sake of comparison, the MAE values for duration prediction were taken after they were averaged across activities by the researcher.

In the joint models, the cascaded LSTM obtained an MAE of 21.75. Hybrid LSTM obtained an MAE of 21.86.

In the non-joint models, the cascaded LSTM obtained an MAE of 32.58, while hybrid LSTM obtained an MAE of 52.20.

3 GRU Models

3.1 *Stock Prediction*

Minh et al. [7] proposed a two-stream GRU that used a Stock2Vec embedding and three technical indicators to predict stock trends.

TGRU operated on the stock dataset, while the Stock2Vec model was trained for sentiment analysis from stock market news sites and documents. The technical indicators were used to analyze whether news affected stock price soon or after a short delay.

The stock prediction was done in the following stages: document preprocessing, document labeling, Stock2Vec embedding, technical indicators calculation, TGRU neural network modeling.

A two-gated GRU was modeled to predict stock pricing and was modified to study the forward and backward context of labeled stock news, much like bi-directional RNN.

Stock prices of the S&P 500 from October 2006 to November 2013 were used, and the TGRU prediction model had an overall accuracy of 66.32%.

Guizhu et al. [8] proposed a GRU model to obtain stock index predictions that swapped out a softmax activation layer for an SVM.

HSI, DAX, and S&P 500 indices were used as the input data for training and testing both the softmax and SVM-enabled GRU models. Daily data ranging from the years 1991–2017 was considered.

When predicting HSI, the GRU-SVM model had an accuracy of 52%. DAX prediction saw GRU-SVM and GRU-Softmax achieve the same accuracy of around 51%. S&P 500 prediction saw GRU-SVM obtain a 51.7% accuracy.

4 Hybrid LSTM Models

4.1 Solar Power Prediction

Woonghee et al. [9] proposed a system to predict solar power obtained over consecutive future days using an LSTM with two CNNs, as a single CNN with a very large filter would significantly increase computation cost. A double CNN (D-CNN) with different filter sizes was used instead.

The data collected was from 71 photovoltaic inverters in 14 sites in South Korea. The data contained information of ten cities spanning a time period from 29 Feb 2012 to 6 Jan 2016. The records contained time, temperature, irradiation, power generation, inverter ID, site ID. Weather data was obtained from the Korea Meteorological Administration. Temperature, humidity, and wind speed were noted at per-hour time intervals.

Comparisons were performed with another implementation consisting of an autoencoder and an LSTM and against other algorithms as well. Accuracy was evaluated with different window sizes (1, 2, 4, and 6 h) and with and without weather data.

With weather data, CNN + LSTM with a 1-h window achieved the lowest MAPE, RMSE, and MAE values, i.e. 13.42%, 0.0987, and 0.0506, respectively, among other CNN + LSTM configurations with varying window sizes.

Without weather data, CNN + LSTM with a 1-h window obtained the lowest MAPE, RMSE, and MAE values of 19.57%, 0.1409, and 0.0585, respectively, among other CNN + LSTM configurations with varying window sizes.

Table 3 MSE values of TC-LSTM on American stock and SSE 50 Index data [10]

	American stock data	SSE 50 Index data
Prediction time span (days)	MSE	MSE
1	0.9770	0.03078
3	1.3958	0.11752
5	2.0321	0.17620
7	3.1685	0.23488
30	7.2412	1.10512
60	12.5943	2.41511

4.2 Stock Price Prediction

Xukuan et al. [10] proposed a stock price prediction model that combined a regular LSTM with a 1D CNN to form a time convolutional LSTM (TC-LSTM).

Two datasets were used to test the model's accuracy. The first dataset had 50 stocks' daily opening prices, from ten sectors from 2007 to 2016 [11]. The second dataset contained 50 stocks from the SSE 50 Index from 2008 to 2017.

Predictions were done for 1, 3, 7, 5, 30, and 60 days. The MSE values obtained per different prediction time spans of data for the different stock data chosen are shown in Table 3.

4.3 Wind Speed Prediction

Jie et al. [12] proposed a method for wind speed prediction to optimally schedule and control energy generation and conversion. A unique model called EnsemLSTM was proposed which used an ensemble of nonlinear-learning and deep learning powered time-series prediction based on support vector regression machine (SVRM), LSTMs, and an EO (extremal optimization) algorithm.

The proposed EnsemLSTM contained six separate LSTMs. The output of the LSTMs was fed into the SVRM, which learned the nonlinear relationship of the six LSTMs predictors like solving multivariate regression problem. The output of the EnsemLSTM was an SVRM forecast combined with EO, for final wind prediction.

The dataset was obtained from a wind farm in Inner Mongolia, China. The aim of the prediction model was to forecast ahead in the short term, utmost 10-min and 1-h of wind speed.

EnsemLSTM for short-term prediction achieved an MAE of 1.1410, an RMSE of 1.5335, and a MAPE of 17.1076%.

EnsemLSTM for utmost short-term prediction achieved an MAE of 0.5746, an RMSE of 0.7552, and a MAPE of 5.4167%.

5 Analysis (Tables 4, 5, and 6)

6 Conclusions

6.1 Based on Forecasting Area

Hotel Reservation The time-based LSTM (avg) performed the best, with the lowest MSE and RMSE values in its forecasting area.

Power Load Since both the methodologies in this area used different performance metrics, a direct comparison was not possible. However, the load forecasting LSTM obtained a MAPE value of 44.25% which would suggest that its accuracy be 55.75%.

Financial Market The direction prediction model performed the best, as it took the most amount of days as input and still obtained a very low RMSE value.

Human Activity Duration The cascaded joint LSTM performed the best as it obtained the lowest MAE value from the other methodologies presented in the area.

Stock Prediction There was no clear performance leader as this area consisted of GRUs and hybrid LSTMs, each using their own performance metric.

Solar Power The CNN + LSTM (1 h, weather data) methodology performed the best as it obtained low values across various performance metrics (barring MSE, which was non-existent).

Wind Speed EnsemLSTM (utmost short-term) showed higher performance among the 2 cases of EnsemLSTM, based on the lower error values obtained. Although it must be noted that EnsemLSTM (utmost short-term) performed prediction for the next 10 min, while EnsemLSTM (short-term) predicted an hour ahead.

6.2 Based on Performance Metric

MSE The regular time-based LSTM (avg) methodology obtained the lowest MSE value.

RMSE The regular time-based LSTM (avg) methodology obtained the lowest RMSE value.

MAPE The regular LSTM ModAugNet-c (KOSPI 200) methodology obtained the lowest MAPE value.

MAE The hybrid LSTM CNN + LSTM (1 h, weather data) methodology obtained the lowest MAE value.

Accuracy The GRU TGRU methodology obtained the highest accuracy.

Table 4 Regular LSTM results

Forecast area	Proposed methodology	Dataset span (days)	MSE	RMSE	MAPE (%)	MAE
Hotel reservation	Time-based LSTM (avg)	1877	**0.0004**	**0.0196**	–	–
	Time-based LSTM (Boston)	798	0.0004	0.0199	–	–
	Time-rate-based LSTM (avg)	1877	0.0007	0.0252		
	Time-rate-based LSTM (Boston)	798	0.0006	0.0244	–	–
Power load	Load forecasting LSTM (2 steps, indiv. forecast)	92	–	–	44.39	–
	Load forecasting LSTM (6 steps, indiv. forecast)	92	–	–	44.31	–
	Load forecasting LSTM (12 steps, indiv. forecast)	92	–	–	44.06	–
	Load forecasting LSTM (2 steps, aggreg. forecast)	92	–	–	8.18	–
	Load forecasting LSTM (6 steps, aggreg. forecast)	92	–	–	8.39	–
	Load forecasting LSTM (12 steps, aggreg. forecast)	92	–	–	8.64	–
	Load forecasting LSTM (2 steps, predict aggreg.)	92	–	–	9.14	–
	Load forecasting LSTM (6 steps, predict aggreg.)	92	–	–	8.95	–
	Load forecasting LSTM (12 steps, predict aggreg.)	92	–	–	8.58	–
	PDRNN	90	–	0.4505	–	0.2510
Financial market	Direction prediction model	6489	–	0.0206	–	–
	ModAugNet-f (S&P 500)	4414	1665.31	–	2.0089	33.849
	ModAugNet-c (S&P 500)	4414	342.48	–	1.0759	12.058
	ModAugNet-f (KOSPI 200)	4388	14.45	–	1.2397	2.832
	ModAugNet-c (KOSPI 200)	4388	7.56	–	**1.0077**	**1.975**
Human activity duration	Cascaded joint LSTM	1830	–	–	–	21.75
	Hybrid joint LSTM	1830	–	–	–	21.86
	Cascaded non-joint LSTM	1830	–	–	–	32.58
	Hybrid non-joint LSTM	1830	–	–	–	52.20

The bold values in Tables 4 represent the values with least error (desirable attribute) among all the other values present within the columns in those tables.

Table 5 Hybrid LSTM results

Forecast area	Proposed methodology	Dataset span (days)	MSE	RMSE	MAPE (%)	MAE
Solar power	CNN + LSTM (1 h, weather data)	1407	–	**0.0987**	13.42	**0.0506**
	CNN + LSTM (1 h, no weather data)	1407	–	0.1409	19.57	0.0585
Stock prediction	TC-LSTM 1 day (U.S. stock data)	2512	0.9770	–	–	–
	TC-LSTM 3 days (U.S. stock data)	2512	1.3958	–	–	–
	TC-LSTM 5 days (U.S. stock data)	2512	2.0321	–	–	–
	TC-LSTM 7 days (U.S. stock data)	2512	3.1685	–	–	–
	TC-LSTM 30 days (U.S. stock data)	2512	7.2412	–	–	–
	TC-LSTM 60 days (U.S. stock data)	2512	12.5943	–	–	–
	TC-LSTM 1 day (SSE 50 Index data)	2526	**0.03078**	–	–	–
	TC-LSTM 3 days(SSE 50 Index data)	2526	0.11752	–	–	–
	TC-LSTM 5 days (SSE 50 Index data)	2526	0.17620	–	–	–
	TC-LSTM 7 days (SSE 50 Index data)	2526	0.23488	–	–	–
	TC-LSTM 30 days (SSE 50 Index data)	2526	1.10512	–	–	–
	TC-LSTM 60 days (SSE 50 Index data)	2526	2.41511	–	–	–
Wind speed	EnsemLSTM (short-term)	29	–	1.5335	17.1076	1.1410
	EnsemLSTM (utmost short-term)	5	–	0.7552	**5.4167**	0.5746

The bold values in Tables 5 represent the values with least error (desirable attribute) among all the other values present within the columns in those tables.

Table 6 GRU results

Forecast area	Proposed methodology	Dataset span (days)	Accuracy (%)
Stock prediction	TGRU	1800	**66.32**
	GRU-SVM (HSI)	6423	52
	GRU-SVM (S&P 500)	6551	51.7
	GRU-SVM (DAX)	6580	~51

The bold value in Table 6 represents the value with highest accuracy (desirable attribute) among all the other values present in that column in the table.

From the above conclusions, it is observed that LSTMs are versatile and can adapt to almost any forecasting area if it contained time-series data. In most cases, LSTMs performed well, as performance lead was taken by LSTMs across most forecasting areas (barring few areas where results between RNN variations were inconclusive) and performance metrics.

References

1. Wang, J., Duggasani, A.: Forecasting hotel reservations with long short-term memory-based recurrent neural networks. Int. J. Data Sci. Anal. **9**(1), 77–94 (2020). https://doi.org/10.1007/s41060-018-0162-6
2. Kong, W., Dong, Z.Y., Jia, Y., Hill, D.J., Xu, Y., Zhang, Y.: Short-term residential load forecasting based on LSTM recurrent neural network. IEEE Trans. Smart Grid. **10**(1), 841–851 (2019). https://doi.org/10.1109/TSG.2017.2753802
3. Shi, H., Xu, M.: Ran Li: deep learning for household load forecasting—a novel pooling deep RNN. IEEE Trans. Smart Grid. **9**(5), 5271–5280 (2018). https://doi.org/10.1109/TSG.2017.2686012
4. Fischer, T.: Christopher Krauss: deep learning with long short-term memory networks for financial market predictions. Eur. J. Oper. Res. **270**(2), 654–669 (2018). https://doi.org/10.1016/j.ejor.2017.11.054
5. Baek, Y., Kim, H.Y.: ModAugNet: a new forecasting framework for stock market index value with an overfitting prevention LSTM module and a prediction LSTM module. Expert Syst. Appl. **113**, 457–480 (2018). https://doi.org/10.1016/j.eswa.2018.07.019
6. Krishna, K., Jain, D., Mehta, S.V., Choudhary, S.: An LSTM based system for prediction of human activities with durations. Proc. ACM Interact. Mob. Wearable Ubiquitous Technol. **1**(4), 1–31 (2018). https://doi.org/10.1145/3161201
7. Minh Dang, L., Sadeghi-Niaraki, A., Huynh, H.D., Min, K., Moon, H.: Deep learning approach for short-term stock trends prediction based on two-stream gated recurrent unit network. IEEE Access. **6**, 55392–55404 (2018). https://doi.org/10.1109/ACCESS.2018.2868970
8. Shen, G., Tan, Q., Zhang, H., Zeng, P., Xu, J.: Deep learning with gated recurrent unit networks for financial sequence predictions. Procedia Comput. Sci. **131**, 895–903 (2018). https://doi.org/10.1016/j.procs.2018.04.298
9. Lee, W., Kim, K., Park, J., Kim, J., Kim, Y.: Forecasting solar power using long-short term memory and convolutional neural networks. IEEE Access. **6**, 73068–73080 (2018). https://doi.org/10.1109/ACCESS.2018.2883330

10. Zhan, X., Li, Y., Li, R., Xiwu, G., Habimana, O., Wang, H.: Stock Price prediction using time convolution long short-term memory network. In: Liu, W., Giunchiglia, F., Yang, B. (eds.) Knowledge Science, Engineering and Management. KSEM 2018, vol. 11061, pp. 461–468. Springer, Cham (2018)
11. Zhang, L., Aggarwal, C., Qi, G.-J.: Stock Price prediction via discovering multi-frequency trading patterns. In: Proceedings of the 23rd ACM SIGKDD International Conference on Knowledge Discovery and Data Mining, pp. 2141–2149. ACM, Halifax, NS (2017)
12. Chen, J., Zeng, G.-Q., Zhou, W., Wei, D., Lu, K.-D.: Wind speed forecasting using nonlinear-learning ensemble of deep learning time series prediction and extremal optimization. Energy Convers. Manag. **165**, 681–695 (2018)

SMaRT: A Framework for Social Media Based Recommender for Tourism

Shini Renjith, A. Sreekumar, and M. Jathavedan

1 Introduction

Like any other service industry, travel and tourism is also extensively making use of recommender systems to enhance various business operations. Its application ranges in multiple areas like selection of destinations, choice of attractions, decision of routes, mode of transport, decision on accommodations and restaurants, etc. Travel recommender systems undergone considerable enhancements since its inception in conjunction with the developments in technology—started with virtual travel guides [1–4] and evolved to contextualized personal travel recommenders [5–9]. Of late, social media has transformed as an important source of data for modern recommender systems in almost every industry with no exception to the travel domain.

The usage of social media is witnessing exponential growth in this era and as a result, the availability of social media data is also increasing in every domain. In the travel and tourism industry, quite a lot of social media data is getting generated on a daily basis in the form of travel reviews, blog posts, testimonials, messages in travel forums, etc. Traditional travel recommender systems built using collaborative or hybrid algorithms lack scalability with the huge size and dimensions associated with social media big data. The key challenge with social media data is the need

S. Renjith (✉)
Department of Computer Applications, Cochin University of Science and Technology, Kochi, Kerala, India

Department of Computer Science and Engineering, Mar Baselios College of Engineering and Technology, Thiruvananthapuram, Kerala, India

A. Sreekumar · M. Jathavedan
Department of Computer Applications, Cochin University of Science and Technology, Kochi, Kerala, India

M. Palesi et al. (eds.), *Second International Conference on Networks and Advances in Computational Technologies*, Transactions on Computational Science and Computational Intelligence, https://doi.org/10.1007/978-3-030-49500-8_26

of high computing power and the time required for processing it. Application of clustering algorithms [10] is considered as a solution to restrict the data volume as it helps to segregate the most relevant information that need to be processed. In similar lines, the adoption of an appropriate dimensionality reduction approach [11] is required to deal with the curse of dimensionality issue associated with social media big data.

Section 2 of this paper briefs on the related literature in this area and Sect. 3 briefs on the proposed architecture. Section 4 discusses about the implementation aspects and Sect. 5 summarizes the paper along with details of planned enhancements in consideration.

2 Background and Related Works

With the advent of big data, almost every industry started analyzing large volumes of historic data to enable data-driven decision-making process in various operations so as to improve enterprise level efficiencies and quality of services and products offered. Jagadish et al. [12] provided a very good articulation of various challenges associated with big data aspects like heterogeneity, inconsistency, incompleteness, visualizations, timeliness, privacy, etc. in recommender systems. MapReduce has evolved as a popular framework to deal with big data context in association with Hadoop cluster. Grolinger et al. [13] explained the key challenges associated with big data processing using MapReduce. Najafabadi et al. [14] explained how deep learning can tackle data analytics problems and the enhancements required for better performance. L'Heureux et al. [15] highlighted the cause–effect relationship of various constraints in machine learning with big data against its four dimensions—volume, velocity, variety, and veracity.

From the early researches in the area of recommender systems, travel and tourism industry is a key focus area. Tourism recommenders evolved from static virtual travel guides to the contextualized personal travel recommenders over the period of time. Modern travel recommenders started leveraging social media as an important source of input data. One example of such an attempt is the personalized travel sequence recommendation model proposed by Jiang et al. [16] which utilized diverse data from multiple big data sources like travelogues and community contributed photos in Flickr. M. Figueredo et al. [17] proposed a solution to infer traveler preferences from photos on social media to come up with travel recommendations.

The major issue in dealing with social media data is its huge volume that needs to be processed. The choice of an appropriate clustering algorithm is essential to deal with this scenario. Shirkhorshidi et al. [18] published a theoretical analysis of multiple clustering algorithms used in big data context The authors [19–21] cover more conceptual studies on the application of clustering algorithms with big data.

There are multiple empirical studies available that deals with the performance of clustering algorithms [22–28]. Similarly, there is a wide range of literature that can be found on the theoretical and application aspects of various dimensionality reduction techniques. In general, it can be classified as linear and non-linear approaches. The key linear approaches include principal component analysis (PCA) [29, 30] and independent component analysis (ICA) [31], whereas the non-linear category includes t-distributed stochastic neighbor embedding (t-SNE) [32, 33], locally linear embedding (LLE) [34, 35], self-organizing maps (SOM) [36], ISOMAP [37, 38], autoencoder [39–41], etc.

No clustering algorithm can be considered superior in their performance and the clustering quality and performance may vary based on the evaluation criteria adopted and dataset being processed. The same is the case with dimensionality reduction techniques as well. So, it is important to choose appropriate clustering algorithm and dimensionality reduction technique based on certain experimental analysis. Various performance aspects of clustering algorithms like clustering quality and turnaround time are experimentally evaluated against different clustering algorithms and dimensionality reduction techniques by Renjith et al. as part of their experiment series [42–45].

3 Proposed Architecture

The proposed framework is a combination of five core processing stages as depicted in Fig. 1. The framework is capable of consuming structured and/or unstructured social media text data from various sources as input in order to generate personalized recommendations for a specific user based on individual as well as societal traits. Subsequent sections explain each of these processing stages in detail.

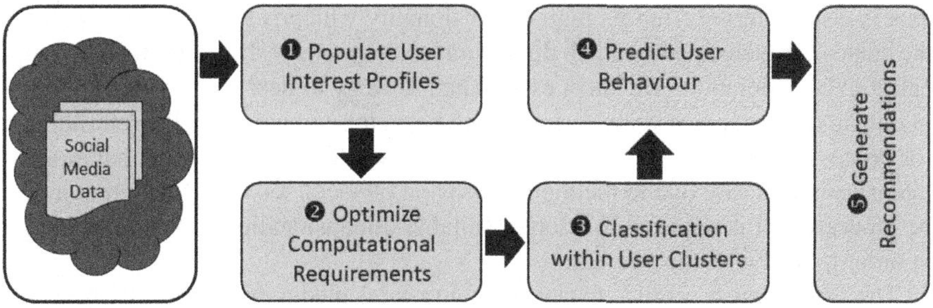

Fig. 1 High level architecture of SMaRT framework

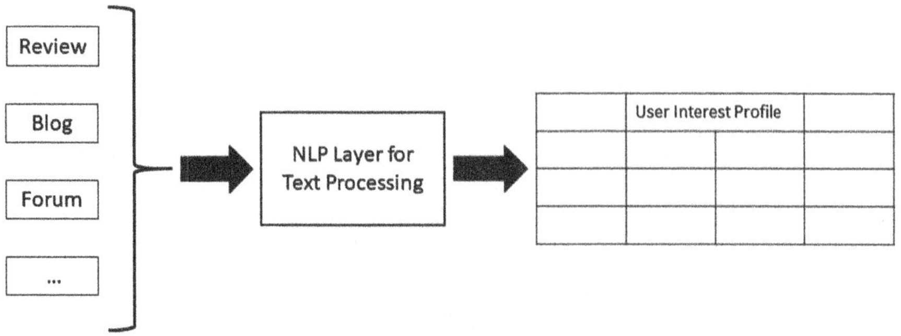

Fig. 2 Stage 1—Population of user interest profiles

3.1 Population of User Interest Profiles

The most vital step in social media based recommendation generation is the extraction of user interest information from the huge volume of input data available. The prominent social media data sources in tourism domain include travel reviews, blogs, forums, etc. A natural language processing (NLP) layer acts as the core of this processing stage. It helps in extracting the underlying user interest information from the text data in the form of a matrix where each row represents the interest of a unique user and each column of the matrix represents unique interest information. Figure 2 represents the architectural depiction of this processing stage.

3.2 Optimization of Computational Requirements

In addition to the inherent challenges associated with huge volume, social media big data is also suffering from its high dimensionality which is typically referred to as the curse of dimensionality. Each dimension corresponds to a unique attribute of the entity in consideration. The most established strategy to deal with the huge volume of data in data analytics is clustering which helps to segregate only the relevant data for processing. However, the clustering process itself can become computationally intensive in this case due to the high dimension of social media big data required to be processed. It is required to adopt a suitable dimensionality reduction technique in order to deal with this scenario.

The user interest profiles extracted from social media data using the proposed framework will contain as much number of records equivalent to the total number of social media users contributed with reviews, blog posts, forum updates, etc. The dimensionality of the user interest profile also will be high corresponding to the total number of unique attributes found across reviews. As depicted in Fig. 3, this processing stage will help in identifying subsets of users that can be considered for subsequent processing. The selection of clustering algorithm and dimensionality

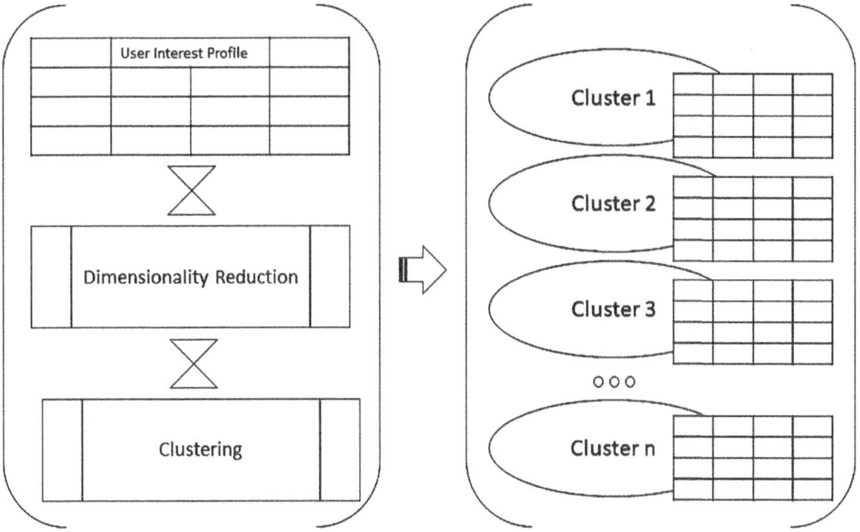

Fig. 3 Stage 2—Optimization of computational requirements by application of dimensionality reduction and clustering

reduction technique has to be done by evaluating the candidate solutions against the dataset in consideration. More details on the strategy to be followed and parameters to be considered during the selection of clustering algorithm and dimensionality reduction approach can be found in [29–32].

3.3 Identification of Similar Users Within the Cluster

Unlike previous phases of the framework, the third phase onwards will function on real-time basis. In order to generate the recommendation for a specific user, the framework will detect the corresponding cluster and apply a classification algorithm within the cluster to determine similar users. The framework proposes to use a tourism specific taxonomy lookup at this stage to enable more accurate classification. The architectural depiction of this phase is provided as Fig. 4.

3.4 Prediction of User Behavior

The next phase aims in predicting the personalized user behavior. In this phase, there are two key considerations—the user's individual preferences available as part of the user interest profile and the collaborative predictions generated by looking at the preferences of similar users within the cluster. Individual preferences represent

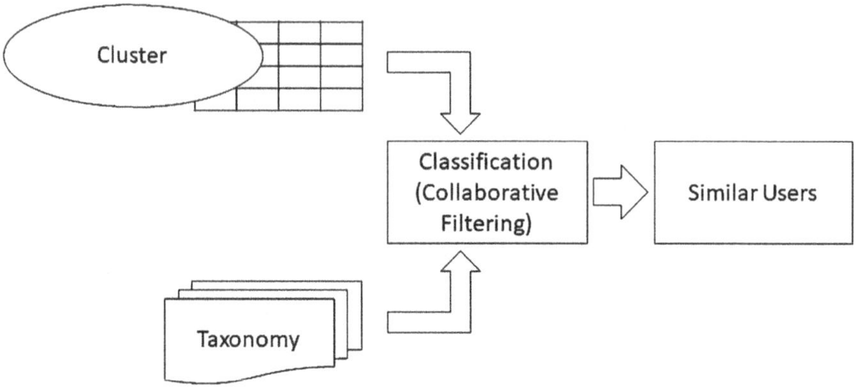

Fig. 4 Stage 3—Classification of users with cluster

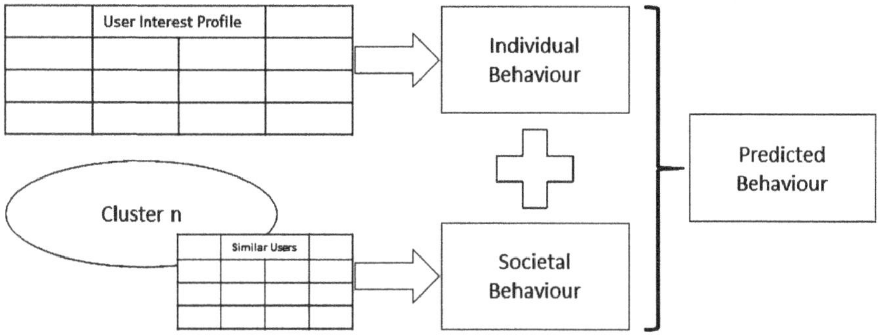

Fig. 5 Stage 4—Prediction of user behavior

the past traits of the user and the user is expected to exhibit the same interests in his/her future transactions as well. Similar users are the ones who exhibited similar transactional traits as the user in consideration. So, if there is an action performed by similar users in the past but not by the user in consideration, still there is a high probability for the user to exhibit that transaction in the future. The outcome from this phase is the predicted behavior by the user in consideration which is the cumulation of individual and societal behaviors as shown in Fig. 5.

3.5 Generation of Recommendations

The fifth and final phase of the framework is depicted in Fig. 6 which deals with the generation of recommendations. There are three key considerations in this stage and the first and foremost one is the predicted user behavior derived from the fourth phase. Other attributes include the list of target actions and constraints

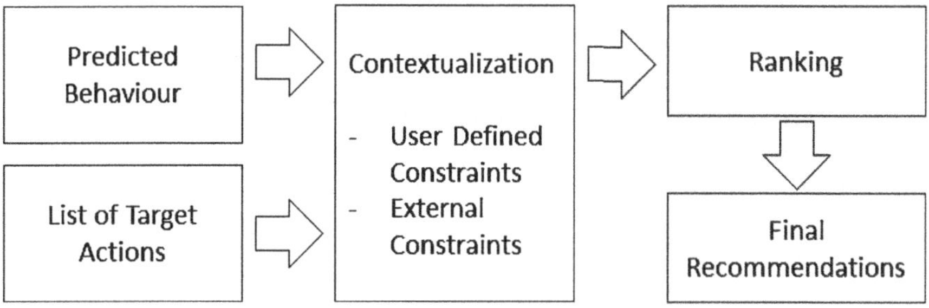

Fig. 6 Stage 5—Generation of recommendations

set by the user or environment. In the context of travel recommender, the target action can be actions like visiting a point of interest, selecting an accommodation option, choosing a restaurant, deciding a commutation option, etc. Constraints are parameters that influence contextualization which can be either set by the user (like budget constraints, physical ability, health conditions, etc.) or by external parameters (like climatic conditions, political scenarios, restrictions imposed by authorities, etc.). Recommendations are generated using predicted behavior and then filtered by applying constraints and ranked before presenting to the end user.

4 Implementation Aspects

This work attempts to propose a generic framework with five defined phases to deal with social media big data in the context of tourism recommendation generation. The proposed framework is modularized in such a way that researchers can plug and play each of the module and try different combinations of algorithms and technologies as required. Please refer to Table 1 for the technology stack used in the SMaRT proof of concept as well as the key differentiators of the proposed framework. The key assumption is that the user of this framework has access to social media data either directly or via an interface like API or web service.

The first two phases of the framework can be considered as offline or batch processing stages, while the rest of the phases has to be implemented as real-time stages. The first phase of the framework deals with preprocessing of data for consumption in subsequent phases. The architecture is created with the consideration that it receives textual data and NLP layer can process the same so as to populate user interest profiles. If the input data is non-textual (for example, social media images), the NLP layer can be skipped, and a relevant data processing layer can be introduced in this phase. This phase as such can be bypassed, in case user interest profiles can be made available directly to the framework.

The second phase can be considered as the machine learning and data mining layer of the framework. It is designed to implement the performance optimization

Table 1 Technology stacks used for the SMaRT proof of concept and key differentiators

Processing stage as defined in SMaRT	Technology stack used for proof of concept	Key differentiators
1. Population of user interest profiles	Java, RabbitMQ, MongoDB	• Modularized • Loosely coupled • Self-contained • Technology agnostic • Reusable
2. Optimization of computational requirements	Apache Spark, Python, R	
3. Identification of similar users within the cluster	Java, R, MongoDB	
4. Prediction of user behavior	Java, MongoDB	
5. Generation of recommendations	Java, JSP, Apache Tomcat	

measures to improve the computational efficiency while processing data in large volumes and with high dimension. Depending on the nature of dataset being populated, a relevant clustering and dimensionality reduction strategy need to be formulated. Though the proposed framework is technology agnostic, it is recommended to choose the best suiting data science technology for implementing this phase to yield the best results.

The third phase is the classification stage where similar users need to be identified by using a collaborative filtering algorithm. If demographic information or user preferences are available for reference, the same can be leveraged for improving the classification accuracy using a hybrid approach. The framework assumes the availability of a tourism specific taxonomy lookup so as to enable more accurate classification.

The next two phases deal with user behavior prediction and recommendation generation as in the case of standard tourism recommenders. It also deals with the contextualization and ranking of recommendations being generated. One important aspect of contextualization is the response to external constraints. This phase requires to consume third party services or APIs in order to gather such information so as to take appropriate actions.

5 Conclusion

Social media data has evolved as an eminent source of information for various industry segments like e-commerce, telecom, insurance, education, hospitality, tourism, marketing, advertising, etc. Based on the estimates by statista.com, a leading online statistics portal, there will be around 2.77 billion social media users around the globe in 2019, against the 2.46 billion in 2017. This will be more than 70% of the total internet population. Though social media data can be considered as a true and real-time reflection of societal inclinations, the huge volume of it restricts conventional recommender systems from directly leveraging it.

This paper proposes SMaRT, a generic framework to deal with social media data in the context of tourism industry. It is proposed to have five non-cohesive technology agnostic modules to constitute the overall architecture of the framework. Though the current focus is on tourism industry, the same framework can be extended to any service or product industry where social media data representing user interests can be made available.

As the immediate next step, it is planned to build and evaluate a full-fledge tourism recommender using the SMaRT framework using travel reviews collated from Tripadvisor as the input data. Also, as an extension to this architectural framework, it is planned to come up with a customizable user interface which can function as a simple front end for the researchers who want to leverage the SMaRT framework.

References

1. Abowd, G., Atkeson, C., Hong, J., Long, S., Kooper, R., Pinkerton, M.: Cyberguide: a mobile context-aware tour guide. Wirel. Netw. **3**, 421–433 (1997). https://doi.org/10.1023/a:1019194325861
2. Davies, N., Cheverst, K., Mitchell, K., Friday, A.: 'Caches in the air': disseminating tourist information in the GUIDE system. In: Proceedings WMCSA'99. Second IEEE Workshop on Mobile Computing Systems and Applications. IEEE (1999). https://doi.org/10.1109/mcsa.1999.749273
3. Cheverst, K., Davies, N., Mitchell, K., Friday, A., Efstratiou, C.: Developing a context-aware electronic tourist guide. In: Proceedings of the SIGCHI Conference on Human Factors in Computing Systems - CHI '00. ACM, New York (2000). https://doi.org/10.1145/332040.332047
4. Malaka, R., Zipf, A.: DEEP MAP: challenging IT research in the framework of a tourist information system. Inf. Commun. Technol. Tour. **2000**, 15–27 (2000). https://doi.org/10.1007/978-3-7091-6291-0_2
5. Renjith, S., Anjali, C.: A personalized travel recommender model based on content-based prediction and collaborative recommendation. Int. J. Comput. Sci. Mob. Comput. **ICMIC13**, 66–73 (2013)
6. Renjith, S., Anjali, C.: A personalized mobile travel recommender system using hybrid algorithm. In: 2014 First International Conference on Computational Systems and Communications (ICCSC). IEEE (2014). https://doi.org/10.1109/compsc.2014.7032612
7. Braunhofer, M., Ricci, F.: Selective contextual information acquisition in travel recommender systems. Inf. Technol. Tour. **17**, 5–29 (2017). https://doi.org/10.1007/s40558-017-0075-6
8. Li, C., Chen, H., Chen, R., Hsieh, H.: On route planning by inferring visiting time, modeling user preferences, and mining representative trip patterns. Knowl. Inf. Syst. **56**, 581–611 (2017). https://doi.org/10.1007/s10115-017-1106-5
9. Hsueh, Y., Huang, H.: Personalized itinerary recommendation with time constraints using GPS datasets. Knowl. Inf. Syst. **60**(1), 523–544 (2019). https://doi.org/10.1007/s10115-018-1217-7
10. Cattell, R.: The description of personality: basic traits resolved into clusters. J. Abnorm. Soc. Psychol. **38**, 476–506 (1943). https://doi.org/10.1037/H0054116

11. Pudil, P., Novovičová, J.: Novel methods for feature subset selection with respect to problem knowledge. In: Feature Extraction, Construction and Selection, pp. 101–116. Springer, New York (1998). https://doi.org/10.1007/978-1-4615-5725-8_7
12. Jagadish, H., Gehrke, J., Labrinidis, A., Papakonstantinou, Y., Patel, J., Ramakrishnan, R., Shahabi, C.: Big data and its technical challenges. Commun. ACM. **57**, 86–94 (2014). https://doi.org/10.1145/2611567
13. Grolinger, K., Hayes, M., Higashino, W., L'Heureux, A., Allison, D., Capretz, M.: Challenges for MapReduce in big data. In: 2014 IEEE World Congress on Services. IEEE (2014). https://doi.org/10.1109/services.2014.41
14. Najafabadi, M., Villanustre, F., Khoshgoftaar, T., Seliya, N., Wald, R., Muharemagic, E.: Deep learning applications and challenges in big data analytics. J. Big Data. **2**, 1 (2015). https://doi.org/10.1186/s40537-014-0007-7
15. L'Heureux, A., Grolinger, K., Elyamany, H., Capretz, M.: Machine learning with big data: challenges and approaches. IEEE Access. **5**, 7776–7797 (2017). https://doi.org/10.1109/access.2017.2696365
16. Jiang, S., Qian, X., Mei, T., Fu, Y.: Personalized travel sequence recommendation on multi-source big social media. IEEE Trans. Big Data. **2**, 43–56 (2016). https://doi.org/10.1109/tbdata.2016.2541160
17. Figueredo, M., Ribeiro, J., Cacho, N., Thome, A., Cacho, A., Lopes, F., Araujo, V.: From photos to travel itinerary: a tourism recommender system for smart tourism destination. In: 2018 IEEE Fourth International Conference on Big Data Computing Service and Applications (BigDataService). IEEE (2018). https://doi.org/10.1109/BigDataService.2018.00021
18. Shirkhorshidi, A., Aghabozorgi, S., Wah, T., Herawan, T.: Big data clustering: a review. In: The 14th International Conference on Computational Science and its Applications – ICCSA 2014, pp. 707–720. IEEE (2014). https://doi.org/10.1007/978-3-319-09156-3_49
19. Sajana, T., Sheela Rani, C., Narayana, K.: A survey on clustering techniques for big data mining. Indian J. Sci. Technol. **9**(3), 1–12 (2016). https://doi.org/10.17485/IJST/2016/V9I3/75971
20. Ajin, V., Kumar, L.: Big data and clustering algorithms. In: 2016 International Conference on Research Advances in Integrated Navigation Systems (RAINS), pp. 101–106. IEEE (2016). https://doi.org/10.1109/RAINS.2016.7764405
21. Dave, M., Gianey, H.: Different clustering algorithms for big data analytics: a review. In: 2016 International Conference System Modeling & Advancement in Research Trends (SMART), pp. 328–333. IEEE (2016). https://doi.org/10.1109/SYSMART.2016.7894544
22. Wei, C., Lee, Y., Hsu, C.: Empirical comparison of fast partitioning-based clustering algorithms for large data sets. Expert Syst. Appl. **24**(4), 351–363 (2003). https://doi.org/10.1016/S0957-4174(02)00185-9
23. Zhang, B.: Comparison of the performance of center-based clustering algorithms. In: Advances in Knowledge Discovery and Data Mining, PAKDD 2003, Lecture Notes in Computer Science, vol. 2637, pp. 63–74. Springer, Berlin (2003). https://doi.org/10.1007/3-540-36175-8_7
24. Wang, X., Hamilton, H.: A comparative study of two density-based spatial clustering algorithms for very large datasets. In: Advances in Artificial Intelligence, AI 2005, Lecture Notes in Computer Science, vol. 3501, pp. 120–132. Springer, Berlin (2005). https://doi.org/10.1007/11424918_14
25. Poonam, Dutta, M.: Performance analysis of clustering methods for outlier detection. In: 2012 Second International Conference on Advanced Computing & Communication Technologies (ACCT 2012), pp. 89–95. IEEE (2012). https://doi.org/10.1109/ACCT.2012.84
26. Fahad, A., Alshatri, N., Tari, Z., Alamri, A., Khalil, I., Zomaya, A., Foufou, S., Bouras, A.: A survey of clustering algorithms for big data: taxonomy and empirical analysis. IEEE Trans. Emerg. Top. Comput. **2**(3), 267–279 (2014). https://doi.org/10.1109/TETC.2014.2330519
27. Jung, Y., Kang, M., Heo, J.: Clustering performance comparison using k-means and expectation maximization algorithms. Biotechnol. Biotechnol. Equip. **28**(2), S44–S48 (2014). https://doi.org/10.1080/13102818.2014.949045

28. Bhatnagar, V., Majhi, R., Jena, P.: Comparative performance evaluation of clustering algorithms for grouping manufacturing firms. Arab. J. Sci. Eng. **43**(8), 4071–4083 (2017). https://doi.org/10.1007/S13369-017-2788-4
29. Hotelling, H.: Analysis of a complex of statistical variables into principal components. J. Educ. Psychol. **24**(6), 417–441 (1933). https://doi.org/10.1037/H0071325
30. Abdi, H., Williams, L.: Principal component analysis. Wiley Interdiscip. Rev. Comput. Stat. **2**(4), 433–459 (2010). https://doi.org/10.1002/wics.101
31. Isomura, T., Toyoizumi, T.: A local learning rule for independent component analysis. Sci. Rep. **6**, 28073 (2016). https://doi.org/10.1038/srep28073
32. Maaten, L., Hinton, G.: Visualizing data using t-SNE. J. Mach. Learn. Res. **9**(Nov), 2579–2605 (2008)
33. Maaten, L.: Accelerating t-SNE using tree-based algorithms. J. Mach. Learn. Res. **15**, 3221–3245 (2014)
34. Roweis, S., Saul, L.: Nonlinear dimensionality reduction by locally linear embedding. Science. **290**(5500), 2323–2326 (2000). https://doi.org/10.1126/science.290.5500.2323
35. Ridder, D., Kouropteva, O., Okun, O., Pietikäinen, M., Duin, R.: Supervised locally linear embedding. In: Artificial Neural Networks and Neural Information Processing - ICANN/ICONIP 2003, pp. 333–341. Springer, Berlin (2003). https://doi.org/10.1007/3-540-44989-2_40
36. Kohonen, T.: Exploration of very large databases by self-organizing maps. In: International Conference on Neural Networks (ICNN'97), vol. 1, pp. PL1–PL6. IEEE (1997). https://doi.org/10.1109/ICNN.1997.611622
37. Tenenbaum, J.: A global geometric framework for nonlinear dimensionality reduction. Science. **290**(5500), 2319–2323 (2000). https://doi.org/10.1126/science.290.5500.2319
38. De'ath, G.: Extended dissimilarity: a method of robust estimation of ecological distances from high beta diversity data. Plant Ecol. **144**, 191–199 (1999). https://doi.org/10.1023/A:1009763730207
39. Liou, C., Huang, J., Yang, W.: Modeling word perception using the Elman network. Neurocomputing. **71**(16-18), 3150–3157 (2008). https://doi.org/10.1016/J.NEUCOM.2008.04.030
40. Hinton, G.: Reducing the dimensionality of data with neural networks. Science. **313**(5786), 504–507 (2006). https://doi.org/10.1126/science.1127647
41. Wang, Y., Yao, H., Zhao, S.: Auto-encoder based dimensionality reduction. Neurocomputing. **184**, 232–242 (2016). https://doi.org/10.1016/j.neucom.2015.08.104
42. Renjith, S., Sreekumar, A., Jathavedan, M.: Evaluation of partitioning clustering algorithms for processing social media data in tourism domain. In: 2018 IEEE Recent Advances in Intelligent Computational Systems (RAICS), pp. 127–131. IEEE (2018). https://doi.org/10.1109/RAICS.2018.8635080
43. Renjith, S., Sreekumar, A., Jathavedan, M.: Performance evaluation of clustering algorithms for varying cardinality and dimensionality of data sets. Materials Today: Proceedings. **27**(1), 627–633 (2020). https://doi.org/10.1016/j.matpr.2020.01.110
44. Renjith, S., Sreekumar, A., Jathavedan, M.: Pragmatic evaluation of the impact of dimensionality reduction in the performance of clustering algorithms. In: Advances in Electrical and Computer Technologies, Lecture Notes in Electrical Engineering, vol. 672. Springer, Singapore (2020). https://doi.org/10.1007/978-981-15-5558-9_45
45. Renjith, S., Sreekumar, A., Jathavedan, M.: A comparative analysis of clustering quality based on internal validation indices for dimensionally reduced social media data. In: Advances in Artificial Intelligence and Data Engineering, Advances in Intelligent Systems and Computing, Vol. 1133, pp. 1047–1065. Springer, Singapore (2020). https://doi.org/10.1007/978-981-15-3514-7_78

Spotting a Phishing URL: A Machine Learning Approach

S. Kanthimathi, Saumya Sachdev, and Shashank Shekhar

1 Introduction

Phishing is a type of attack that involves tricking the user into giving up private information because they believe that the web page asking for details is legitimate. This can be very dangerous for obvious reasons. Often people do not realize that they have been attacked until after it is too late. Successfully deceiving even one person in a corporation can compromise the entire system of the company. Log in credentials obtained through phishing can lead to huge data breaches and end up in the hands of hackers. Thus preventing such attacks is crucial in today's data-intensive world.

According to research by Trend Micro, 91% of cyber-attack that results in a data breach in corporate firms begin with a phishing attack. The first phishing attack came up in 1980s and started to increase with the increase in Internet usage. As people became aware of such attacks, they tried building models to tackle phishing attacks. A common way came up as to match the website against a database containing a list of phishing websites. But this did not produce convincible results and hence appealed for a robust system. Various technical algorithms are available to prevent hacking of personal information with the use of phishing attacks.

Some of them include browsers alerting users to fraudulent websites, augmenting password logins, filtering out phishing mails, monitoring and taking down, etc.

There are a lot of agencies working against phishing and maintain a regular database of the phishing URLs. Popular ones include PhishTank, OpenPhish etc.

S. Kanthimathi (✉) · S. Sachdev · S. Shekhar
PES University Electronic City Campus, Bangalore, India
e-mail: kanthimathi@pes.edu

M. Palesi et al. (eds.), *Second International Conference on Networks and Advances in Computational Technologies*, Transactions on Computational Science and Computational Intelligence, https://doi.org/10.1007/978-3-030-49500-8_27

Although they are maintaining the updated list, there are high chances that some website that is developed recently for phishing purpose will not be enlisted and thus poses potential threat to a specific or wide range of users depending on the motive of the creator. In order to tackle situations like these, we need a system that uses machine learning to identify potential phishing websites.

It is necessary for a system to classify phishing or suspicious URLs out of all browsed URLs. This can help avoid social engineering attacks as a user will be notified if the link they are clicking at is malicious or not. Since closest proximity to user will be via browser extension that will use machine learning model at back end to classify whether a website is phishing or safe, we are proposing a model that will do the same.

2 Related Work

In Abdelhamid et al.'s study [1], various machine learning algorithms including Bayesian network, C4.5 decision tree, SVM, AdaBoost, eDRI, OneRule, conjunctive rule and RIDOR are compared using their effect on phishing activities. The dataset used was gathered from PhishTank and MillerSmiles repositories. Decision tree, Bayesian network and SVM gave the best results. One disadvantage of this technique is that the models made from machine learning algorithms were hard to understand by end users. In Zuhair and Selamat's study [2], more efficient phishing classification models are made by using machine learning techniques and boosting its induction power via factors like feature and feature category, dataset size, active learning and adaptable modelling. The resulting assembly uses multiple modules working offline and/or online to learn actively and adapt to identifying phishing variants from the web data available around it.

In Marchal et al.'s study [3], a system to detect phishing URLs relying primarily on URL's lexical analysis was proposed. This approach is based on the intra-relatedness of parts of URLs. They use search engines to extract features from a URL based on the intra-relatedness of components of URL and its popularity on website ranking pages. The features were then used in a classification algorithm on a dataset of about one lakh URLs that were a combination of phishing and legitimate pages. This experiment yielded a very high classification accuracy of 94.91% and also a low false positive rate of 1.44%. In Manoj Kumar and Alekhya's study [4], fuzzy logic evaluates the degree of phishing in web pages. It also tries to implement genetic algorithm to deal with fraudulent websites using fuzzy logic technique. The phishing characteristics used were not completely reliable as the presence of certain characteristics was not enough to label the website as phishing site. The results were found to ambiguous in some cases.

In Zhang et al.'s study [5], several semantic features like word embedding and word2vec are used to capture both semantic and syntactic information of words. The disadvantage of this method is that it is highly dependent on language. Word embedding of words, whose language model has been learned, can be obtained. In Yahya Dae et al.'s study [6], Markov chains are used to construct N-grams from URLs after pre-processing. Three classifiers: J4.8, SVM and LR are used to evaluate the 12 N-gram features that are extracted from the URLs. As a result, the phishing classifier presented is light weight and not time consuming. Error rate can be reduced by adding most effective and lightweight bag of word features. In Sanglerdsinlapachai and Rungsawang's study [7], various features from page information, like source codes or online visual appearance, and external resources from third parties are used to define a concept called domain top-page similarity and use it to identify whether the page is phishing or legitimate.

3 Proposed Scheme

In this paper we propose a methodology to identify whether a given URL is malicious or not. Phishing is one of the most common methods of illegally obtaining a user's data without s/he knowing about it. Hence it is the necessity of the hour to create accurate models that can correctly classify whether a given URL is trustworthy or not. Machine learning models are used for classification like Support Vector Machines (SVM), Random Forest, Neural Networks, Naïve Bayes Classifier etc.

The problem of identifying a malicious URL can be thought of as a simple binary classification where 1 indicates Safe URL and -1 indicates Malicious URL. The dataset used for the purposes of this paper is available as a UCI repository: Phishing Websites Data Set [8]. The dataset has 11,055 instances of URLs each with 30 attributes. All the attributes in the dataset are categorical in nature and encoded using one hot encoding with either three or two values depending on the number of classes the attribute has. In this paper we compare three classification methods: namely, SVM, Random Forest and Artificial Neural Networks. The best performing of the three is stored in a browser plugin and used to predict whether a URL clicked by the user is malicious or not. The flow diagram shows the methodology followed in this paper (Fig. 1).

1. The browser plugin is always running in the background. As soon as the user clicks on a URL, the plugin extracts it and starts processing it.
2. The attribute values from the new URL are extracted by the plugin and sent to the stored model for further processing.
3. The plugin stores an already trained model of the best performing model. The extracted attributes are plugged into the model.
4. If the URL is classified as malicious by the model, a pop-up warns the user of the same; otherwise, the plugin waits for the next URL.

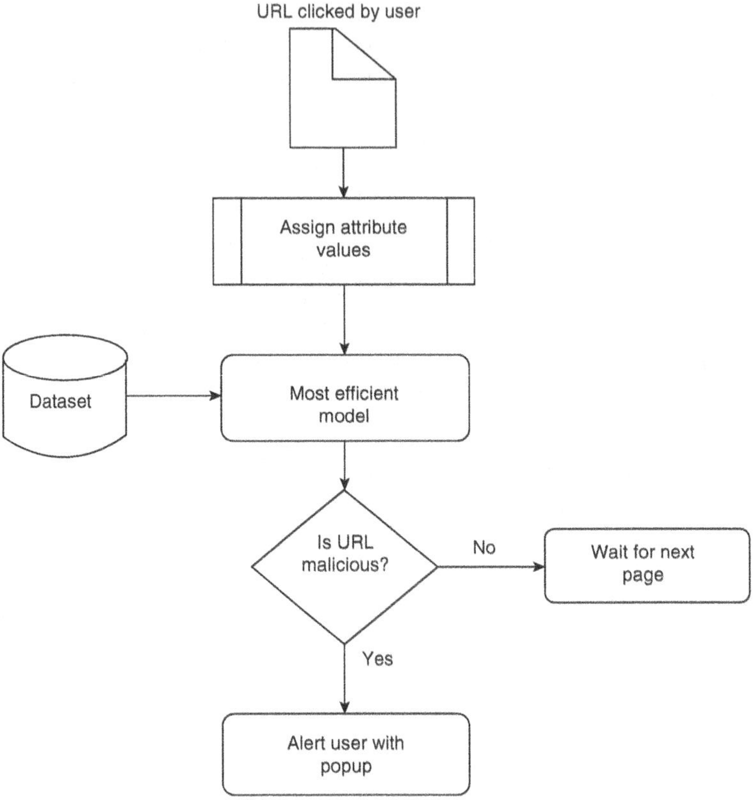

Fig. 1 Flow diagram for proposed scheme

4 Comparison of Models

Support Vector Machines, Random Forest and Artificial Neural Network can all be used for classification purposes. However, their performance depends on the dataset being used. The three models were chosen for this paper because they are accurate and comparatively faster. Fast algorithm is necessary for the plugin as the user needs to be alerted as soon as possible if the URL is intended for harm. Below is a generic comparison of the three models.

1. SVM: It is primarily used for binary classification problems. SVM constructs a hyperplane that splits the input space into two parts. The core task behind SVM is to identify the best hyperplane that can exist for the given dataset. The best hyperplane has the maximum margin from data point. SVM is best suited for data that has more dimensions than the number of samples. It is not an ensemble learning algorithm and is not suitable for dataset with missing values. The training time can be quite long if the dataset is large. SVM is also not capable

of handling too much noise in the dataset. SVM is appropriate for this problem because it is a binary classification problem but not the best as it not fast enough for a browser extension.

2. Random Forest: It is one of the most powerful machine learning algorithms. It works by constructing a large number of decision trees whose results are combined in order to predict the required result. It can be used for both classification and regression. RF is especially robust in cases of a large dataset with high dimensionality. It is a type of ensemble learning algorithm called Bootstrap Aggregator. RF is capable of dealing with noise and missing values in the dataset. RF can sometimes feel like a black box model as you have very little control on what the model does. Random Forest is perfect for this application as decision trees are fast and do not necessarily require values of all attributes.
3. Artificial Neural Networks—ANNs are a type of machine learning algorithms that try to replicate the workings of a brain. It is based on a collection of artificial neurons that are arranged in layers connected to each other through weights. The training of the algorithm includes adjusting the weights to best predict class. They work very well for large datasets and are computationally more expensive than traditional machine learning algorithms. They can outperform nearly every machine learning algorithm but are prone to overfitting. They can also handle noisy or missing values in datasets. The biggest disadvantage of using ANNs is that they are complete black box solution, and hence for this application it is not the most ideal.

5 Implementation

The dataset used in the models below has a collection of 11,055 URLs. Each of them have 30 attributes like length of the URL, whether the URL uses any shortening services, is the web page being redirected, if the website uses SSL protocol, how long is the domain registration length, usage of standard ports etc. [9]. The dataset has 6157 instances of safe/non-phishing URLs and 4898 instances of phishing URLs.

The above-mentioned three models were implemented using the Python library Scikit-learn. The SVM model used Radial Basis Function (RBF) as the kernel method for pattern analysis. To optimize the values C and gamma parameters, the hyper-parameters, grid search was used. SVM is ideal for instances where the number of attributes is more than the number of test cases. In random forest we used 1000 base decision trees and achieved accuracy close to 97%. There is no restriction on the depth of the decision tree. The splitting criterion for each decision tree is Gini Index. Confusion matrix and accuracy are used to evaluate the model. We have used sklearn's multilayer perceptron (MLP), which is a type of feed forward network. In our model we have used default three layers, which are input, hidden and output layer. Each node of input nodes is a neuron that uses a rectified linear unit (reLu) activation function. MLP utilizes a supervised learning technique called backpropagation for training. The ANN model has 30 input nodes for the 30 parameters in the dataset, three nodes in the hidden layer and two nodes in the output node.

After implementing the models, we observed that Random Forest was performing better than others. The following table depicts the results obtained.

Model	Accuracy
SVM	96%
ANN	92%
Random Forest	97%

The classification report for Random Forest classifier is shown below:

	Precision	Recall	F1-score	Support
Phishing	0.97	0.95	0.96	1183
Safe	0.96	0.98	0.97	1581

Because of the high accuracy and its ability to handle categorical values very well, we decided to implement Random Forest. Random Forest provides the highest accuracy as it does not depend on a single tree for classification but rather depends on the prediction by the maximum number of trees. We made a browser extension that would extract the URL and send it to the back-end server. The browser extension was made using JavaScript, HTML and CSS, JQuery and Django. The browser extension works by producing a pop-up window when the user clicks on the icon. The features from a new URL are extracted from a python script, which runs on the Django server, and fed to the classifier. Our browser extension was able to identify legitimate and phishing websites to a great extent. However, we do seek a scope of improvement as running the model induces a slight delay on the user counterpart.

6 Conclusion

Based on the classification models and chrome extension, we saw that although the features we extracted were helpful in the identification of phishing sites, but the performance can be improved by evolving the features as the new phishing attackers are evolving the websites they used to attack. As a future work, we are reserving the task to identify additional features and improve the computation time of the extension.

References

1. Abdelhamid, N., Thabtah, F., Abdeljaber, H.: Phishing detection: a recent intelligent machine learning comparison based on models content and features. In: 2017 IEEE International Conference on Intelligence and Security Informatics (ISI), Beijing, pp. 72–77. IEEE (2017)
2. Zuhair, H., Selamat, A.: Phishing classification models: issues and perspectives. In: 2017 IEEE Conference on Open Systems (ICOS), Miri, pp. 26–31. IEEE (2017)
3. Marchal, S., Franois, J., State, R., Engel, T.: PhishStorm: detecting phishing with streaming analytics. IEEE Trans. Netw. Serv. Manag. **11**(4), 458–471 (2014)
4. Manoj Kumar, K.N., Alekhya, K.: Detecting phishing websites using fuzzy logic. Int. J. Adv. Res. Comput. Eng. Technol. **5**(10), 2413–2417 (2016)
5. Zhang, X., Zeng, Y., Jin, X., Yan, Z., Geng, G.: Boosting the phishing detection performance by semantic analysis. In: 2017 IEEE International Conference on Big Data (Big Data), Boston, MA, pp. 1063–1070. IEEE (2017)
6. Yahya Dae, A., Badlishah Ahmad, R., Jacob, Y.: Lexical based method for phishing URLs detection. J. Theor. Appl. Inf. Technol. **88**(3), 54–60 (2016)
7. Sanglerdsinlapachai, N., Rungsawang, A.: Using domain top-page similarity feature in machine learning-based web phishing detection. In: 2010 Third International Conference on Knowledge Discovery and Data Mining, Phuket, pp. 187–190. IEEE (2010)
8. Phishing Websites Data Set: https://archive.ics.uci.edu/ml/datasets/phishing+websites
9. Phishing Website Features.docx: https://archive.ics.uci.edu/ml/machine-learning-databases/00327/

Index

M. Palesi et al. (eds.), *Second International Conference on Networks and Advances in Computational Technologies*, Transactions on Computational Science and Computational Intelligence, https://doi.org/10.1007/978-3-030-49500-8

The manufacturer's authorised representative in the EU is Springer Nature Customer Service Centre GmbH, Europaplatz 3, 69115 Heidelberg, Germany. If you have any concerns regarding our products, please contact ProductSafety@springernature.com

Printed and bound by CPI Group (UK) Ltd, Croydon, CR0 4YY
30/09/2026
02225693-0006